Industrial Engineering and Operations Management

Industrial Engineering and Operations Management

Edited by **Courtney Hoover**

Willford Press
New York

Published by Willford Press,
118-35 Queens Blvd., Suite 400,
Forest Hills, NY 11375, USA
www.willfordpress.com

Industrial Engineering and Operations Management
Edited by Courtney Hoover

International Standard Book Number: 978-1-68285-103-6 (Hardback)

Printed in China.

Contents

Preface

Industrial engineering is a multidisciplinary scientific field that aims to integrate the engineering theories and concepts and management techniques to solve & manage complex processes and ensure effective production across various industries. It involves development and analysis of various complex systems comprising of people, energy, information systems and equipment. The field of industrial engineering incorporates concepts from management science, manufacturing engineering, operations research, etc. This book assimilates diverse aspects of industrial engineering and operations management like quality control, process engineering, supply chain management and logistics, econometrics, etc. Researches and case studies included in this book are compiled by internationally acclaimed experts and researchers from various parts of the globe that make this book a truly international effort. It aims to serve as a resource guide for experts and students alike and contribute to the overall growth of the discipline.

Various studies have approached the subject by analyzing it with a single perspective, but the present book provides diverse methodologies and techniques to address this field. This book contains theories and applications needed for understanding the subject from different perspectives. The aim is to keep the readers informed about the progress in the field; therefore, the contributions were carefully examined to compile novel researches by specialists from across the globe.

Indeed, the job of the editor is the most crucial and challenging in compiling all chapters into a single book. In the end, I would extend my sincere thanks to the chapter authors for their profound work. I am also thankful for the support provided by my family and colleagues during the compilation of this book.

Editor

1

Reliability optimization of binary state non-repairable systems: A state of the art survey

Roya Soltani*

School of Industrial Engineering, Iran University of Science and Technology, Tehran, Iran

CHRONICLE

Keywords:
Reliability Allocation
Redundancy Allocation
Reliability-Redundancy Allocation
Stochastic Uncertainty
Fuzzy Uncertainty
Interval Uncertainty
Robust optimization
Mathematical programming
Exact methods
Heuristic and Meta-heuristic

ABSTRACT

The purpose of this paper is to discuss the state of the art on models and methods for reliability optimization problems (ROPs) including reliability allocation, redundancy allocation and reliability-redundancy allocation. There are literally few surveys for reviewing the literature of the ROPs. Tillman et al. (1980) classified the related papers by system structure, problem type, and solution methods, separately. In another work, Tzafestas (1980) reviewed system reliability optimization models and the optimization techniques. Yearout (1986) reviewed the literature related to standby redundancy. Kuo (2000) studied the system reliability optimization based on system structure and solution methods. Kuo and Prasad (2004) overviewed system reliability optimization methods. Later, Kuo (2007) reviewed recent advances in optimal reliability allocation problems. The present study adds to the previous literature surveys and focuses mainly on papers after year 2000 but with a quick review on the previous works so that the readers become familiar with the existing approaches. This research investigates the literature from system structure, system performance, uncertainty state and solution approach standpoints, simultaneously.

Acronyms

ROP: Reliability Optimization Problem
RAP: Redundancy Allocation Problem
RRAP: Reliability- Redundancy Allocation Problem
GA: Generic Algorithm
SA: Simulated Annealing
GP: Geometric Programming
HBMO: Honey Bee Mating Optimization
ABC: Artificial Bee Colony
ACO: Ant Colony Optimization
DC: Degrade Ceiling
PSO: Particle Swarm Optimization
IA: Immune Algorithm
DE: Differential Evolution
HS: Harmony Search
VND: Variable Neighborhood Descent
VNS: Variable Neighborhood Search
EGHS: Effective Global Harmony Search
NGHS: Novel Global Harmony Search
SOMA: Self-Organizing Migrating Algorithm
BBMOPSO: Bare- Bones Multi-Objective Particle Swarm Optimization
DSAMOPSO: Dynamic Self-Adaptive Multi-Objective Particle Swarm Optimization
AUGMECON: Augmented Epsilon Constraint
CTV-MOPSO: Customized Time-Variant Multi- Objective Particle Swarm Optimization
UMOSA: Ulungu Multi-Objective Simulated Annealing
LMBB: Lagrange Multiplier and the Branch-and-Bound
ICA: Imperialist Competitive Algorithm
FLC: Fuzzy Logic Controller
TS: Tabu Search
IIMOM: Intelligent Interactive Multi-Objective Optimization Method
GDA: Great Deluge Algorithm
CEA: Cellular Evolutionary Approach
NSGA-II: Non-dominated Sorting Genetic Algorithm
EVM : Expected Value Model
CCP : Chance-Constrained Programming (Redundancy CCP)
DCP: Dependent Chance Programming
MC: Monte Carlo
MCS: Monte Carlo Simulation
RSM: Response Surface Methodology
BBD: Box-Behnken design
OSSO: Orthogonal Simplified Swarm Optimization
SPEA2: Strength Pareto Evolutionary Genetic Algorithm
MOHGA: Multi-objective hierarchical genetic algorithm
SMOSA: Suppapitnarm Multi-Objective Simulated Annealing
PSA: Pareto Simulated Annealing
PDMOSA: Pareto Domination based Multi-Objective Simulated Annealing
WMOSA: Weight based Multi-Objective Simulated Annealing
LHS: Latin Hypercube Sampling
NN: Neural Network

* Corresponding author.

1. Introduction

To be more competitive in the market, the reliabilities of systems or products need to be optimized by system designers. This goal is accomplished by applying suitable reliability optimization techniques, which includes three main categories, i.e. the reliability allocation, the redundancy allocation and the reliability-redundancy allocation as presented in Fig. 1. Redundancy can be in system level, subsystem level (Module) and component level.

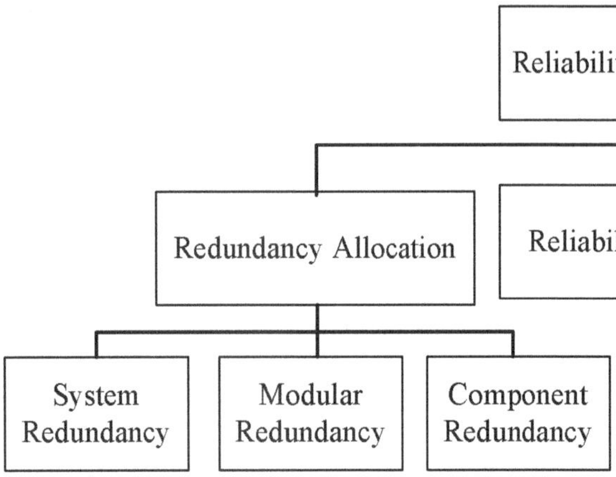

Fig. 1. Reliability optimization problems

In the reliability allocation problems, the reliability levels of the components is determined such that the consumption of a resource under a reliability constraint is minimized while the redundancy allocation problem generally involves the selection of components and redundancy levels to maximize the system reliability given various system-level's constraints. The reliability-redundancy allocation, which is the combination of two aforesaid problems, determines the reliability levels and the redundancy levels, simultaneously.

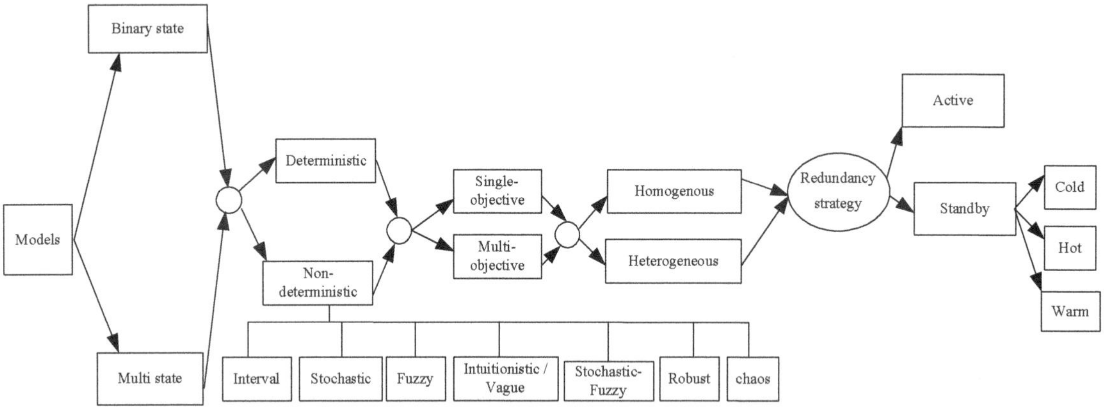

Fig. 2. RAP models

Fig. 2 presents taxonomy of redundancy allocation problems. By starting from the model and going through each path, a new model is resulted. The RAP is first classified and separated based on whether the problem is binary state or multi state. In the former, the components are in two states of functioning and non-functioning whilst in the latter, the states between them are also considered. Each of these categories can be studied with deterministic or non-deterministic parameters, which involve uncertainties such as stochastic uncertainty, interval uncertainty, fuzzy uncertainty, intuitionistic fuzzy or vague set, stochastic-fuzzy uncertainty, perturbation in a defined uncertainty set and chaos. Each problem can be modelled with a single objective or multiple objectives each of which with homogenous and heterogeneous components. The redundancy strategy can be either active or standby. The standby redundancy in turn is categorized into cold, hot and warm. The models have been formulated with different system structures and with different performance measures such as reliability, availability, mean time to failure (average life), percentile life, etc. For more information about the application of percentile life and reliability as performance measures readers are referred to the work by Kim and Kuo (2003). Note that fault-tolerance mechanisms, which are the ability of a system to continue performing its intended function in spite of faults, are not investigated in this research.

Fig. 3 classifies the solution methods for RAPs. Mathematical programming methods, which include exact and approximation methods are the most suitable for small-sized problems. Chern (1992) proves that RAP in series system is an NP-Hard problem. As the system configuration becomes complex and other simplified assumptions are released the NP-Hardness of the problem increases. Therefore, heuristic and meta-heuristic approaches have been proposed to deal with RAPs.

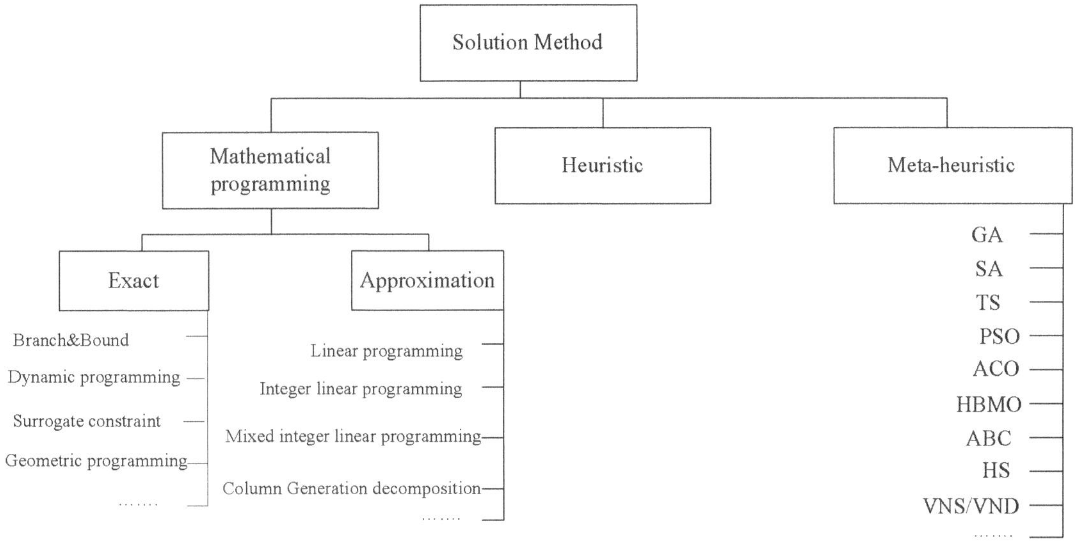

Fig. 3. RAP solution methods

The rest of the paper is organized as follows. In section 2 the RAP is investigated which include deterministic and non-deterministic models. Section 3 is devoted to reliability allocation. Section 3 surveys the reliability-redundancy allocation problems. Finally, concluding remarks have been provided in section 4.

2. Redundancy allocation

As presented in Fig.1, redundancy can be in system level (i.e. system redundancy), in subsystem level (i.e. modular redundancy) or in component level (i.e. component redundancy). It is a well-known principle to the design engineers that active redundancy at the component level is better than that at the system level in usual stochastic ordering (Nanda & Hazra, 2013; Barlow & Proschan, 1975). In the following subsections, modular and component redundancies are investigated regarding deterministic or non-deterministic parameters. Redundancy can be in active or standby forms. Yearout et al. (1986) and Li and Ding (2013) respectively reviewed standby and active redundancies in reliability.

Basic assumptions for RAP are as follows:

1) The availability of the components is unlimited;

2) Failures of individual components are independent;

3) All failed components will not damage the system and are not repaired.

4) No preventive maintenance is considered.

In the followings, other assumptions such as redundancy strategy, being deterministic or non-deterministic, etc. are mentioned. Furthermore, wherever a basic assumption is violated, it is mentioned in a column named other considerations.

2.1 Component redundancy

2.1.1 Deterministic models

Deterministic models assume that all parameters including component reliability, cost, weight, volume, etc. are known, precisely. The RAP is restricted by available system cost, weight, volume, etc. In the field of system-reliability, it has been reported that various exact/mathematical programming and heuristic/meta-heuristic techniques have been used to cope with RAPs. For review on the work before year 2000 readers are referred to the works by Tillman et al. (1980), Kuo (2000), Kuo and Prasad (2004) and Kuo (2007). In the area of single objective models with active strategy, Fyffe et al. (1968) were the first who proposed a model for RAP where system's reliability is maximized subject to constraints on cost and weight. They used a Lagrangian multiplier in the objective function to handle multiple constraints.

Nakagawa and Miyazaki (1981) showed that, sometimes, the use of a Lagrangian multiplier becomes inefficient. Therefore, they proposed the *surrogate constraint* approach. Other exact and mathematical programming techniques include *DP* (e.g. Bellman & Dreyfus, 1958; Fyffe et al., 1968; Jianping, 1996; Misra, 1971), Integer programming (e.g. Ghare, 1969; Misra & Sharma, 1991), Partial bound enumeration (e.g. Jianping & Xishen , 1997), Branch and bound (e.g. Bulfin & Liu,1985; Sup & Kwon,1999), GP (e.g. Federowicz & Mazumdar,*1968;* Govil, 1983*;* Misra & Sharma, 1973), Lagrange Multiplier (Govil & Agarwala,1983), etc.

Table 1 lists the related works in the literature after year 2000. Note that the references are not arranged in chronological order but they are arranged based on the solution method. As it is observed, most works have been carried out in active redundancy area except the work by Amari (2010) and Tannous et al. (2011) which consider both active and/or warm standby redundancy strategies and seems like more researches are needed with respect to redundancy strategy.

In the 1970s, Gopal et al. (1978) (GAG), and Nakagawa and Nakashima (1977) (NAN) introduced iterative heuristics for redundancy allocation. In the 1980s, the methods were improved by Kohda and Inoue (1982) (KI), and Shi (1987) (SHI). Kim and Yum (1993) (KY), and Jianping (1996) (JP), concentrated on improving solutions from a local optimum. Other heuristics include the work by Dinghua (1987), Nakagawa and Miyazaki (1981). Meta-heuristic approaches for single objective problems before year 2000 include GA (Coit & Smith, 1996), penalty guided GA (Coit & Smith, 1996), hybrid NN-GA (Coit & Smith, 1996), SA (Ravi et al., 1997), Ant system (Liang & Smith, 1999), etc.

Table 1 (a)

Exact and Mathematical programming approaches for RAP (Single Objective Problems)

Reference	System configuration	Uncertainty state	Strategy	System performance	Solution method	Other considerations
Prasad et al. (2001)	Series- parallel	Deterministic	Active	Percentile life	Lexicographic search	An upper bound is used to reduce the search space
Ng & Sancho (2001)	Series- parallel/ overspeed protection system	Deterministic	Active	System reliability	Hybrid DP/depth first search	
Yalaoui et al. (2005)	Series- parallel	Deterministic	Active	System cost	DP	
Onishi et al. (2007)	Series- parallel	Deterministic	Active	System reliability	Improved surrogate constraint	
Djerdjour and Rekab (2001)	Series- parallel	Deterministic	Active	System cost	Branch and bound	A heuristic for finding a good lower bound is presented along with some implicit enumeration criteria
Ramirez-Marquez et al. (2004)	Series- parallel	Deterministic	Active	Max-Min reliability	Mixed integer linear programming	
Lee et al. (2003)	Series- parallel	Deterministic	Active	Max-Min reliability	Comparison of Min-Max approach and NN	
Hsieh (2003)	Series- parallel	Deterministic	Active	System reliability	Simple linear approximation	Multiple component choices, Constraints must be linear and separable
Hsieh (2002)	Series- parallel	Deterministic	Active	System reliability	Two-Phase Linear Programming	The method provides a good lower bound for branch-and-bound algorithm or artificial methods
Billionnet (2008)	Series- parallel	Deterministic	Active	System reliability	Integer linear programming	
Nourelfath & Nahas (2003)	Series	Deterministic	-	System reliability	Quantized hopfield networks	For each component, there are different technologies available
Amari (2010)	*k*-out-of-*n*	deterministic	Active/warm standby	System reliability	linear programming (LP)-based branch-and-bound	Arbitrary general Distributions (such as Weibull or Gamma)
Tannous et al. (2011)	Series-parallel	deterministic	Warm standby	System reliability	GA and exact integer programming for reformulated model	SBDD for driving the objective function , exponential failure distribution

Table 1 (b)

Exact and Mathematical programming approaches for RAP (Multi Objective Problems)

Reference	System configuration	Uncertainty state	Strategy	System performance	Solution method	Other considerations
Mahapatra (2009)	Series- parallel	Deterministic	Active	System reliability and Entropy	Global criterion method	Non-linear cost constraint
Khalili-Damghani & Amiri(2012)	Series-parallel	Deterministic	Active	System reliability and cost	Epsilon constraint along with DEA (to prune the pareto set)	
Sadjadi et al. (2014)	Series-parallel	Deterministic	Active and cold standby	System reliability, cost and weigh	Compromise programming	Switch reliability is imperfect, Earlang time to failure distribution
Soltani et al. (2015)	Series-parallel	Deterministic	Active and cold standby	System reliability, cost and entropy/weight entropy	Compromise programming	Switch reliability is imperfect, Earlang time to failure distribution
Cao et al. (2013)	Series-parallel	Deterministic	Active	System reliability, cost and weight	Decomposition approach	
Khalili-Damghani et al. (2013)	Series-parallel	Deterministic	Active	System reliability, cost and weight	A decision support system based on TOPSIS, modified ε-constraint and DEA	

Table 2 lists the related works in the literature after year 2000. Active and cold standby redundancy strategies have been studied more and more researches are needed regarding the warm standby redundancy.

Table 2 (a)
Heuristic and Meta-heuristic approaches for RAP (Single Objective Problems)

Reference	System configuration	Uncertainty state	Strategy	System performance	Solution method	Other considerations
Ha (2004)	Bridge, Series-parallel	Deterministic	Active	System reliability	Multi-path heuristics called tree and scanning	There are no other restrictions such as separable, linear, and convex.
You & Chen (2005)	Series-parallel	Deterministic	Active	System reliability	Heuristic (based on greedy method and GA)	Assumption of Fyffe (1968) hold.
Sadjadi & Soltani (2009)	Series-parallel	Deterministic	Active	System reliability	Heuristic and GA	Allowing one component type
Coit & Liu (2000)	k-out-of-n	Deterministic	Active	System reliability	Standard algorithm for zero-one IP	Exponential time to failure, allowing one component type
Kulturel-Konak et al. (2003)	k-out-of-n	Deterministic	Active	System reliability	TS	Exponential time to failure, allowing one component type
Ryoo (2005)	Complex systems	Deterministic	Active	System reliability	Robust meta-heuristics (TS)	
Kim et al. (2004)	Series-parallel	Deterministic	Active	System reliability	SA	
Ravi (2004)	Complex/series-parallel	Deterministic	Active	System reliability	GDA	
Rocco (2000)	Complex	Deterministic	Active	System reliability	CEA	
Sadjadi & Soltani (2012)	Series-parallel	Deterministic	Active	System reliability	Heuristic and HBMO	Mix of components
Hsieh & Yeh (2012)	Series-parallel	Deterministic	Active	System reliability	Penalty guided bees search	Mix of components
Liang & Smith (2004)	k-out-of-n	Deterministic	Active	System reliability	ACO	
Nahas & Nourelfath (2005)	Series	Deterministic	Active	System reliability	Ant system with local search	One technology can be used for each component
Ahmadizar & Soltanpanah (2011)	series	Deterministic	Active	System reliability	ACO	One technology can be used for each component
Nahas et al. (2007)	k-out-of-n	Deterministic	Active	System reliability	Coupling ACO and DC	
Beji et al. (2010)	Series-parallel	Deterministic	Active	System reliability	Hybrid PSO with local search	An adaptive penalty function was proposed.
Chen & You (2005)	Series-parallel	Deterministic	Active	System reliability	IA	Multiple component choices
Nahas & Thien-My (2010)	Series-parallel	Deterministic	Active	System reliability	HS	
Liang & Wu (2005)	Series-parallel	Deterministic	Active	System reliability	VND	
Liang & Chen (2007)	Series-parallel	Deterministic	Active	System reliability	VNS	
Coit (2001)	Series-parallel	Deterministic	Cold standby	System reliability	Integer programming	Multiple component choices for each subsystem and imperfect switching
Coit (2003)	Series-parallel	Deterministic	Active and Cold standby	System reliability	Integer programming	Assumption of Coit (2001) hold.
Tavakkoli-Moghaddam et al. (2008)	Series-parallel	Deterministic	Active and Cold standby	System reliability	GA	Assumptions of Coit (2003) hold.
Tavakkoli-Moghaddam & Safari (2007)	Series-parallel	Deterministic	Active and cold standby	System reliability	-	Mixing components is allowed.
Safari & Tavakkoli-Moghaddam (2010)	Series-parallel	Deterministic	Active and cold standby	System reliability	Memetic algorithm	No component mixing is allowed.
Karimi et al, (2011)	Series-parallel	Deterministic	Cold standby	System reliability	GA and SA	Exponential and hypo-exponential time to failure distribution
Chambari et al. (2013)	Series-parallel	Deterministic	Active and cold standby	System reliability	SA	Assumptions of Coit (2003) hold.
Safaei et al. (2012)	Series-parallel	Deterministic	Active	System reliability	Annealing-based PSO	Assumptions of Hsieh (2002) hold.
Najafi et al. (2013)	Series-parallel	Deterministic	Active	Mean time to failure	Tuned SA and GA	Mixing components is not allowed
Yeh (2014)	Series-parallel	Deterministic	Active	System reliability	OSSO	Mixing components is allowed.
Soltani et al. (2013)	Series-parallel	Deterministic	Active	System reliability	Heuristic and HBMO	Mixing components is allowed. Price discounting are considered.

Table 2 (b)
Heuristic and Meta-heuristic approaches for RAP (Multi Objective Problems)

Reference	System configuration	Uncertainty state	Strategy	System performance	Solution method	Other considerations
Salazar et al. (2006)	Complex, series-parallel, life-support system in a space capsule	Deterministic	Active	System reliability and cost	NSGA-II	Three types of problems including reliability allocation, RAP and RRAP are considered.
Coit & Konak (2006)	Series-parallel	Deterministic	Active	Subsystem reliability	Multiple weighted objective heuristic	
Zhao et al. (2007)	Series-parallel, Series-parallel with k-out-of-n:G subsystem (gearbox reliability optimization)	Deterministic	Active	Subsystem cost and weight	multi-objective ant colony system	
Taboada et al. (2007)	Series-parallel	Deterministic	Active	System reliability, cost and weight	NSGA-II	Pareto set is pruned through data clustering (k-means)
Taboada & Coit (2008)	Series-parallel	Deterministic	Active	System reliability, cost and weight	Multiple objective GA (MOEA-DAP)	
Kulturel-Konak et al. (2008)	Series-parallel	Deterministic/stochastic reliability	Active	System reliability, cost and weight/risk	Multinomial TS with MC to Prune Pareto optimal set)	Multiple prioritized objectives were considered.
Liang & Lo (2010)	Series-parallel	Deterministic	Active	System reliability and cost/weight	MOVNS	Problem of Fyffe et al. (1968) and Salazar et al. (2006) were considered.
Suman (2003)	Series-parallel , life support system in a capsule	Deterministic	Active	System reliability, cost and weight	SMOSA, UMOSA, PSA, PDMOSA and WMOSA	
EbrahimNezhad et al. (2011)	Series-parallel	Deterministic	Active and Cold Standby	System reliability and profit	NSGA-II	Erlang time to failure
Safari (2012)	Series-parallel	Deterministic	Active and Cold Standby	System reliability and cost	NSGA-II	Erlang time to failure
Chambari et al. (2012)	Series-parallel	Deterministic	Active and Cold Standby	System reliability and cost	NSGA-II	Erlang time to failure
EbrahimNezhad et al. (2012)	Series-parallel	Deterministic	Active and Cold Standby	System reliability and profit	NSGA-II and Memetic algorithm	Erlang time to failure
Khalili-Damghani et al. (2013)	Series-parallel	Deterministic	Active	System reliability, cost and weight	DSAMOPSO, AUGMECON, NSGA-II, CTV-MOPSO	
Zhang et al. (2014)	Series-parallel	Deterministic	Active	System reliability, cost and weight	BBMOPSO followed by k-Means and Hierarchical clustering	Equivalent components
Azizmohammadi et al. (2013)	Series-parallel	Deterministic	Active and standby	System reliability (max-min), cost and volume	HMOICA (hybrid ICA and GA)	
Shelokar et al. (2002)	Series-parallel, complex, bridge	Deterministic	Active	System reliability and cost	Ant algorithm	

2.1.2 Non-deterministic models

Non-deterministic models assume that at least one of the design parameters is not known, precisely. Regarding the level of access to the historical data the uncertainty can be considered in the following forms.

2.1.2.1 Stochastic Uncertainty

In the stochastic uncertainty either the distribution of the data is specified or mean and standard deviation of the data are known. One way to formulate these problems is to maximize the expected value of system reliability (e.g. Rubinstein et al., 1997). However, this method is not suitable for designers who are risk-averse. Therefore, both expected value and variance of the system reliability (e.g. Marseguerra et al., 2005) or a function of both such as coefficient of variation (e.g. Tekiner & Coit, 2011) is taken into consideration. In case where both expected value and variance of the system reliability are considered the problem is treated as a multi-objective problem. Table 3 lists the RAP with respect to stochastic uncertainty.

Table 3
Stochastic uncertainty

Reference	System configuration	Uncertainty state	Strategy	System performance	Solution method	Other considerations
Coit & Smith (2002)	Series-parallel	Random scale parameter for weibull distribution	Active	Lower percentile of system time to failure	GA	The scale parameter doesn't change with time
Coit & Wattanapongsakorn (2004)	Series-parallel, bridge, fault tolerant distributed system	Stochastic component reliability	Active	System reliability and variance	Stochastic optimization	Risk is considered in decision making (risk-averse)
Yeh (2003)	Series-parallel, Network	Stochastic component reliability	Active	System cost	MCS-RSM	MCS is used to find the estimated reliability then BBD of RSM is used to find the reliability function
Marseguerra et al. (2005)	Network	Stochastic component reliability	Active	Expected and variance of system reliability	GA and MC	Multiple choices of component types
Tekiner & Coit (2011)	Series-parallel	Stochastic component reliability	Active	Coefficient of variation	Neighborhood search, and linear integer programming	With and without component mixing
Rubinstein et al. (1997)	Series-parallel	Stochastic component reliability	Active	Expected values of system reliability	Simulation and GA	Cost is a separable nonlinear function
Reddy et al. (2011)	Network	Stochastic component reliability	Active	System reliability	Simulation method	Identical components
Zhao & Liu (2003)	Series-parallel , complex(bridge),communication network, lifetime-support system, series	Random lifetime, stochastic cost	Active/ standby	Mean system-lifetime,α -system lifetime, or system reliability	Stochastic programming technique solved by stochastic simulation, NN and GA	There is only one type of elements available, switch reliability is perfect
Yadavalli et al. (2007)	Series-parallel	Resource Chance constraint	Active	System reliability	Branch and bound	Uniform, normal, log-normal distributions
Li & Hu (2008)	Series-parallel	Random lifetimes	Active/ standby	Survival function	Stochastic comparison	
Zafiropoulos & Dialynas (2004)	Power electronic device	Component failure rate uncertainty	-	System structure considering System reliability and cost	SA	Triangular and normal distributions. The failure rates are simulated using LHS method

2.1.2.2 Interval Uncertainty

In the interval uncertainty it is assumed that just the information about the lower and upper bounds of the interval is known. Yokota et al. (1995) are the first who formulated the optimal design problem of system reliability as a nonlinear integer programming problem with interval coefficients. They transformed the problem into a bi-criteria 0-1 nonlinear programming problem without interval coefficients, and solve it directly through GA. In general, there are some ways to formulate interval reliability problems, which are maximization of the left bound and/or center of the interval (Yokota et al., 1996; Taguchi et al., 1998), nonlinear goal programming formulation (Taguchi et al., 1997), etc. There are some methods to deal with such problems such as GA (Taguchi & Gen, 1997; Yokota et al., 1996), hybrid GA and SA (Taguchi et al., 1998), Improved GA (Taguchi & Yokota, 1999), etc. Table 4 lists the related works after year 2000.

Table 4(a)
Redundancy allocation under interval uncertainty (Single objective Problems)

Reference	System configuration	Uncertainty state	Stretegy	System performance	Solution method	Other considerations
Gupta et al. (2009)	Series-parallel	Interval component reliability	Active	System reliability	Advanced GA with interval fitness function	Interval valued mission time, identical components
Bhunia et al. (2010)	Series-parallel	Interval reliability and stochastic cost	Active	System reliability	Real-coded GA	Resource constraints are chance constraints
Sahoo et al. (2010)	Hierarchical series-parallel, parallel-series, complex, k-out-of-n	Interval reliability	Active	System reliability	GA	
Taguchi & Yokota (2011)	Series-parallel	Interval reliability	Active	System reliability	Hybrid GA, SA and FLC	
Sahoo et al. (2013)	Bridge	Interval reliability, cost and amount of resources	Active	System reliability	GA	Low level (component) redundancy and high level (system) redundancy are considered.

Table 4(b)
Redundancy allocation under interval uncertainty (Multi-objective Problems)

Reference	System configuration	Uncertainty state	Redundancy strategy	System performance	Solution method	Other considerations
Sahoo et al. (2012)	Series-parallel	Interval reliability and cost	Active	System reliability and cost	GA,Tchebycheff, weighted Tchebycheff and lexicographic weighted Tchebycheff	Identical components
Bhunia & Sahoo (2012)	Series-parallel	Interval reliability and cost	Active	System reliability and cost	GA, Global criterion method, Tchebycheff and weighted Tchebycheff	Identical components
Roy et al. (2014)	Series-parallel (break lining manufacturing plant)	Interval component reliability and cost and system entropy	Active	System reliability and cost with entropy constraint	Entropy based region reducing GA	System cost depends on reliability of the components

2.1.2.3 Fuzzy Uncertainty

The concept of fuzzy reliability has been introduced and formulated in the book by Cai (1996). One form of fuzzy uncertainty is having flexible goals and constraints (e.g. Park, 1987). Another form is having imprecise data for components reliabilities/lifetimes, cost, weight, etc. Cheng and Mon (1993) used the α-cut of a triangular fuzzy number to get the interval and find the fuzzy reliability of the series system and the parallel system.

Table 5
Fuzzy uncertainty

Reference	System configuration	Uncertainty state	Strategy	System performance	Solution method	Other considerations
Ravi et al. (2000)	Series-parallel	Fuzzy goals and constraints	Active	System reliability, cost, weight and volume	A global optimization meta-heuristics (Threshold accepting)	Linear membership functions have been assumed for all the goals/objectives.
Zhao & Liu (2005)	Series-parallel	Fuzzy lifetimes	Standby	Expected system lifetime, α-system lifetime, system reliability (EVM, CCP, DCP)	Integrated fuzzy simulation, NN and GA	
Hou & Wu (2006)	Series-parallel	Fuzzy reliability	Active	System reliability	Fuzzy simulation-based GA	Nonlinear chance constrained programing models and three goal programing models based on possibility and credibility measures
Mahapatra & ROY (2011)	Series-parallel	Fuzzy reliability, cost and weight	Active	System reliability	Fuzzy parametric geometric programming	
Yao et al. (2008)	Series, parallel	Triangular fuzzy numbers	Active	System reliability	signed distance method to defuzzify	
Lee et al. (2012)	Parallel	Level (λ,ρ) interval-valued fuzzy numbers	Active	System reliability	signed distance method to defuzzify	
Han et al. (2006)	Flexible Manufacturing Cells	Triangular fuzzy numbers	-	System reliability	Fuzzy fault tree	
Jameel & Radhi (2014)	Series-parallel	Fuzzy reliability and flexible constraints	Active	System reliability	Penalty function mixed with Nelder and Mend's algorithm	
			Multi-Objective			
Sasaki & Gen (2003)	Series-parallel	Fuzzy objectives	Active	System reliability and cost	Hybrid GA	With Generalized upper bounding (GUB) structure
Chen & Liu(2011)	Series-parallel	Type-2 fuzzy lifetime	Standby	System lifetime and cost	Fuzzy Goal programming and Approximation approach based PSO	Switching device of the standby system is perfect

Chen (1994) used fuzzy numbers to find the fuzzy reliability of both the series system and the parallel system. Mon and Cheng (1994) used the α -cut of fuzzy number to derive a non-linear program of the fuzzy system in both the series and parallel cases. Singer (1990) considered the fuzzy reliability of both series and parallel systems using an approximation of a fuzzy binary operation $\otimes$ with two L-R type fuzzy numbers. Chen (1996) presented a method for fuzzy system reliability analysis based on fuzzy time series and the α-cuts operations of fuzzy numbers. Hong et al. (1997) analyzed fuzzy system reliability by the use of *t*-norm based convolution of fuzzy arithmetic operation, where the reliability of each component is represented by L-R type fuzzy numbers. Table 5 lists other works after year 2000.

2.1.2.4 Intuitionistic fuzzy and vague sets

Kumar et al. (2011) extended the concept of fuzzy set by idea of triangular intuitionistic fuzzy sets (IFS) and proposed a general procedure to construct the membership function and non-membership function of the reliability function using intuitionistic fuzzy failure rate. The major advantage of using intuitionistic fuzzy sets over fuzzy sets is that intuitionistic fuzzy sets separate the positive and the negative evidence for the membership of an element in a set (Mahapatra & Roy, 2009).Table 6 lists the works related to intuitionistic fuzzy and vague sets.

Table 6
Intuitionistic fuzzy and vague sets

Reference	System configuration	Uncertainty state	Redundancy strategy	System performance	Solution method	Other considerations
Kumar et al. (2006)	Series, parallel, Marine power plant	Interval valued trapezoidal vague sets	Active	System reliability	A method for analyzing the fuzzy system reliability	An interval is considered for the grade of membership and non- membership of an element (interval valued vague set)
Chen (2003)	Series, parallel, series-parallel	Triangular vague set for components reliabilities	Active	System reliability	A method based on the vague set theory	A real number in [0, 1] is considered for the grade of membership and non-membership of an element
Kumar et al. (2007)	Series, parallel	LR type interval valued triangular vague set for component reliability	Active	System reliability	Tw (the weakest t norm) based arithmetic operation	The results are exact
Kumar & Yadav (2012)	Series, parallel	Intuitionistic fuzzy failure rate	Active	System reliability	Non-linear programming techniques	Different types of intuitionistic fuzzy failure rate are considered.
Mahapatra & Roy (2009)	Series, parallel, electric network model of dark room	Triangular intuitionistic fuzzy reliability	Active	System reliability	(α, β)-cut	
Pandey et al. (2011)	Series, parallel, web server	Triangular intuitionistic fuzzy reliability	Active	System reliability	A method based on IFS theory	
Mahapatra & Roy (2014)	Complex system	Intuitionistic fuzzy cost	Active	System reliability	Intuitionistic fuzzy optimization method	

2.1.2.5. Fuzzy-random Uncertainty

Due to subjective judgment, imprecise human knowledge and perception in capturing statistical data, the real data of lifetimes in many systems are both random and fuzzy in nature. Table 7 lists the related work to fuzzy-random uncertainty.

Table 7
Fuzzy-random uncertainty

Reference	System configuration	Uncertainty state	Redundancy strategy	System performance	Solution method	Other considerations
Zhao & Liu (2004)	Multistage systems and communication network systems	Random-fuzzy lifetimes	Standby	Expected system lifetime, (α,β)-system lifetime and system reliability	Integrated random fuzzy simulation, NN and GA	
Wang & Watada (2009)	Parallel-series	Random-fuzzy lifetimes	Active	System reliability	Hybrid binary PSO	
Wang & Watada (2009)	Parallel-series	Random-fuzzy lifetimes	Active	System reliability/cost (Mean chance measure)	Fuzzy random simulation and GA	Compute the reliability with convex and non-convex lifetimes
Wang et al. (2012)	Series-parallel (some subsystems with gradual failure (e.g. single helical gear reducer) and some with sudden failure)	Random-fuzzy lifetime	Active	System reliability	Saddlepoint Approximation (SAP)- simulation for gradual failure and statistics-based reliability analysis for components with sudden failure	There is a time-dependent degradation process
Nematian et al. (2008)	Series-parallel	Random-fuzzy lifetimes	Active /Standby	Expected value of system lifetime	Integer programming	

2.1.2.6 Robust optimization

Robust approaches are based on the idea that the resulted robust solution would be immune against different realizations of uncertain parameters. There are two main robustness measures: one based on regret (e.g. Feizollahi & Modarres, 2012; Soltani et al., 2013) and the other one based on the realized performance (Feizollahi et al., 2014; Soltani & Sadjadi, 2014). Table 8 lists the existing works in the literature. Amongst the works, just one work, i.e. the work by Soltani et al. (2013) considers cold standby strategy and more researches are needed with respect to redundancy strategy. The author has currently an under review work considering the redundancy strategy choices. Developing other robust optimization techniques is of interest and the author is currently working on this subject matter.

Table 8
Robust RAP

Reference	System configuration	Uncertainty state	Redundancy strategy	System performance	Solution method
Feizollahi & Modarres (2012)	Series- parallel	Interval uncertainty	Active	Min-Max regret of System reliability	MIP and Benders decomposition
Soltani et al. (2013)	Series- parallel	Interval uncertainty	Cold standby	Min-Max regret of System reliability	Benders decomposition, GA and Enumeration method
Soltani & Sadjadi (2014)	Series- parallel	Fuzzy uncertainty	Active	Expected value and variance of the system reliability	Branch and cut
Feizollahi et al. (2014)	Series- parallel	Budgeted uncertainty	Active	System reliability	MIP and Benders decomposition

2.1.2.7 Chaos uncertainty

Chaos means the oscillations, which seem random but in fact they are generated by the deterministic nonlinear model. The idea of chaos theory application in the reliability modeling appears in Zou and Li (2001). The integration of fuzzy reliability model with the phase portraits of variables creates preconditions for the construction of phase portrait reflecting the system reliability dynamics (Rotshtein et al., 2012). To the best of our knowledge, in the area of RAP just one work for chaos uncertainty has been reported as presented in Table 9 and more researches in this area would be interesting.

Table 9
Chaos uncertainty for RAP

Reference	System configuration	Uncertainty state	Redundancy strategy	System performance	Solution method	Other considerations
Rotshtein et al. (2012)	Series-parallel	Deterministic	Active	System reliability	GA	Integration of fuzzy logic and chaos theory

2.2 Multi-level redundancy (Modular redundancy)

As mentioned earlier, redundancy at the component level is more effective than redundancy at the system level. This is true under some specific assumptions but Boland and EL-Neweihi (1995) showed that this was not true in the case of redundancy with non-identical spare parts. Modular redundancy can be more effective than component redundancy. In other words, providing redundancy at high levels such as modules or subsystems can be more economical than providing redundancy at low level of components. Multi-level redundancy allocation problems (MRAP) have been addressed, considering the redundancy at the subsystem level (modular redundancy). Another form of this problem is multiple multi-level redundancy allocation problem (MMRAP) in which all available items for redundancy (system, module and component) can be simultaneously chosen. During the optimization process, the values of design variables expressing redundancy are interdependent, since they are hierarchical, i.e., the reliability of an upper level unit depends on the reliability of its lower level units. In the hierarchical structure of a reliability system, the system level is the top most level and the component level is the lowest. Subsystem or module levels are located between the top level and the bottom level. Table 10 lists the related works to multi-level or modular redundancy. To the best of our knowledge, just one work, i.e. the work by Pourdarvish and Ramezani (2013), considers cold standby redundancy. Therefore, more researches on redundancy strategy are needed. More researches are also required in the area of multi-objective multi-level redundancy. Non-deterministic multi-level redundancy is an interesting area too.

Table 10
Multi-level redundancy

Reference	System configuration	Uncertainty state	Redundancy strategy	System performance	Solution method	Other considerations
Yun & Kim (2004)	Multi-level series	Deterministic	Active	System reliability	Heuristic and GA	Only one unit can be used in a direct line
Yun et al. (2007)	Multiple multi-level series (hierarchical structure)	Deterministic	Active	System reliability	Simple GA (SGA)	If an item is used, all its sibling items should be used or its function should be satisfied by the corresponding child items.
Jang & Kim (2011)	Multiple multi-level	Deterministic	Active	System reliability	TS	TS of memory-based mechanisms that balances intensification with diversification via the short-term and long-term memory
Kumar et al. (2009a)	Multilevel series and series–parallel	Deterministic	Active	System reliability	Hierarchical GA	
Yeh (2009)	Multiple multi-level	Deterministic	Active	System reliability	Two stage discrete PSO (2DPSO)	Assumptions of Yun et al. (2007) hold.
Wang et al. (2010a)	Multilevel redundancy	Deterministic	Active	System reliability	Memetic algorithm with breadth-first crossover and breadth-first mutation	
Wang et al. (2010b)	Multilevel redundancy	Deterministic	Active	System reliability	PSO	
He et al (2013)	Multilevel redundancy	Deterministic	Active	System reliability	Hybrid GA based on the two dimensional redundancy encoding mechanism	
Pourdarvish & Ramezani (2013)	Multilevel redundancy	Deterministic	Cold standby	System reliability	MA and Hierarchical GA	
			Multi-objective Problems			
Kumar et al. (2009 b)	Multilevel redundancy	Deterministic	Active	System reliability and cost	MOHGA including SPEA2 and NSGA-II	

2. Reliability Allocation

The reliability allocation problems assume fix or known redundancy levels and they aim at determining the reliability levels. There are some methods for determining reliability levels such as Hooke and Jeeves (H-J) pattern search (Tillman et al., 1977), GAG2 (Gopal et al., 1980), to name a few. For more study on these methods, the readers can view the work by Tillman (1985). Table 11 lists the reliability allocation models existed in the literature.

Table 11(a)
Reliability allocation (Single Objective Problems)

Reference	System configuration	Uncertainty state	System performance	Solution method	Other considerations
Park (1987))	Two-component series	Deterministic	System reliability	Fuzzy non-linear programming	
Ravi et al. (2000)	Complex	Fuzzy goals and constraints	System cost	A global optimization meta-heuristics (Threshold accepting)	
Mahapatra & Mahapatra (2010)	Series	Fuzzy reliability and cost	System reliability/ System cost	Fuzzy parametric geometric programming	Objective and constraint are posynomial functions
Mahapatra & Roy (2012)	Bridge system	Triangular fuzzy number for reliability	System reliability	α -cut	
Elegbede et al. (2003)	Series-parallel	Deterministic	System cost	ECAY (exact)	
Yalaoui et al. (2005)	Parallel-series	Deterministic	System cost	Resolution method	Different technologies which have the same function are used in parallel, convex cost functions
Sriramdas et al. (2014)	Series (transceiver)	Trapezoidal fuzzy numbers	System reliability	Approximation method based on linear programming for trapezoidal fuzzy numbers	Centroid method of defuzzification is used to evaluate weighting factor

Table 11(b)
Reliability allocation (Multi Objective Problems)

Reference	System configuration	Uncertainty state	System performance	Solution method	Other considerations
Sakawa & Yano (1985)	Series (standby)	L-R type fuzzy numbers for DM's judgment	System reliability, Cost, weight, volume and the product of weight and volume	Fuzzy sequential proxy optimization technique (FSPOT)	Reliability of switch is one, the state of components are independent of each other
Huang et al. (2005)	Series-parallel	Deterministic	System reliability, Cost and weight	IIMOM and ANN	
Mahapatra & Roy (2006)	Series, complex	Generalized fuzzy set	System reliability, Cost	Fuzzy multi-objective mathematical programming	
Mahapatra & Roy (2006)	Series	Crisp/fuzzy cost and flexible constraints	System reliability/ Cost	Fuzzy parametric geometric programming	
Mahapatra et al. (2010)	LCD display system	Deterministic	System reliability and Cost	Intuitionistic fuzzy programming	

As observed, the problem of reliability allocation has gained less attention amongst researchers in comparison with RAP or RRAP. The uncertainty type considered in the literature for this kind of problem is of fuzzy uncertainty. Considering other forms of uncertainty can be an interesting research area.

3. Reliability- Redundancy Allocation

The reliability-redundancy problem was initially introduced by Misra and Ljubojevic (1973). They treated the number of redundancies as real variables and solved the problem as a reliability allocation problem by using Lagrange multipliers. **A** trial and error approach was used afterward to obtain the integer solution for the number of redundancies. The problem has been studied on four typical system configurations such as series, series-parallel, complex and over-speed protection system for a gas turbine. The aim is to determine reliability levels and redundancy levels to maximize system reliability subject to constraints on cost, weight and volume usually in nonlinear forms. The techniques for solving RRAP mainly combine the methods for reliability allocation and the methods for redundancy allocation. Tillman et al. (1977) were among the first to solve RRAP using *THK* method, which is based on a combination of Hooke-Jeeves (H-J) pattern search (Hooke & Jeeves, 1961) for reliability allocation and the Aggrawal-Gupta-Misra (A-G-M) greedy heuristic (Aggrawal et al., 1975) for redundancy allocation in a series-parallel system. Other combinations include H-J&GAG (Gopal et al., 1978), H-J&S-V (Sharama & Venkateswaran, 1971), H-J&N-N (Nakagawa & Nakashima, 1977), H-J&K-I (Kohda & Inoue, 1982), GAG2 (Gopal et al., 1980) and *GAG, GAG2&S-V, GAG2&NN, GAG2&K-I*, where the first heuristic finds reliability levels and the second heuristic finds redundancy levels. Besides, Xu et al. (1990) proposed XKL method, which uses a heuristic to find redundancy levels and an analytical approach to find reliability levels. Jacobson and Arora (1996) proposed a Simplex search for solving RRAP. Mathematical programming methods include *LMBB* method (Kuo et al., 1987) which combines the branch-and-bound and Lagrange multipliers methods, and surrogate constraint (Hikita et al., 1992). Meta-heuristic approaches include GA (Yokota et al., 1996; Hsieh et al., 1998) and the evolutionary algorithm developed by Prasad and Kuo (1997). After year 2000 with the advent of the new heuristic and meta-heuristics, methods such as Column Generation, IA, SA, PSO, HS, ICA, cuckoo search separately or in hybrid forms have been implemented to this problem. Regarding the method for multi-objective reliability-redundancy problems before year 2000 we can refer to goal programming and goal attainment formulations (e.g. Dhingra, 1992; Rao & Dhingra, 1992). Table 12 summarizes the existing work after year 2000. Note that the table is sorted based on the solution method. To the best of our knowledge, amongst the paper investigated, just two works have studied cold standby strategy (Ardakan & Hamadani, 2014; Liu, 2006) and a few works are in non-deterministic area. Considering other redundancy strategies and other forms of uncertainties can be a fruitful area of research.

Table 12(a)
Reliability- Redundancy Allocation (Single Objective Problems)

Reference	System configuration	Uncertainty state	Redundancy strategy	System performance	Solution method
Sun et al. (2001)	Bridge, network	Deterministic	Active	System reliability/cost	Convexification method (exact)
Lee et al. (2002a), (2002b)		Deterministic	Active	System reliability	Hybrid NN-GA with FLC
Ha & Kuo (2005)	Series, bridge	Deterministic	Active	System reliability	Scaling method based on multi-path iterative heuristic
Liu (2006)	Series-parallel, k-out-of-n:G	Deterministic	Active/standby	System reliability/cost	A combination method including Hooke–Jeeves pattern search and dynamic programming, applying dominating sequences
Kim et al. (2006)	Series, series-parallel, complex (bridge)	Deterministic	Active	System reliability	SA
Mori et al. (2007)	Series-parallel	Deterministic	Active	System reliability	Hybrid SA-GA
Chen (2006)	Series, series-parallel, complex and over-speed protection system	Deterministic	Active	System reliability	IA
Hsieh & You (2011)	Series, series-parallel, complex and over-speed protection system	Deterministic	Active	System reliability	IA based two-phase
Coelho (2009)	Over-speed protection system for a gas turbine	Deterministic	Active	System reliability	DE based on chaotic sequences using Lozi's map
Coelho et al. (2009)	Over-speed protection system for a gas turbine	Deterministic	Active	System reliability	HS with DE
Zou et al. (2010)	Complex (bridge), Over-speed protection system for a gas turbine, large scale system	Deterministic	Active	System reliability	NGHS
Zou et al. (2011)	Complex (bridge), Over-speed protection system for a gas turbine	Deterministic	Active	System reliability	EGHS (hybrid HS-PSO)
Wang & Li (2012)	Series, series-parallel, complex and over-speed protection system	Deterministic	Active	System reliability	DE and HS
Yeh & Hsieh(2011)	Series, series-parallel, complex and over-speed protection system	Deterministic	Active	System reliability	ABC
Garg et al. (2013)	Series, series-parallel, complex, over-speed protection system, Life support system in a space capsule	Deterministic	Active	System reliability/system cost	Two phase algorithm including ABC
Afonso et al. (2013)	Series, series-parallel, complex and over-speed protection system	Deterministic	Active	System reliability	Modified ICA
Coelho (2009b)	Series, series-parallel and over-speed protection system	Deterministic	Active	System reliability	SOMA with a Gaussian operator
Zia & Coit (2010)	Series-parallel	Deterministic	Active	System reliability	Column Generation
Coelho (2009)	Complex (bridge), over-speed protection system	Deterministic	Active	System reliability	PSO-GC
Wu et al. (2011)	Series, series-parallel, complex, over-speed protection system, large scale system	Deterministic	Active	System reliability	Improved PSO (IPSO)
Garg & Sharma (2013)	Pharmaceutical Plant	Deterministic	Active	System reliability	PSO
Tan et al. (2013)	series, series-parallel, complex (bridge), Over-speed protection	Deterministic	Active	System reliability	DE-PSO with chaotic local search
Zhang et al. (2013)	series, series-parallel, complex (bridge), Over-speed protection, large scale	Deterministic	Active	System reliability	Hybrid HS/ LXPM –IPSO with Golden search (GS)
Valian & Valian (2013)	series, series-parallel, complex (bridge), Over-speed protection, large scale system	Deterministic	Active	System reliability	Cuckoo search by Lévy flights
Valian et al. (2013)	series, series-parallel, complex (bridge), Over-speed protection, large scale system	Deterministic	Active	System reliability	Improved cuckoo search
Kanagaraj et al. (2013)	series, series-parallel, complex (bridge), Over-speed protection	Deterministic	Active	System reliability	Hybrid cuckoo search-GA
Ardakan & Hamadani (2014)	series, series-parallel, complex (bridge)	Deterministic	Cold standby	System reliability	Modified GA
Mahato et al. (2012)	Complex bridge system, over speed protection system for a gas turbine	Interval reliability	Active	System reliability	GA

Table 12(b)
Reliability- Redundancy Allocation (Multi Objective Problems)

Reference	System configuration	Uncertainty state	Redundancy strategy	System performance	Solution method
Huang (2004)	Over-speed protection system for a gas turbine	Deterministic	Active	Reliability, cost and weight	Interactive physical programming
Ebrahimipour & Sheikhalishahi (2012)	Series-bridge	Triangular fuzzy reliability	Active	System reliability and cost considering price discounting	MOPSO
Sheikhalishahi et al. (2013)	series, series–parallel, and complex (bridge)	Deterministic	Active	System reliability, cost, weight and volume	hybrid GA-PSO
Mutingi & Mapfaira (2013)	Series	Fuzzy goals and constraints	Active	System reliability and cost	Fuzzy multi-objective GA
Garg & Sharma (2013)	Series–parallel (pharmaceutical plant)	Intuitionistic fuzzy reliability and flexible constraints	Active	System reliability and cost	MOPSO
Garg et al. (2014)	Series-parallel, Complex Bridge, Over-speed protection	Triangular fuzzy reliability	Active	System reliability and cost	PSO
Garg et al. (2014)	Series-parallel, Complex	Intuitionistic fuzzy reliability	Active	System reliability and cost	Intuitionistic fuzzy programming technique and PSO

4. Concluding remarks

In the present research, three types of reliability optimization methods including redundancy allocation, reliability allocation and reliability-redundancy allocation have been investigated from different perspectives, which have been collected in a table format. Since the area of the reliability optimization problems is vast and so many works have been carried out in this area, the focus of this study has been on binary state non-repairable systems and the problems have been categorized with respect to different criteria. Scholars and researchers who wish to study on each of the mentioned area, need to flashback to the initial works and see the changing trend of the subjects from the scratch so that they can perceive the need for developing new models and methods in reliability. For this purpose, for each area the author has tried to have a quick review on early works and for each classification some of the prominent works in the literature have been introduced so that the interested readers can refer them and find other related papers to study in depth. This study has classified the literature in a novel way and at the end of each classification research guides for further researches have been proposed to help researchers and seems like the same is needed for multistate and/or repairable systems.

References

Abouei Ardakan, M., & Zeinal Hamadani, A. (2014). Reliability–redundancy allocation problem with cold-standby redundancy strategy. *Simulation Modelling Practice and Theory*, *42*, 107-118.

Afonso, L. D., Mariani, V. C., & dos Santos Coelho, L. (2013). Modified imperialist competitive algorithm based on attraction and repulsion concepts for reliability-redundancy optimization. *Expert Systems with Applications*, *40*(9), 3794-3802.

Aggarwal, K. K., Gupta, J. S., & Misra, K. B. (1975). A new heuristic criterion for solving a redundancy optimization problem. *Reliability, IEEE Transactions on*, *24*(1), 86-87.

Ahmadizar, F., & Soltanpanah, H. (2011). Reliability optimization of a series system with multiple-choice and budget constraints using an efficient ant colony approach. *Expert systems with Applications*, *38*(4), 3640-3646.

Amari, S. V., & Dill, G. (2010, January). Redundancy optimization problem with warm-standby redundancy. In *Reliability and Maintainability Symposium (RAMS), 2010 Proceedings-Annual* (pp. 1-6). IEEE.

Azizmohammadi, R., Amiri, M., Tavakkoli-Moghaddam, R., & Mohammadi, M. (2013). Solving a Redundancy Allocation Problem by a Hybrid Multi-objective Imperialist Competitive Algorithm. *International Journal of Engineering (1025-2495)*, *26*(9).

Barlow, R. E., & Proschan, F. (1975). *Statistical theory of reliability and life testing: probability models*. Florida State University Tallahassee.

Beji, N., Jarboui, B., Eddaly, M., & Chabchoub, H. (2010). A hybrid particle swarm optimization algorithm for the redundancy allocation problem. *Journal of Computational Science*, *1*(3), 159-167.

Bellman, R., & Dreyfus, S. (1958). Dynamic programming and the reliability of multicomponent devices. *Operations Research*, *6*(2), 200-206.

Bhunia, A. K., Sahoo, L., & Roy, D. (2010). Reliability stochastic optimization for a series system with interval component reliability via genetic algorithm.*Applied Mathematics and Computation*, *216*(3), 929-939.

Bhunia, A. K., & Sahoo, L. (2012). Optimization of Constrained Multi-objective Reliability Problems with Interval Valued Reliability of Components via Genetic Algorithm. *Indian Journal of Industrial and Applied Mathematics*, *3*(1s), 25-44.

Billionnet, A. (2008). Redundancy allocation for series-parallel systems using integer linear programming. *Reliability, IEEE Transactions on*, *57*(3), 507-516.

Boland, P. J., & El-Neweihi, E. (1995). Component redundancy vs system redundancy in the hazard rate ordering. *Reliability, IEEE Transactions on*,*44*(4), 614-619.

Bulfin, R. L., & Liu, C. Y. (1985). Optimal allocation of redundant components for large systems. *Reliability, IEEE Transactions on*, *34*(3), 241-247.

Cai, K. Y. (1996). *Introduction to fuzzy reliability*. Springer.

Cao, D., Murat, A., & Chinnam, R. B. (2013). Efficient exact optimization of multi-objective redundancy allocation problems in series-parallel systems.*Reliability Engineering & System Safety*, *111*, 154-163.

Chambari, A., Rahmati, S. H. A., & Najafi, A. A. (2012). A bi-objective model to optimize reliability and cost of system with a choice of redundancy strategies.*Computers & Industrial Engineering*, *63*(1), 109-119.

Chambari, A., Najafi, A. A., Rahmati, S. H. A., & Karimi, A. (2013). An efficient simulated annealing algorithm for the redundancy allocation problem with a choice of redundancy strategies. *Reliability Engineering & System Safety*, *119*, 158-164.

Chen, S. M. (1996). New method for fuzzy system reliability analysis.*Cybernetics & Systems*, *27*(4), 385-401.

Chen, S. M. (2003). Analyzing fuzzy system reliability using vague set theory.*International Journal of Applied Science and Engineering*, *1*(1), 82-88.

Chen, S. M. (1994). Fuzzy system reliability analysis using fuzzy number arithmetic operations. *Fuzzy Sets and Systems*, *64*(1), 31-38.

Chen, T. C., & You, P. S. (2005). Immune algorithms-based approach for redundant reliability problems with multiple component choices. *Computers in Industry*, *56*(2), 195-205.

Chen, Y., & Liu, Y. (2011). Standby redundancy optimization with type-2 fuzzy lifetimes. In *Advances in Swarm Intelligence* (pp. 329-337). Springer Berlin Heidelberg.

Chen, T. C. (2006). IAs based approach for reliability redundancy allocation problems. *Applied Mathematics and Computation*, *182*(2), 1556-1567.

Cheng, C. H., & Mon, D. L. (1993). Fuzzy system reliability analysis by interval of confidence. *Fuzzy Sets and Systems*, *56*(1), 29-35.

Chern, M. S. (1992). On the computational complexity of reliability redundancy allocation in a series system. *Operations Research Letters*, *11*(5), 309-315.

Coelho, L. D. S., de Andrade Bernert, D. L., & Mariani, V. C. (2009, October). A harmony search algorithm combined with differential operator applied to reliability-redundancy optimization. In *Systems, Man and Cybernetics, 2009. SMC 2009. IEEE International Conference on* (pp. 5215-5219). IEEE.

Coelho, L. D. S. (2009). An efficient particle swarm approach for mixed-integer programming in reliability–redundancy optimization applications. *Reliability Engineering & System Safety*, *94*(4), 830-837.

Coelho, L. D. S. (2009). Reliability–redundancy optimization by means of a chaotic differential evolution approach. *Chaos, Solitons & Fractals*, *41*(2), 594-602.

Coelho, L. D. S. (2009). Self-organizing migrating strategies applied to reliability-redundancy optimization of systems. *Reliability, IEEE Transactions on*, *58*(3), 501-510.

Coit, D. W., & Smith, A. E. (1996). Reliability optimization of series-parallel systems using a genetic algorithm. *Reliability, IEEE Transactions on*, *45*(2), 254-260.

Coit, D. W., & LIU, J. C. (2000). System reliability optimization with k-out-of-n subsystems. *International Journal of Reliability, Quality and Safety Engineering*,*7*(02), 129-142.

Coit, D. W., & Smith, A. E. (2002). Genetic algorithm to maximize a lower-bound for system time-to-failure with uncertain component Weibull parameters.*Computers & Industrial Engineering*, *41*(4), 423-440.

Coit, D. W., & Smith, A. E. (1996). Reliability optimization of series-parallel systems using a genetic algorithm. *Reliability, IEEE Transactions on*, *45*(2), 254-260.

Coit, D. W., & Smith, A. E. (1996). Solving the redundancy allocation problem using a combined neural network/genetic algorithm approach. *Computers & Operations Research*, *23*(6), 515-526.

Coit, D. W., & Konak, A. (2006). Multiple weighted objectives heuristic for the redundancy allocation problem. *Reliability, IEEE Transactions on*, *55*(3), 551-558.

Coit, D. W. (2001). Cold-standby redundancy optimization for nonrepairable systems. *IIE Transactions*, *33*(6), 471-478.

Coit, D. W. (2003). Maximization of system reliability with a choice of redundancy strategies. *IIE transactions*, *35*(6), 535-543.

Coit, D. W., & Smith, A. E. (1996). Penalty guided genetic search for reliability design optimization. *Computers & industrial engineering*, *30*(4), 895-904.

Coit, D. W., Jin, T., & Wattanapongsakorn, N. (2004). System optimization with component reliability estimation uncertainty: a multi-criteria approach. *Reliability, IEEE Transactions on*, *53*(3), 369-380.

Dhingra, A. K. (1992). Optimal apportionment of reliability and redundancy in series systems under multiple objectives. *Reliability, IEEE Transactions on*,*41*(4), 576-582.

Dinghua, S., (1987). A new heuristic algorithm for constrained redundancy-optimization in complex systems, *IEEE Trans* on Reliability, 36, 621-623.

Djerdjour, M., & Rekab, K. (2001). A branch and bound algorithm for designing reliable systems at a minimum cost. *Applied Mathematics and Computation*,*118*(2), 247-259.

EbrahimNezhad Moghadam Rashti, M., Vishkaei, B. M., & Esmaeilpour, R. (2012). A Memtic genetic algorithm for a redundancy allocation problem.*International Journal of Applied Operational Research-An Open Access Journal*,*2*(2), 49-57.

EbrahimNezhad, M., Vishkaei, B. M., Pasandideh, H. R., & Safari, J. (2011). Increasing the Reliability and the Profit in a Redundancy Allocation Problem. *International Journal of Applied Operational Research*, *1*(2), 57-64.

Ebrahimipour, V., Haeri, A., Sheikhalishahi, M., Asadzadeh, S. M. (2012) Application of Multi-Objective Particle Swarm Optimization to Solve a Fuzzy Multi-Objective Reliability Redundancy Allocation Problem. Journal of Safety Engineering, 1(2), 26-38.

Elegbede, A. C., Chu, C., Adjallah, K. H., & Yalaoui, F. (2003). Reliability allocation through cost minimization. *Reliability, IEEE Transactions on*, *52*(1), 106-111.

Federowicz, A. J., & Mazumdar, M. (1968). Use of geometric programming to maximize reliability achieved by redundancy. *Operations Research*, *16*(5), 948-954.

Feizollahi, M. J., & Modarres, M. (2012). The robust deviation redundancy allocation problem with interval component reliabilities. *Reliability, IEEE Transactions on*, *61*(4), 957-965.

Feizollahi, M. J., Ahmed, S., & Modarres, M. (2014)The Robust Redundancy Allocation Problem in Series-Parallel Systems With Budgeted Uncertainty. *Reliability, IEEE Transactions on*, 8(1), 239-250.

Fyffe, D. E., Hines, W. W., & Lee, N. K. (1968). System reliability allocation and a computational algorithm, *IEEE Transactions on Reliability*, R-17, 64-69.

Garg, H., Rani, M., & Sharma, S. P. (2013). An efficient two phase approach for solving reliability–redundancy allocation problem using artificial bee colony technique. *Computers & Operations Research*, *40*(12), 2961-2969.

Garg, H. A. R. I. S. H., & Sharma, S. P. (2013). Reliability redundancy allocation problem of pharmaceutical plant. *Journal of Engineering Science and Technology*, *8*(2), 190-198.

Garg, H., & Sharma, S. P. (2013). Multi-objective reliability-redundancy allocation problem using particle swarm optimization. *Computers & Industrial Engineering*, *64*(1), 247-255.

Garg, H., Rani, M., Sharma, S. P., & Vishwakarma, Y. (2014). Bi-objective optimization of the reliability-redundancy allocation problem for series-parallel system. *Journal of Manufacturing Systems*, *In Press,* Available online.

Garg, H., Rani, M., Sharma, S. P., & Vishwakarma, Y. (2014). Intuitionistic fuzzy optimization technique for solving multi-objective reliability optimization problems in interval environment. *Expert Systems with Applications*, *41*(7), 3157-3167.

Ghare, P.M. and Taylor, R.E. (1969) Optimal redundancy for reliability in series system. *Operations Research*, **17**, 838–847.

Gopal, K., Aggarwal, K. K., & Gupta, J. S. (1978). An improved algorithm for reliability optimization. *Reliability, IEEE Transactions on*, *27*(5), 325-328.

Gopal, K., Aggarwal K. K., & Gupta J. S. (1980). A new method for solving reliability optimization problem, *IEEE Transactions on Reliability*, R-29, 36-38.

Govil, K. K. (1983). Geometric programming method for optimal reliability allocation for a series system subject to cost constraint. *Microelectronics Reliability*, *23*(5), 783-784.

Govil, K. K., & Agarwala, R. A. (1983). Lagrange multiplier method for optimal reliability allocation in a series system. *Reliability Engineering*, *6*(3), 181-190.

Gupta, R. K., Bhunia, A. K., & Roy, D. (2009). A GA based penalty function technique for solving constrained redundancy allocation problem of series system with interval valued reliability of components. *Journal of Computational and Applied Mathematics*, *232*(2), 275-284.

Ha, C. (2004). *Reliability-yield allocation for semiconductor integrated circuits: modeling and optimization* (Doctoral dissertation, Texas A&M University).

Ha, C., & Kuo, W. (2005). Multi-path approach for reliability-redundancy allocation using a scaling method. *Journal of Heuristics*, *11*(3), 201-217.

Han, B. T., Zhang, C. B., Sun, C. S., & Xu, C. J. (2006). Reliability analysis of flexible manufacturing cells based on triangular fuzzy number. *Communications in Statistics—Theory and Methods*, *35*(10), 1897-1907.

He, P., Wu, K., Xu, J., Wen, J., & Jiang, Z. (2013). Multilevel redundancy allocation using two dimensional arrays encoding and hybrid genetic algorithm.*Computers & Industrial Engineering*, *64*(1), 69-83.

Hikita, M., Nakagawa, Y., Nakashima, K., & Narihisa, H. (1992). Reliability optimization of systems by a surrogate-constraints algorithm. *Reliability, IEEE Transactions on*, *41*(3), 473-480.

Hou, F., & Wu, Q. (2006). Genetic algorithm-based redundancy optimization problems in fuzzy framework. *Communications in Statistics—Theory and Methods*, *35*(10), 1931-1941.

Hong, D. H., & Do, H. Y. (1997). Fuzzy system reliability analysis by the use of Tw (the weakest t-norm) on fuzzy number arithmetic operations. *Fuzzy Sets and Systems*, *90*(3), 307-316.

Hooke, R., & Jeeves, T. A. (1961). ``Direct Search"Solution of Numerical and Statistical Problems. *Journal of the ACM (JACM)*, *8*(2), 212-229.

Hsieh, T. J., & Yeh, W. C. (2012). Penalty guided bees search for redundancy allocation problems with a mix of components in series–parallel systems.*Computers & Operations Research*, *39*(11), 2688-2704.

Hsieh, Y. C., Chen, T. C., & Bricker, D. L. (1998). Genetic algorithms for reliability design problems. *Microelectronics Reliability*, *38*(10), 1599-1605.

Hsieh, Y. C. (2003). A linear approximation for redundant reliability problems with multiple component choices. *Computers & industrial engineering*, *44*(1), 91-103.

Hsieh, Y. C., & You, P. S. (2011). An effective immune based two-phase approach for the optimal reliability–redundancy allocation problem. *Applied Mathematics and Computation*, *218*(4), 1297-1307.

Hsieh, Y. C. (2002). A two-phase linear programming approach for redundancy allocation problems. *The Yugoslav Journal of Operations Research ISSN: 0354-0243 EISSN: 2334-6043*, *12*(2).

Huang, H. Z., TIAN, Z. G., & GU, Y. K. (2004). Reliability and redundancy apportionment optimization using interactive physical programming. *International Journal of Reliability, Quality And Safety Engineering*, *11*(03), 213-222.

Huang, H. Z., Tian, Z., & Zuo, M. J. (2005). Intelligent interactive multiobjective optimization method and its application to reliability optimization. *IIE Transactions*, *37*(11), 983-993.

Jacobson, D. W., & Arora, S. R. (1996, January). Simultaneous allocation of reliability & redundancy using simplex search. In *Reliability and Maintainability Symposium, 1996 Proceedings. International Symposium on Product Quality and Integrity., Annual* (pp. 243-250). IEEE.

Jameel, A. F., Radhi, A. Z. (2014). Solution of Fuzzy Redundancy Reliability Optimization Problem by Modified Penalty Function Method. International Journal of Modern Mathematical Sciences, 10(1), 60-74.

Jang, K. W., & Kim, J. H. (2011). A tabu search for multiple multi-level redundancy allocation problem in series-parallel systems. *International Journal of Industrial Engineering*, *18*(3).

Jianping, L. (1996). A bound heuristic algorithm for solving reliability redundancy optimization. *Microelectronics Reliability*, *36*(3), 335-339.

Jianping, L. I., & Xishen, J. (1997). A new partial bound enumeration technique for solving reliability redundancy optimization. *Microelectronics Reliability*, *37*(2), 237-242.

Kanagaraj, G., Ponnambalam, S. G., & Jawahar, N. (2013). A hybrid cuckoo search and genetic algorithm for reliability–redundancy allocation problems. *Computers & Industrial Engineering*, *66*(4), 1115-1124.

Khalili-Damghani, K., Abtahi, A. R., & Tavana, M. (2013). A Decision Support System for Solving Multi-Objective Redundancy Allocation Problems. *Quality and Reliability Engineering International*.

Khalili-Damghani, K., & Amiri, M. (2012). Solving binary-state multi-objective reliability redundancy allocation series-parallel problem using efficient epsilon-constraint, multi-start partial bound enumeration algorithm, and DEA. *Reliability Engineering & System Safety*, *103*, 35-44.

Khalili-Damghani, K., Abtahi, A. R., & Tavana, M. (2013). A new multi-objective particle swarm optimization method for solving reliability redundancy allocation problems. *Reliability Engineering & System Safety*, *111*, 58-75.

Kim, J. H., & Yum, B. J. (1993). A heuristic method for solving redundancy optimization problems in complex systems. *Reliability, IEEE Transactions on*,*42*(4), 572-578.

Kim, H. G., Bae, C. O., & Park, D. J. (2006). Reliability-redundancy optimization using simulated annealing algorithms. *Journal of Quality in Maintenance Engineering*, *12*(4), 354-363.

Kim, H., Bae, C., and Park, S., (2004). Simulated Annealing Algorithm for Redundancy Optimization with Multiple Component Choices, *In Advanced Reliability Modeling, Proceedings of the Asian International Workshop, World Scientific*, 237–244.

Kim, K. O., & Kuo, W. (2003). Percentile life and reliability as performance measures in optimal system design. *IIE Transactions*, *35*(12), 1133-1142.

Kohda, T., & Inoue K. (1982). A reliability optimization method for complex system with the criterion local optimality, *IEEE Transactions on Reliability*, R-31, 109-111.

Kulturel-Konak, S., Coit, D. W., & Baheranwala, F. (2008). Pruned Pareto-optimal sets for the system redundancy allocation problem based on multiple prioritized objectives. *Journal of Heuristics*, *14*(4), 335-357.

Kulturel-Konak, S., Smith, A. E., & Coit, D. W. (2003). Efficiently solving the redundancy allocation problem using tabu search. *IIE transactions*, *35*(6), 515-526.

Kumar, R., Izui, K., Yoshimura, M., & Nishiwaki, S. (2009). Optimal multilevel redundancy allocation in series and series–parallel systems. *Computers & Industrial Engineering*, *57*(1), 169-180.

Kumar, R., Izui, K., Yoshimura, M., & Nishiwaki, S. (2009). Multi-objective hierarchical genetic algorithms for multilevel redundancy allocation optimization.*Reliability Engineering & System Safety*, *94*(4), 891-904.

Kumar, A., Yadav, S. P., & Kumar, S. (2006). Fuzzy reliability of a marine power plant using interval valued vague sets. *International Journal of Applied Science and Engineering*, *4*(1), 71-82.

Kumar, A., Yadav, S. P., & Kumar, S. (2007). Fuzzy system reliability analysis usingTw (the weakest t -norm) based arithmetic operations on L–R type interval valued vague sets. *International Journal of Quality & Reliability Management*, *24*(8), 846-860.

Kumar, M., Yadav, S. P., & Kumar, S. (2011). A new approach for analysing the fuzzy system reliability using intuitionistic fuzzy number. *International Journal of Industrial and Systems Engineering*, *8*(2), 135-156.

Kumar, M., & Yadav, S. P. (2012). A novel approach for analyzing fuzzy system reliability using different types of intuitionistic fuzzy failure rates of components.*ISA transactions*, *51*(2), 288-297.

Kuo, W., & Wan, R. (2007). Recent advances in optimal reliability allocation. In *Computational Intelligence in Reliability Engineering* (pp. 1-36). Springer Berlin Heidelberg.

Kuo, W., & Prasad, V. R. (2000). An annotated overview of system-reliability optimization. *Reliability, IEEE Transactions on*, *49*(2), 176-187.

Kuo, W., Lin, H. H., Xu, Z., & Zhang, W. (1987). Reliability optimization with the Lagrange-multiplier and branch-and-bound technique. *Reliability, IEEE Transactions on*, *36*(5), 624-630.

Kuo, W., & Prasad, R. (2004). System Reliability Optimization: An Overview. In *Mathematical Reliability: An Expository Perspective* (pp. 31-54). Springer US.

Lee, C.Y., Gen M., & Kuo W. (2002a). Reliability optimization design using hybridized genetic algorithm with a neural network technique. *IEICE Transactions on Fundamentals of Electronics, Communications and Computer Sciences*: E84-A: 627-637.

Lee, C.Y., Gen, M., & Tsujimura Y. (2002b). Reliability optimization design using hybrid NN-GA with fuzzy logic controller. *IEICE Transactions on Fundamentals of Electronics, Communications and Computer Sciences*, *85*(2), 432-446.

Lee, H., Kuo, W., & Ha, C. (2003). Comparison of max-min approach and NN method for reliability optimization of series-parallel system. *Journal of Systems Science and Systems Engineering*, *12*(1), 39-48.

Lee, H.-M., Fuh, C.-F., & Su J.-S., (2012). Fuzzy parallel system reliability analysis based on level (λ, ρ) interval-valued fuzzy numbers. *International Journal of Innovative Computing, Information and Control*, 8(8), 5703-5713.

Li, X., & Hu, X. (2008). Some new stochastic comparisons for redundancy allocations in series and parallel systems. *Statistics & Probability Letters*,*78*(18), 3388-3394.

Li, X., & Ding, W. (2013). On Allocation of Active Redundancies to Systems: A Brief Review. In *Stochastic Orders in Reliability and Risk* (pp. 235-254). Springer New York.

Liang, Y. C., & Lo, M. H. (2010). Multi-objective redundancy allocation optimization using a variable neighborhood search algorithm. *Journal of Heuristics*, *16*(3), 511-535.

Liang, Y. C., & Smith, A. E. (2004). An ant colony optimization algorithm for the redundancy allocation problem (RAP). *Reliability, IEEE Transactions on*, *53*(3), 417-423.

Liang, Y.C., & Wu., C.C., (2005). A Variable Neighborhood Descent Algorithm for the Redundancy Allocation Problem. *Ind Eng Manage Syst* , 4(1): 109– 116.

Liang, Y. C., & Chen, Y. C. (2007). Redundancy allocation of series-parallel systems using a variable neighborhood search algorithm. *Reliability Engineering & System Safety*, *92*(3), 323-331.

Liang, Y. C., & Smith, A. E. (1999). An ant system approach to redundancy allocation. In *Evolutionary Computation, 1999. CEC 99. Proceedings of the 1999 Congress on* (Vol. 2). IEEE.

Liu, G. S. (2006). A combination method for reliability-redundancy optimization.*Engineering Optimization*, *38*(04), 485-499.

Mahapatra, G. S., & Roy, T. K. (2011). Optimal Redundancy Allocation in Series-Parallel System using Generalized Fuzzy Number. *Tamsui Oxford Journal of Information and Mathematical Sciences*, *27*(1), 1-20.

Mahapatra, G. S., & Roy, T. K. (2009). Reliability Evaluation using Triangular Intuitionistic Fuzzy Numbers Arithmetic Operations. *Proceedings of World Academy of Science: Engineering & Technology, 50*.

Mahapatra, B. S., & Mahapatra, G. S. (2010). Reliability and cost analysis of series system models using fuzzy parametric geometric programming. *Fuzzy Information and Engineering, 2*(4), 399-411.

Mahapatra G. S., & Roy T. K., (2006). Optimal Fuzzy Reliability for a series system with Cost Constraint using Fuzzy Geometric Programming, *Tamsui Oxford Journal of Management Sciences*, 22(2): 53-63.

Mahapatra, G. S., & Roy, T. K. (2006). Fuzzy multi-objective mathematical programming on reliability optimization model. *Applied Mathematics and Computation, 174*(1), 643-659.

Mahapatra, G. S., & Roy, T. K. (2012). Reliability evaluation of bridge system with fuzzy reliability of components using interval nonlinear programming. *Electronic Journal of Applied Statistical Analysis, 5*(2) : 151 – 163.

Mahapatra, G. S., Mitra, M., & Roy, T. K. (2010). Intuitionistic fuzzy multi-objective mathematical programming on reliability optimization model.*International Journal of Fuzzy Systems, 12*(3), 259-266.

Mahapatra, G. S. (2009). Reliability Optimization of Entropy Based Series-Parallel System Using Global Criterion Method. *Intelligent Information Management, 1*(3), 145-149.

Mahapatra, G. S., & Roy, T. K. (2012). Reliability optimization of complex system using intiotionistic fuzzy optimization technique. International journal of industrial and systems engineering, 16(3), 279-295.

Mahato, S. K., Sahoo, L., & Bhunia, A. K. (2012). Reliability-redundancy optimization problem with interval valued reliabilities of components via genetic algorithm. *J Inf Comput Sci, 7*(4), 284-295.

Marseguerra, M., Zio, E., Podofillini, L., & Coit, D. W. (2005). Optimal design of reliable network systems in presence of uncertainty. *Reliability, IEEE Transactions on, 54*(2), 243-253.

Misra, K. B., & Sharma, U. (1991). An efficient algorithm to solve integer-programming problems arising in system-reliability design. *Reliability, IEEE Transactions on, 40*(1), 81-91.

Misra, K. B., & Sharma, J. (1973). A new geometric programming formulation for a reliability problem†. *International Journal of Control, 18*(3), 497-503.

Misra, K. B. (1971). Dynamic programming formulation of the redundancy allocation problem. *International Journal of Mathamatical Education in Science and Technology, 2*(3), 207-215.

Misra, K. B., & Ljubojevic, M. D. (1973). Optimal reliability design of a system: a new look. *Reliability, IEEE Transactions on, 22*(5), 255-258.

Mon, D. L., & Cheng, C. H. (1994). Fuzzy system reliability analysis for components with different membership functions. *Fuzzy Sets and Systems,64*(2), 145-157.

Mori, B. D., de Castro, H. F., & Cavalca, K. L. (2007). Development of hybrid algorithm based on simulated annealing and genetic algorithm to reliability redundancy optimization. *International Journal of Quality & Reliability Management, 24*(9), 972-987.

Nahas, N., Nourelfath, M., & Ait-Kadi, D. (2007). Coupling ant colony and the degraded ceiling algorithm for the redundancy allocation problem of series–parallel systems. *Reliability Engineering & System Safety, 92*(2), 211-222.

Nahas, N., & Nourelfath, M. (2005). Ant system for reliability optimization of a series system with multiple-choice and budget constraints. *Reliability Engineering & System Safety, 87*(1), 1-12.

Nahas, N., & Thien-My, D. (2010). Harmony search algorithm: application to the redundancy optimization problem. *Engineering Optimization, 42*(9), 845-861.

Najafi, A. A., Karimi, H., Chambari, A., & Azimi, F. (2013). Two metaheuristics for solving the reliability redundancy allocation problem to maximize mean time to failure of a series–parallel system. *Scientia Iranica, 20*(3), 832-838.

Nakagawa, Y., & Nakashima, K. (1977). A heuristic method for determining optimal reliability allocation. *Reliability, IEEE Transactions on, 26*(3), 156-161.

Nakagawa, Y., & Miyazaki, S. (1981). An experimental comparison of the heuristic methods for solving reliability optimization problems. *Reliability, IEEE Transactions on*, *30*(2), 181-184.

Nakagawa, Y., & Miyazaki, S. (1981). Surrogate constraints algorithm for reliability optimization problems with two constraints. *Reliability, IEEE Transactions on*, *30*(2), 175-180.

Nanda, A. K., & Hazra, N. K. (2013). Some results on active redundancy at component level versus system level. *Operations Research Letters*, *41*(3), 241-245.

Ng, K. Y., & Sancho, N. G. F. (2001). A hybrid 'dynamic programming/depth-first search'algorithm, with an application to redundancy allocation. *IIE Transactions*, *33*(12), 1047-1058.

Nematian, J., Eshghi, K., & Eshragh-Jahromi, A. (2008). Fuzzy reliability optimization models for redundant systems. *Journal of Industrial Engineering International Islamic Azad University*, *4*(7), 1-9.

Nourelfath, M., & Nahas, N. (2003). Quantized hopfield networks for reliability optimization. *Reliability Engineering & System Safety*, *81*(2), 191-196.

Onishi, J., Kimura, S., James, R. J., & Nakagawa, Y. (2007). Solving the redundancy allocation problem with a mix of components using the improved surrogate constraint method. *Reliability, IEEE Transactions on*, *56*(1), 94-101.

Park, K. S. (1987). Fuzzy apportionment of system reliability. *Reliability, IEEE Transactions on*, *36*(1), 129-132.

Pandey, D., Tyagi, S. K., & Kumar, V. (2011). Reliability Analysis of a Series and Parallel Network using Triangular Intuitionistic Fuzzy Sets. *Applicat. Appl. Math.-An Int. J*, *6*(11), 1845-1855.

Pourdarvish, A., & Ramezani, Z. (2013). Cold standby redundancy allocation in a multi-level series system by memetic algorithm. *International Journal of Reliability, Quality and Safety Engineering*, *20*(03).

Prasad, V. R. , & Kuo, W., (1997). An evolutionary algorithm for reliability-redundancy allocation problem, *Technical report, Department of Industrial Engineering,* Texas A&M University, College Station, TX.

Prasad, V. R., Kuo, W., & Kim, K. O. (2001). Maximization of a percentile life of a series system through component redundancy allocation. *IIE Transactions,33*(12), 1071-1079.

Ramirez-Marquez, J. E., Coit, D. W., & Konak, A. (2004). Redundancy allocation for series-parallel systems using a max-min approach. *IIE Transactions*, *36*(9), 891-898.

Rao, S. S., & Dhingra, A. K. (1992). Reliability and redundancy apportionment using crisp and fuzzy multiobjective optimization approaches. *Reliability Engineering & System Safety*, *37*(3), 253-261.

Ravi, V. (2004). Optimization of complex system reliability by a modified great deluge algorithm. *Asia-Pacific Journal of Operational Research*, *21*(04), 487-497.

Ravi, V., Reddy, P. J., & Zimmermann, H. J. (2000). Fuzzy global optimization of complex system reliability. *Fuzzy Systems, IEEE Transactions on*, *8*(3), 241-248.

Ravi, V., Murty, B. S. N., & Reddy, P. J. (1997). Nonequilibrium simulated-annealing algorithm applied to reliability optimization of complex systems.*Reliability, IEEE Transactions on*, *46*(2), 233-239.

Reddy, Y. S. K. R. C. N. M. R., Reddy, C. N. M., & Reddy, K. H. C. (2011). Stochastic redundancy allocation problem using simulation. *Journal of Engineering and Technology Research*, *3*(10), 287-291.

Rocco, C. M., Miller, A. J., Moreno, J. A., & Carrasquero, N. (2000). A cellular evolutionary approach applied to reliability optimization of complex systems. In*Reliability and Maintainability Symposium, 2000. Proceedings. Annual* (pp. 210-215). IEEE.

Rotshtein, A., Katielnikov, D., & Pustylnik, L. (2012). Reliability Modeling and Optimization Using Fuzzy Logic and Chaos Theory. *International Journal of Quality, Statistics, and Reliability, 2012*.

Roy, P., Mahapatra, B. S., Mahapatra, G. S., & Roy, P. K. (2014). Entropy based region reducing genetic algorithm for reliability redundancy allocation in interval environment. *Expert Systems with Applications*, In Press, Available online.

Rubinstein, R. Y., Levitin, G., Lisnianski, A., & Ben-Haim, H. (1997). Redundancy optimization of static series-parallel reliability models under uncertainty. *Reliability, IEEE Transactions on*, *46*(4), 503-511.

Ryoo H.S. (2005). Robust meta-heuristic algorithm for redundancy optimization in large-scale complex systems. *Annals of Operations Research*, 133: 209-228.

Sadjadi, S. J., & Soltani, R. (2012). Alternative design redundancy allocation using an efficient heuristic and a honey bee mating algorithm. *Expert Systems with Applications*, *39*(1), 990-999.

Sadjadi, S. J., & Soltani, R. (2009). An efficient heuristic versus a robust hybrid meta-heuristic for general framework of serial–parallel redundancy problem.*Reliability Engineering & System Safety*, *94*(11), 1703-1710.

Sadjadi, S. J., Tofigh, A. A., & Soltani, R. (2014). A New Nonlinear Multi-objective Redundancy Allocation Model with the Choice of Redundancy Strategy Solved by Compromise Programming Approach. *International Journal of Engineering (1025-2495)*, *27*(4).

Safaei, N., Tavakkoli-Moghaddam, R., & Kiassat, C. (2012). Annealing-based particle swarm optimization to solve the redundant reliability problem with multiple component choices. *Applied Soft Computing*, *12*(11), 3462-3471.

Safari, J., & Tavakkoli-Moghaddam, R. (2010). A redundancy allocation problem with the choice of redundancy strategies by a memetic algorithm. *Journal of Industrial Engineering International*, *6*(11), 6-16.

Safari, J. (2012). Multi-objective reliability optimization of series-parallel systems with a choice of redundancy strategies. *Reliability Engineering & System Safety*, *108*, 10-20.

Sahoo, L., Bhunia, A. K., & Roy, D. (2010). A Genetic algorithm based reliability redundancy optimization for interval valued reliabilities of components. *Journal of Applied Quantitative Methods*, *5*(2), 270–287.

Sahoo, L., Bhunia, A. K., & Roy, D. (2013). Reliability optimization with high and low level redundancies in interval environment via genetic algorithm. *International Journal of System Assurance Engineering and Management*, 1-11.

Sahoo, L., Bhunia, A. K., & Kapur, P. K. (2012). Genetic algorithm based multi-objective reliability optimization in interval environment. *Computers & Industrial Engineering*, *62*(1), 152-160.

Sakawa, M., & Yano, H., (1985). Interactive multi-objective reliability design of a standby system by the fuzzy sequential proxy optimization technique (FSPOT), *International Journal of System Science*, 16 (2), 177-195.

Salazar, D., Rocco, C. M., & Galván, B. J. (2006). Optimization of constrained multiple-objective reliability problems using evolutionary algorithms. *Reliability Engineering & System Safety*, *91*(9), 1057-1070.

Sasaki, M., & Gen, M. (2003). Fuzzy multiple objective optimal system design by hybrid genetic algorithm. *Applied Soft Computing*, *2*(3), 189-196.

Sharama, J., & Venkateswaran, K.V., (1971). A direct method for maximizing the system reliability. *IEEE Trans*. On Reliability, R-20, 256–259.

Sheikhalishahi, M., Ebrahimipour, V., Shiri, H., Zaman, H., & Jeihoonian, M. (2013). A hybrid GA–PSO approach for reliability optimization in redundancy allocation problem. *The International Journal of Advanced Manufacturing Technology*, *68*(1-4), 317-338.

Shelokar, P. S., Jayaraman, V. K., & Kulkarni, B. D. (2002). Ant algorithm for single and multiobjective reliability optimization problems. *Quality and Reliability Engineering International*, *18*(6), 497-514.

Shi, D. H. (1987) A new heuristic algorithm for constrained redundancy optimization in complex system, IEEE Trans. Reliability, R-36, 621-623.

Singer, D. (1990). A fuzzy set approach to fault tree and reliability analysis.*Fuzzy sets and systems*, *34*(2), 145-155.

Soltani, R., & Sadjadi, S. J. (2014). Reliability optimization through robust redundancy allocation models with choice of component type under fuzziness. *Proceedings of the Institution of Mechanical Engineers, Part O: Journal of Risk and Reliability*, 1748006X14527075.

Soltani , R., Sadjadi, S. J., & Tavakkoli-Moghaddam, R., (2015). Entropy based redundancy allocation in serial-parallel systems with choices of a redundancy strategy and component type: A multi-objective model. *Applied mathematics and information sciences*, to be appear in Vol. 9 No. 1.

Soltani, R., Sadjadi, S. J., & Tavakkoli-Moghaddam, R. (2013). Robust cold standby redundancy allocation for nonrepairable series–parallel systems through Min-Max regret formulation and Benders' decomposition method.*Proceedings of the Institution of Mechanical Engineers, Part O: Journal of Risk and Reliability*, 1748006X13514962.

Soltani, R., Sadjadi, S. J., & Tofigh, A. A. (2013). A model to enhance the reliability of the serial parallel systems with component mixing. *Applied Mathematical Modelling,* 38,1064–1076.

Sriramdas, V., Chaturvedi, S. K., & Gargama, H. (2014). Fuzzy arithmetic based reliability allocation approach during early design and development. *Expert Systems with Applications*, *41*(7), 3444-3449.

Suman, B. (2003). Simulated annealing-based multiobjective algorithms and their application for system reliability. *Engineering Optimization*, *35*(4), 391-416.

Sun, X. L., McKinnon, K. I. M., & Li, D. (2001). A convexification method for a class of global optimization problems with applications to reliability optimization. *Journal of Global Optimization*, *21*(2), 185-199.

Sup, S. C., & Kwon, C. Y. (1999). Branch-and-bound redundancy optimization for a series system with multiple-choice constraints. *Reliability, IEEE Transactions on*, *48*(2), 108-117.

Taboada, H. A., & Coit, D. W. (2008, May). Development of a multiple objective genetic algorithm for solving reliability design allocation problems. In*Proceedings of the 2008 industrial engineering research conference. Available at: http://ie. rutgers. edu/resource/research_paper/paper_08-004. pdf.*

Taboada, H. A., Baheranwala, F., Coit, D. W., & Wattanapongsakorn, N. (2007). Practical solutions for multi-objective optimization: An application to system reliability design problems. *Reliability Engineering & System Safety*,*92*(3), 314-322.

Taguchi, T., & Yokota, T. A Solution Method for Nonlinear Integer Programming Problem with Interval Coefficients Using Hybrid GA/SA Algorithms. In *Proceedings of the 41st international conference on computers & industrial engineering.*

Taguchi, T., Ida, K., & Gen, M. (1997). Method for solving nonlinear goal programming with interval coefficients using genetic algorithm. *Computers & industrial engineering*, *33*(3), 597-600.

Taguchi, T., & Yokota, T. (1999). Optimal design problem of system reliability with interval coefficient using improved genetic algorithms. *Computers & industrial engineering*, *37*(1), 145-149.

Taguchi, T., Yokota, T., & Gen, M. (1998). Reliability optimal design problem with interval coefficients using hybrid genetic algorithms. *Computers & Industrial Engineering*, *35*(1), 373-376.

Tan, Y., Tan, G. Z., & Deng, S. G. (2013). Hybrid particle swarm optimization with differential evolution and chaotic local search to solve reliability-redundancy allocation problems. *Journal of Central South University*, *20*, 1572-1581.

Tannous, O., Xing, L., & Bechta Dugan, J. (2011, January). Reliability analysis of warm standby systems using sequential BDD. In *Reliability and Maintainability Symposium (RAMS), 2011 Proceedings-Annual* (pp. 1-7). IEEE.

Tavakkoli-Moghaddam, R., & Safari, J. (2007). A new mathematical model for a redundancy allocation problem with mixing components redundant and choice of redundancy strategies. *Applied mathematical sciences*, *45*(1), 2221-2230.

Tavakkoli-Moghaddam, R., Safari, J., & Sassani, F. (2008). Reliability optimization of series-parallel systems with a choice of redundancy strategies using a genetic algorithm. *Reliability Engineering & System Safety*, *93*(4), 550-556.

Tekiner-Mogulkoc, H., & Coit, D. W. (2011). System reliability optimization considering uncertainty: minimization of the coefficient of variation for series-parallel systems. *Reliability, IEEE Transactions on*, *60*(3), 667-674.

Tillman, F. A., Hwang, F. A., & Kuo, W., (1985). Optimization of Sysrem Reliability, *Marcel Dekker*.

Tillman, F.A., Hwang, C.L., & Kuo, W. (1980). An annotated overview of system reliability optimization. 49 , *Marcel Dekker*, N.Y.

Tillman, F. A., Hwang, C. L., & Kuo, W. (1977). Determining component reliability and redundancy for optimum system reliability. *Reliability, IEEE Transactions on*, *26*(3), 162-165.

Tzafestas, S. G. (1980). Optimization of system reliability: a survey of problems and techniques. *International Journal of Systems Science*, *11*(4), 455-486.
Valian, E., Tavakoli, S., Mohanna, S., & Haghi, A. (2013). Improved cuckoo search for reliability optimization problems. *Computers & Industrial Engineering*,*64*(1), 459-468.
Valian, E., & Valian, E. (2013). A cuckoo search algorithm by Lévy flights for solving reliability redundancy allocation problems. *Engineering Optimization*,*45*(11), 1273-1286.
Wang, L., & Li, L. P. (2012). A coevolutionary differential evolution with harmony search for reliability–redundancy optimization. *Expert Systems with Applications*, *39*(5), 5271-5278.
Wang, S., & Watada, J. (2009). Modelling redundancy allocation for a fuzzy random parallel–series system. *Journal of computational and applied mathematics*, *232*(2), 539-557.
Wang, S., & Watada, J. (2009). Reliability optimization of a series-parallel system with fuzzy random lifetimes. *International Journal of Innovative Computing, Information and Control*, *5*(6), 1547-1558.
Wang, Z., Huang, H. Z., Li, Y., Pang, Y., & Xiao, N. C. (2012). An approach to system reliability analysis with fuzzy random variables. *Mechanism and Machine Theory*, *52*, 35-46.
Wang, Z., Tang, K., & Yao, X. (2010). A memetic algorithm for multi-level redundancy allocation. *Reliability, IEEE Transactions on*, *59*(4), 754-765.
Wu, P., Gao, L., Zou, D., & Li, S. (2011). An improved particle swarm optimization algorithm for reliability problems. *ISA transactions*, *50*(1), 71-81.
Xu, Z., Kuo, W., & Lin, H. H. (1990). Optimization limits in improving system reliability. *Reliability, IEEE Transactions on*, *39*(1), 51-60.
Yadavalli, V. S. S., Malada, A., & Charles, V. (2007). Reliability stochastic optimization for an n-stage series system with m chance constraints. *South African Journal of Science*, *103*(11-12), 502-504.
Yalaoui, A., Châtelet, E., & Chu, C. (2005). A new dynamic programming method for reliability & redundancy allocation in a parallel-series system.*Reliability, IEEE Transactions on*, *54*(2), 254-261.
Yalaoui, A., Chu, C., & Châtelet, E. (2005). Reliability allocation problem in a series–parallel system. *Reliability engineering & system safety*, *90*(1), 55-61.
Yao, J. S., Su, J. S., & Shih, T. S. (2008). Fuzzy System Reliability Analysis Using Triangular Fuzzy Numbers Based on Statistical Data. *Journal of Information Science & Engineering*, *24*(5): 1521-1535.
Yearout, R. D., Reddy, P., & Grosh, D. L. (1986). Standby redundancy in reliability-a review. *Reliability, IEEE Transactions on*, *35*(3), 285-292.
Yeh, W. C. (2009). A two-stage discrete particle swarm optimization for the problem of multiple multi-level redundancy allocation in series systems. *Expert Systems with Applications*, *36*(5), 9192-9200.
Yeh, W. C. (2003). A MCS-RSM approach for network reliability to minimise the total cost. *The International Journal of Advanced Manufacturing Technology*, *22*(9-10), 681-688.
Yeh, W. C., & Hsieh, T. J. (2011). Solving reliability redundancy allocation problems using an artificial bee colony algorithm. *Computers & Operations Research*, *38*(11), 1465-1473.
Yeh, W. C. (2014). Orthogonal simplified swarm optimization for the series–parallel redundancy allocation problem with a mix of components. *Knowledge-Based Systems*, *64*, 1-12.
Yokota, T., Gen, M., & Li, Y. X. (1996). Genetic algorithm for non-linear mixed integer programming problems and its applications. *Computers & industrial engineering*, *30*(4), 905-917.
Yokota, T., & Gen M., (1996). A method for solving system reliability design problem with interval coefficients using genetic algorithm. *Journal of Japan Society for Fuzzy Theory and Systems*, 8(2): 378–387, (in Japanese).
Yokota, T., Gen, M., Li, Y., & Kim, C. E., (1996). A Genetic Algorithm for Interval Nonlinear Integer Programming Problem. *Computers and industrial Ensag*, 31(3–4): 913–917.
Yokota, T., Gen, M., Taguchi, T., & Li, Y. (1995). A method for interval 0–1 nonlinear programming problem using a genetic algorithm. *Computers & Industrial Engineering*, *29*(1), 531-535.
You, P. S., & Chen, T. C. (2005). An efficient heuristic for series–parallel redundant reliability problems. *Computers & operations research*, *32*(8), 2117-2127.

Yun, W. Y., Song, Y. M., & Kim, H. G. (2007). Multiple multi-level redundancy allocation in series systems. *Reliability Engineering & System Safety*, *92*(3), 308-313.

Yun, W. Y., & Kim, J. W. (2004). Multi-level redundancy optimization in series systems. *Computers & Industrial Engineering*, *46*(2), 337-346.

Zafiropoulos, E. P., & Dialynas, E. N. (2004). Reliability and cost optimization of electronic devices considering the component failure rate uncertainty. *Reliability Engineering & System Safety*, *84*(3), 271-284.

Zhang, E., Wu, Y., & Chen, Q. (2014). A practical approach for solving multi-objective reliability redundancy allocation problems using extended bare-bones particle swarm optimization. *Reliability Engineering & System Safety*, *127*, 65-76.

Zhang, H., Hu, X., Shao, X., Li, Z., & Wang, Y. (2013). IPSO-based hybrid approaches for reliability-redundancy allocation problems. *Science China Technological Sciences*, *56*(11), 2854-2864.

Zhao, J. H., Liu, Z., & Dao, M. T. (2007). Reliability optimization using multiobjective ant colony system approaches. *Reliability Engineering & System Safety*, *92*(1), 109-120.

Zhao, R., & Liu, B. (2004). Redundancy optimization problems with uncertainty of combining randomness and fuzziness. *European Journal of Operational Research*, *157*(3), 716-735.

Zhao, R., & Liu, B. (2005). Standby redundancy optimization problems with fuzzy lifetimes. *Computers & Industrial Engineering*, *49*(2), 318-338.

Zhao, R., & Liu, B. (2003). Stochastic programming models for general redundancy-optimization problems. *Reliability, IEEE Transactions on*, *52*(2), 181-191.

Zia, L., & Coit, D. W. (2010). Redundancy allocation for series-parallel systems using a column generation approach. *Reliability, IEEE Transactions on*, *59*(4), 706-717.

Zou, D., Gao, L., Li, S., & Wu, J. (2011). An effective global harmony search algorithm for reliability problems. *Expert Systems with Applications*, *38*(4), 4642-4648.

Zou, D., Gao, L., Wu, J., Li, S., & Li, Y. (2010). A novel global harmony search algorithm for reliability problems. *Computers & Industrial Engineering*, *58*(2), 307-316.

Zou, F. Z, & Li, C. X. (2001). Chaotic model for software reliability. *Chinese Journal of Computers*, 24(3), 281–291.

2

Multiple characteristics optimization in machining of GFRP composites using Grey relational analysis

Arun Kumar Parida, Rajesh Kumar Bhuyan and Bharat Chandra Routara*

School of Mechanical Engineering, KIIT, Bhubaneswar, Odisha, India

C H R O N I C L E

Keywords:
ANOVA
Taguchi Method
Turning
Grey Relational Analysis

A B S T R A C T

In the present work, a multi-response optimization method is used to optimize the machining parameters in turning of glass fiber reinforced polymer (GFRP) composites. Parameters like spindle speed (N), feed rate (f) and depth of cut (d) are taken to obtain the responses such as surface roughness (Ra) and material removal rate (MRR). Taguchi's L9 orthogonal array has been used for machining the work-piece. Analysis of variance (ANOVA) has been carried out to check the significant process parameter in a single objective performance characteristic. The multiple performance characteristics have been analysed using Grey relational analysis and an appreciable result has been reported with this approach.

1. Introduction

The use of machining such as turning, drilling, milling, etc. in GFRP composite is increasing as the field of application of composite is quite a lot. However, it is often made difficult to machine the GFRP composites as it brings many undesirable results such as rapid tool wear, a defective surface layer with cracks or delamination, a rough surface finish etc. To avoid this problem, it is necessary to select the appropriate process parameters to get the highest performance for desired dimensional accuracy. Davim and Mata (2004) studied the influencing parameters on surface roughness during turning of GFRP using statistical analysis. They reported the percentage of contribution of cutting parameters such as velocity and feed rate on the surface roughness. Surface roughness of unidirectional GFRP composite was experimented by Işık (2008) on the basis of process parameters such as cutting speed, feed, depth of cut and tool geometry in turning with CERMET cutting tool and found that the surface quality is closely related with speed, feed and tool geometry. Palanikumar (2006) used Taguchi method, which offers a simple and systematic approach to find the optimum cutting parameters for surface roughness. Hence, from the literature (Santhanakrishnan, 1990; Sakuma & Seto, 1983; Takeyama & Iijima, 1988; Sang-Ook et al., 1997) a detailed analysis has been carried out for identifying the important parameters, which affects the machining characteristics. Taguchi method is a scientifically disciplined mechanism for evaluating and implementing the improvement in products, processes, materials and facilities.

* Corresponding author.

These improvements are aimed at improving the desired characteristics and simultaneously reducing the number of defects by studying the key variables controlling the process and optimizing the procedures or design to yield the best results (1996). However, this method is designed to optimize a single performance characteristics and difficult to optimize the multi-responses in complex process (1989). Deng (1989) proposed a Grey relational theory which is useful for solving the problems with poor, insufficient and uncertain information. It can be further effectively adopted for solving the complicated interrelationship among the designated performance characteristics. The grey relational theory provides a better solution in which the information is incomplete or the model is incomplete (1989). It also provides an efficient solution to the uncertainty, multiple inputs and discrete data problem (2004). Using Grey relational analysis, Tzeng and Chen (2006) optimized the process parameters in turning of tool steels with eight numbers of independent variables. They determined the optimum parameters, which maximizes the accuracy and minimizes the surface roughness and dimensional precision. Palanikumar et al. (2006) employed orthogonal array with Grey relational analysis to optimize the multiple performance characteristics in turning of GFRP with cemented carbide (k-10) tool. According to them the Grey relational theory simplifies the optimization procedure. It is also economical and more convenient to predict the effect of different influencing combinations of parameters. Gupta and Kumar (2013) applied successfully the Grey relational theory to optimize the process parameters in multiple performance characteristics such as surface roughness and material removal rate during turning of GFRP with Poly-crystalline diamond cutting tool. Rao et al. (2012) used Grey relational analysis for optimizing the multi- response characteristics to minimize the surface roughness and cutting force and maximize the tool life in machining of Inconel-718 with process parameters such as speed, feed, depth of cut and approach angle. From the above studies, it is found that Grey relational analysis is one of the important techniques for optimizing the multi-response characteristics and it can be applied successfully to any machining process.

In this present work, an experimental investigation has been carried out using a normal speed lathe machine. The experiments were carried out on the basis of Taguchi's L_9 orthogonal array. The important turning parameters were chosen as spindle speed, feed rate and depth of cut. The multiple responses such as surface roughness and material removal rate are analyzed here. The Grey relational analysis is used to optimize multi objective responses. ANOVA is also used to find the highly influencing machining parameters that contribute to high quality product. Finally, confirmation tests are conducted to validate the experimental results.

2. Experimental work

The GFRP work-pieces having 40 mm diameter and 50mm in length are used for machining in a lathe. The GFRP bar was fabricated by dry hand lay-up technique. The hand lay-up technique was chosen as it was ideally suited to manufacture low volume with minimum tooling cost. The experiments were planned using Taguchi's L_9 orthogonal array that helps to reduce the number of experimental runs. The three cutting parameters such as spindle speed, feed rate and depth of cut with three different levels are used for the experimentation. The turning experiments were carried out in an all geared lathe machine whose spindle speed ranges from 200 to 400 rpm and feed ranges from 0.03 to 0.05 mm/rev. The machining of wok-pieces was carried out in a dry condition with cemented carbide tool. The composition of fiber, resin and filler and the machining parameters with their levels are shown in Table 1 and Table 2.

2.1 Material Removal Rate (MRR)

The material removal rate is the volume of material removed per unit time. Volume of material removed is a function of spindle speed, feed rate and depth of cut. Material removal with a higher rate is one of the most important criteria during the turning operation. The *MRR* is calculated using the expression

MRR = $\pi D_i d f N$ in mm^3/min,

where D_i is the initial diameter, d is the depth of cut, f is the feed rate and N is the number of revolutions of spindle per minute.

2.2 Surface Roughness (R_a)

Surface roughness is also another important aspect in machining of GFRP composite. Here the roughness is measured three times using a stylus type surface roughness tester(Taylor Hobson, Sutronic 25) of sampling length 0.8 mm, evaluation length of 4 mm and least count of 0.01 μm and the average surface roughness is listed in Table 2.

Table 1
Specification of epoxy, filler and fiber

Sl. No.	Epoxy + Hardener	Filler	E-glass
01	Araldite LY 556 + HY 951(50 wt %)	Graphite (3 wt %) + Ash clay (3 wt %)	R099 1200 P566 (44 wt %)

Table 2
Machining parameters with their levels

Expt. no	*N*	*f*	*d*	*N* (rpm)	*f* (mm/rev)	*d* (mm)	*MRR* in mm^3/min	Surface roughness R_a in (μm)
01	1	1	1	200	0.03	0.5	319.49	2.60
02	1	2	2	200	0.04	1.0	625.8	2.80
03	1	3	3	200	0.05	1.5	913.26	2.00
04	2	1	2	300	0.03	1.0	1071.28	2.50
05	2	2	3	300	0.04	1.5	1555.08	2.80
06	2	3	1	300	0.05	0.5	493.13	3.40
07	3	1	3	400	0.03	1.5	3392.92	3.52
08	3	2	1	400	0.04	0.5	1074.42	4.88
09	3	3	2	400	0.05	1.0	2111.15	3.44

3. Optimization of individual performance characteristics

3.1 Determination of optimal process parameters for MRR

In this section, L_9 orthogonal array is used to determine the optimal process parameters. Machining results are reported in using S/N ratio and ANOVA analysis. In Taguchi method, there are three performance characteristics such as higher-is-better, nominal-is-better and lower-is-better. Here higher-is-better characteristic is used to find the optimal process parameter for *MRR*. The *MRR* and S/N ratio for *MRR* is listed in Table 3.

Table 3
Experimental results for *MRR* and S/N ratio

Run	*N*	*f*	*d*	MRR	S/N ratio
01	1	1	1	319.49	50.0891
02	1	2	2	625.80	55.9287
03	1	3	3	913.26	59.2119
04	2	1	2	1071.28	60.5981
05	2	2	3	1455.08	63.2577
06	2	3	1	493.13	53.8592
07	3	1	3	2392.92	67.5786
08	3	2	1	1074.42	60.6235
09	3	3	2	2111.15	66.4904

3.2 Analysis of S/N ratio for MRR

As the experimental design is orthogonal, so it is possible to separate out the effect of each process parameter at different levels. From the response Table 4 of mean S/N ratio for *MRR*, the optimal process parameters are obtained such as spindle speed at level-3, feed rate at level-2 and depth of cut at level-3 i.e., v_3-f_2-d_3.

Table 4
Response table of mean S/N ratio for *MRR*

Symbol	Process parameters	Mean S/N ratio				Rank
		Level-1	Level-2	Level-3	Max-Min	
N	Spindle speed	55.08	59.24	64.90	9.82	1
f	Feed rate	59.42	59.94	59.85	0.51	3
d	Depth of cut	54.86	61.01	63.35	8.49	2

Total mean S/N ratio = 59.73 dB

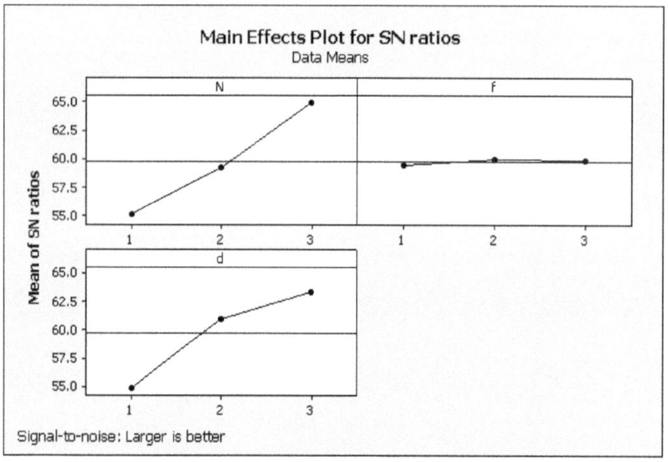

Fig. 1. Mean S/N graph for *MRR*

3.3 ANOVA for MRR

The purpose of the ANOVA is to find the statistical significance of process parameters on the response shown in Table 5. From table, it is found that spindle speed with a P- value less than 0.05 that means it is significant at 95% confidence level. Therefore, spindle speed is the most significant parameter for *MRR*.

Table 5
ANOVA table for *MRR*

Source	DF	SS	MS	F	P
N	2	2414913	1207456	21.13	0.045
f	2	66325	33163	0.58	0.633
d	2	1428931	714465	12.50	0.074
Error	2	114269	57134		
Total	8	4024438			

S = 239.028 R-Sq = 97.16% R-Sq (adj) = 88.64%

3.4 Confirmation test for MRR

Confirmation tests are carried out using the optimal process parameter combination to predict and verify the improvement of the performance characteristics as shown in Table 6. Here the increase of S/N ratio from the initial process parameter to optimal process parameter is 18.0186 dB and based on the confirmation test, *MRR* is increased 7.96 times.

Table 6
Results of confirmation for *MRR*

	Initial process parameter	Optimal process parameter	
		Prediction	Experiment
Level	N_1-f_1-d_1		N_3-f_2-d_3
MRR	319.49		2543.21
S/N ratio(dB)	50.0891	68.7086	68.1076

Improvement of S/N ratio = 18.0186 dB

3.5 Determination of optimal process parameters for surface roughness (R_a)

Similarly, the S/N ratio for surface roughness is calculated. Here lower-is-better characteristic is used to find the optimal process parameter for surface roughness (R_a). The S/N ratio for R_a is listed in Table 7. From the response Table 8 of mean S/N ratio for R_a, the optimal process parameters are obtained such as spindle speed at level-1, feed rate at level-1 and depth of cut at level-3 i.e., N_1-f_1-d_3.

Table 7
Experimental results for surface roughness and S/N ratio

Run	*v*	*f*	*d*	R_a	S/N ratio
01	1	1	1	2.60	-8.2995
02	1	2	2	2.80	-8.9432
03	1	3	3	2.00	-6.0206
04	2	1	2	2.50	-7.9588
05	2	2	3	2.80	-8.9432
06	2	3	1	3.40	-10.6296
07	3	1	3	3.52	-10.9309
08	3	2	1	4.88	-13.7684
09	3	3	2	3.44	-10.7312

3.6 Analysis of S/N ratio for surface roughness (R_a)

Table 8
Response table of mean S/N ratio for R_a

Symbol	Process parameters	Mean S/N ratio				Rank
		Level-1	Level-2	Level-3	Max-Min	
N	Spindle speed	-7.754	-9.177	-11.810	4.056	1
f	Feed rate	-9.063	-10.552	-9.127	1.489	3
d	Depth of cut	-10.899	-9.211	-8.632	2.268	2

Total mean S/N ratio =-9.580 dB

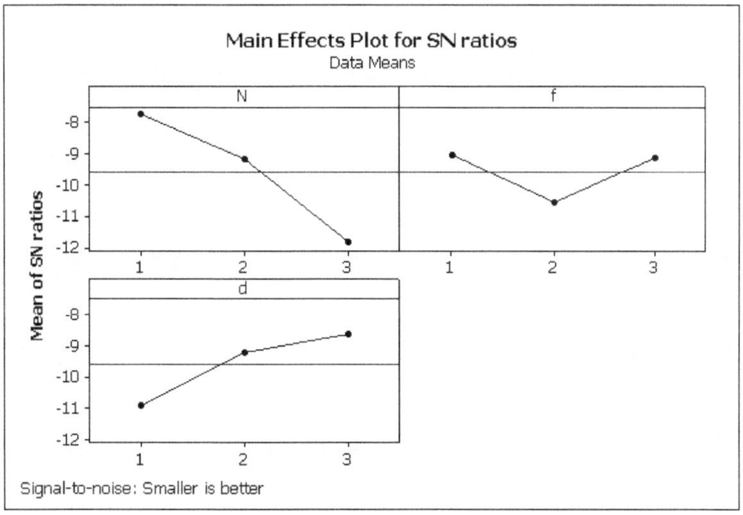

Fig. 2. Mean S/N graph for surface roughness

3.7 ANOVA for surface roughness (R_a)

ANOVA for surface roughness (R_a) is listed in the Table 9. From the table it is clearly found that spindle speed with a P- value less than 0.05 that means it is significant at 95% confidence level. Therefore, spindle speed is the most significant parameter for R_a followed by depth of cut and feed rate.

Table 9
ANOVA table for surface roughness (R_a)

Source	DF	SS	MS	F	P
N	2	3.47369	1.73684	26.46	0.036
f	2	0.68862	0.34431	5.25	0.160
d	2	1.25662	0.62831	9.57	0.095
Error	2	0.13129	0.06564		
Total	8	5.55022			

S = 0.256212 R-Sq = 97.63% R-Sq (adj) = 90.54%

3.8 Confirmation test for surface roughness (R_a)

In the confirmation table for surface roughness (R_a), it is found that the increase of S/N ratio from the initial process parameters to the optimal process parameters is 2.7244 dB which is shown in Table 10. Based on the confirmation test it is also found that the surface roughness is increased 1.36 times.

Table 10
Results of confirmation for surface roughness (R_a)

	Initial process parameter	Optimal process parameters	
		Prediction	Experiment
Level	N_1-f_1-d_1		N_1-f_1-d_3
R_a	2.60		1.9
S/N ratio(dB)	-8.2995	-6.28783	-5.5751

Improvement of S/N ratio=2.7244 dB

4. Grey relational analysis

The grey relational analysis, which is useful for dealing with poor, incomplete and uncertain information can be used to solve complicated inter-relationships among multiple performance characteristics satisfactorily. Following are the steps needed for converting the multi-response characteristics to single response characteristics.

1) Normalize the experimental results of metal removal rate and surface roughness(data preprocessing)
2) Calculate the Grey relational co-efficient.
3) Calculate the Grey relational grade by averaging the Grey relational co-efficient.

In the grey relational analysis, the experimental results are first normalized in the range between zero and unity. This process of normalization is known as the grey relational generation. After then the grey relational coefficient is calculated from the normalized experimental data to express the relationship between the desired and actual experimental data. Then, the overall grey relational grade is calculated by averaging the grey relational coefficient corresponding to each selected process response. The overall evaluation of the multiple process responses are based on the grey relational grade. This method converts a multiple response process optimization problem with the objective function of overall grey relational grade. The corresponding level of parametric combination with highest grey relational grade is considered as the optimum process parameter.

In the grey relational analysis, the normalized data processing for surface roughness (R_a) corresponding to lower-the-better criterion is expressed as

$$x_i(k) = \frac{\max y_i(k) - y_i(k)}{\max y_i(k) - \min y_i(k)} \quad (1)$$

In the turning operation, the material removal rate should be higher. Therefore the normalized data processing for the material removal rate is the higher-the-better performance characteristics considered and is expressed as

$$x_i(k) = \frac{y_i(k) - \min y_i(k)}{\max y_i(k) - \min y_i(k)} \quad (2)$$

Eq. (1) is used for the "lower-the-better" responses and Eq. (2) is used for the "larger-the- better" responses, where $x_i(k)$ is the value after grey relational generation, min $y_i(k)$ is the smallest value of $y_i(k)$ for the k_{th} response, and the max $y_i(k)$ is the largest value of $y_i(k)$ for the k_{th} response. An ideal sequence is $x_o(k)$ where (k= 1, 2 for *MRR* and R_a respectively). The definition of grey relational grade in the course of grey relational analysis is to reveal the relational degree between the nine sequences [$x_o(k)$ and $x_i(k)$, i =1, 2, 3…..9]. The grey relational coefficient $\xi_i(k)$ can be calculated as

$$\xi_i(k) = \frac{\Delta_{\min} + \psi \Delta_{\max}}{\Delta_{oi}(k) + \psi \Delta_{\max}}, \quad (3)$$

where $\Delta_{oi} = x_o(k) - x_i(k)$ =difference of the absolute value between x_o (k) and x_i (k);

Ψ= distinguishing coefficient between zero and one, the purpose of which is to weaken the effect of $\Delta_{\max}$ when it gets too big, and thus enlarges the difference significance of the relational coefficient. In the present case, ψ = 0.5 is used.

After averaging the grey relational coefficients, the grey relational grade γ_i can be calculated as follows:

$$\gamma_i = \frac{1}{n}\sum_{k=1}^{n} \xi_i(k), \quad (4)$$

where *n* = number of process responses. The higher value of grey relational grade is considered as the stronger relational degree between the ideal sequence $x_0(k)$ and the given sequence $x_i(k)$. It has already been mentioned that the ideal sequence $x_0(k)$ is the best process response in the experimental layout. The higher grey relational grade implies that the corresponding parameter combination is closer to the optimal. Thus, the grey relational grade is found out which is shown in Table 11.

Table 11

Grey relational grade table

Expt. No	Responses		Grey relational generation		Evaluation of Δ_{oi}		Grey relational coefficient(ψ=0.5)		Grey relational grade
	MRR	R_a	*MRR*	R_a	*MRR*	R_a	*MRR*	R_a	
01	319.49	2.6	0	0.791667	1	0.208333	0.333333	0.705882	0.519608
02	625.8	2.8	0.099664	0.722222	0.900336	0.277778	0.357057	0.642857	0.499957
03	913.26	2	0.193195	1	0.806805	0	0.382612	1	0.691306
04	1071.28	2.5	0.244609	0.826389	0.755391	0.173611	0.398282	0.742268	0.570275
05	1555.08	2.8	0.402023	0.722222	0.597977	0.277778	0.455383	0.642857	0.54912
06	493.13	3.4	0.056497	0.513889	0.943503	0.486111	0.34638	0.507042	0.426711
07	3392.92	3.52	1	0.472222	0	0.527778	1	0.486486	0.743243
08	1074.42	4.88	0.245631	0	0.754369	1	0.398607	0.333333	0.36597
09	2111.15	3.44	0.582951	0.5	0.417049	0.5	0.545227	0.5	0.522614

4.1 Determination of optimal process parameters for Grey relational grade

Table 12
S/N ratio and grey relational grade of performance characteristics

Run	*N*	*f*	*d*	Grey-relational grade	S/N ratio
01	1	1	1	0.519608	-5.68648
02	1	2	2	0.499957	-6.02135
03	1	3	3	0.691306	-3.20659
04	2	1	2	0.570275	-4.87831
05	2	2	3	0.54912	-5.20665
06	2	3	1	0.426711	-7.39732
07	3	1	3	0.743243	-2.57738
08	3	2	1	0.36597	-8.73109
09	3	3	2	0.522614	-5.63638

4.2 Analysis of S/N ratio for Grey relational grade

Table 13
Response table of mean S/N ratio for Grey relational grade

Symbol	Process parameters	Mean S/N ratio				Rank
		Level-1	Level-2	Level-3	Max-Min	
N	Spindle speed	-4.971	-5.827	-5.648	0.856	3
f	Feed rate	-4.381	-6.653	-5.413	2.272	2
d	Depth of cut	-7.272	-5.512	-3.664	3.608	1

Total mean S/N ratio =-5.482 dB

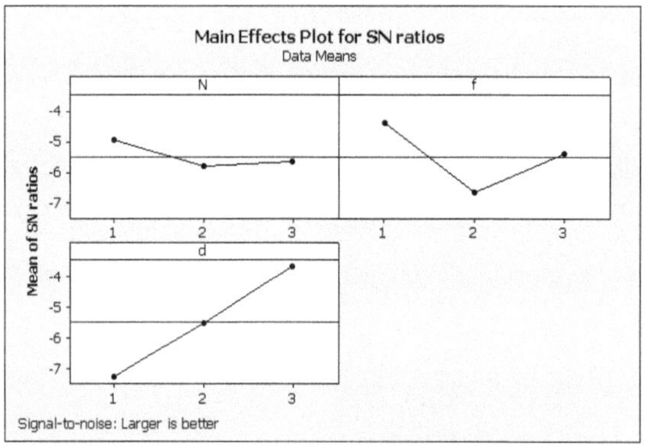

Fig. 3. Grey relational grade graph of multiple performance characteristics

The optimal process parameters are obtained from the response graph as shown in Fig 3. The optimal setting parameters for multiple performance characteristics is spindle speed at level-1, feed rate at level-1 and depth of cut at level-3 i.e. N_1-f_1-d_3.

4.3 ANOVA for Grey relational grade

From the table it is clearly found that depth of cut with a P-value less than 0.05 that means it is significant at 95% confidence level. So depth of cut is the most significant parameter for R_a followed by feed rate and spindle speed for multiple performance characteristics.

Table 14
ANOVA for grey relational grade

Source	DF	SS	MS	F	P
N	2	0.004527	0.002264	4.39	0.185
f	2	0.029193	0.014596	28.32	0.034
d	2	0.075801	0.037900	73.53	0.013
Error	2	0.001031	0.000515		
Total	8	0.110551			

S = 0.0227041 R-Sq = 99.07% R-Sq (adj) = 96.27%

4.4 Confirmation test for Grey relational grade

Table 15
Results for the confirmation tests

	Initial process parameter	Optimal process parameter	
		Prediction	experiment
Level	N_1-f_1-d_1	N_1-f_1-d_3	N_1-f_1-d_3
MRR	319.49		1885.08
R_a	2.6		2.5
Grey relational Grade	0.519608	-2.05095	0.623505

Improvement of Grey relational Grade= 0.10

After the optimum level of machining parameters in multiple performance characteristics is identified, a verification test needs to be carried out in order to check the accuracy of the analysis. Table 15 shows the comparison of estimated grey relational grade with the actual grey relational grade obtained in experiment using the optimal cutting parameters. It is also found that the improvement of grey relational grade from initial process parameter combination to the optimal process parameter combination is 0.10.

5. Conclusions

The use of the Taguchi orthogonal array with grey relational analysis to optimize the machining process with the multiple performance characteristics is reported here. A grey relational analysis of the experimental results of *MRR* and surface roughness are converted to single performance characteristics called the grey relational grade. As a result, optimization of the complicated multiple performance characteristics can be greatly simplified through this approach. It is shown that the performance characteristic of the machining process such as *MRR* and surface roughness are improved together by using this approach. Finally, the suggested multi-response approach using grey relational analysis in combination with Taguchi's robust design methodology is quite capable for any type of optimization problem involving any number of responses.

References

Davim, J. P., & Mata, F. (2004). Influence of cutting parameters on surface roughness in turning glass-fibre-reinforced plastics using statistical analysis. *Industrial Lubrication and Tribology*, 56(5), 270-274.

Deng, J. L. (1989). Introduction to Grey system. *Journal of Grey System, 1, 1–24.*

Gupta, M. & Kumar, S. (2013). Multi-objective optimization of cutting parameters in turning using grey relational analysis. *International Journal of Industrial Engineering Computations,* (4), 547–558.

Işık, B. (2008). Experimental investigations of surface roughness in orthogonal turning of unidirectional glass-fiber reinforced plastic composites. *The International Journal of Advanced Manufacturing Technology*, *37*(1-2), 42-48.

Palanikumar, K. (2006). Cutting parameters optimization for surface roughness in machining of GFRP composites using Taguchi's method. *Journal of Reinforced Plastics and Composites*, *25*(16), 1739-1751.

Palanikumar, K., Karunamoorthy, L., & Karthikeyan, R. (2006). Multiple performance optimization of machining parameters on the machining of GFRP composites using carbide (K10) tool. *Materials and Manufacturing Processes,21*(8), 846-852.

Ross, P. J. (1996). Taguchi Techniques for Quality Engineering, *McGraw-Hill, New York.*

Sadasiva, R., Rajesh V. & Venu Gopal. A. (2012). Taguchi based Grey Relational Analysis to optimize face milling process with multiple performance characteristics. *International Conference on Trends in Industrial and Mechanical Engineering, Dubai, (ICTIME), March 24-25.*

Sakuma, K. & Seto, M. (1983) Tool wear in cutting glass-fiber-reinforced plastics (the relation between fiber orientation and tool wear). *Bulletin of the JSME. 26 (218), 1420–1427.*

Sang-Ook, A., Eun-Sang, L. & Sang-Lai, N. (1997). A study on the cutting characteristics of glass fiber reinforced plastics with respect to tool materials and geometries. *Journal of Materials Processing Technology, 68, 60–67.*

Santhanakrishnan, G. (1990). Investigations on machining of FRP composites and their Tribological Behaviour. *PhD thesis, IIT Madras: Chennai, India.*

Singh, N. P., Raghukandan, K. & Pai, P. C. (2004). Optimization by Grey Relational Analysis of EDM Parameters on Machining Al-10% SICP Composites. *Journal of Materials Processing Technology,* 155–156, 1658–1661.

Takeyama, H., & Iijima, N. (1988). Machinability of glassfiber reinforced plastics and application of ultrasonic machining. *CIRP Annals-Manufacturing Technology, 37*(1), 93-96.

Tzeng, Y. F., & Chen, F. C. (2006). Multi-objective process optimization for turning of tool steels. *International Journal of Machining and Machinability of Materials, 1(1), 76–93.*

3

Optimal dynamic pricing and replenishment policies for deteriorating items

Masoud Rabbani*, **Nadia Pourmohammad Zia and Hamed Rafiei**

School of Industrial Engineering, College of Engineering, University of Tehran, Tehran, Iran

CHRONICLE

Keywords:
Dynamic pricing
Replenishment
Advertisement
Time-dependent deterioration rate

ABSTRACT

Marketing strategies and proper inventory replenishment policies are often incorporated by enterprises to stimulate demand and maximize profit. The aim of this paper is to represent an integrated model for dynamic pricing and inventory control of deteriorating items. To reflect the dynamic characteristic of the problem, the selling price is defined as a time-dependent function of the initial selling price and the discount rate. In this regard, the price is exponentially discounted to compensate negative impact of the deterioration. The planning horizon is assumed to be infinite and the deterioration rate is time-dependent. In addition to price, the demand rate is dependent on advertisement as a powerful marketing tool. Several theoretical results and an iterative solution algorithm are developed to provide the optimal solution. Finally, to show validity of the model and illustrate the solution procedure, numerical results are presented.

1. Introduction

Dynamic pricing is the process of changing price during a period due to variations in stock level, demand level, inventory quality, etc. It is one of the most efficient tools of the revenue management, which enhances ability of the firms in managing demand and making profit. In recent years, this marketing strategy has been widely used by industries (Coy, 2000) due to three main reasons including emergence of a very competitive environment in businesses and industries, increasing access to demand information, and ease of applying price changes due to advent of new technologies. In classic inventory control models, it is assumed that the goods can be held infinitely for future demand, while such assumption does not hold in reality as the majority of the products lose their original value as time passes and for some of them this happens more quickly (Khanlarzade et al., 2014). These products are called deteriorating items. In inventory management problems, deterioration is defined as damage, putrefaction, demolition, evaporation, expiration, and reduction of efficiency and margin value of product, which lead to a decrease in product serviceability (Yang & Wee, 2003). Medicine, fruits and vegetables, seasonal and temporary products, and electronic devices are common examples of deteriorating items. Since deterioration results in a drop in quality and stock-level of the inventory, it imposes additional costs on the inventory system. Therefore, efficient inventory control of deteriorating items is of great importance.

* Corresponding author.

Previously, pricing and inventory management policies were incorporated separately; however integration of these two policies enhances the profit of the firms through coordination of supply and demand. Integrated decision making on inventory management and pricing decisions for deteriorating products is a branch of revenue management, which has attracted many researchers working in inventory management, economy, marketing and operations research areas. In this study an integrated model for dynamic pricing and inventory control of deteriorating products is developed. To tackle the practical conditions of the inventory systems, not only the deterioration rate changes over time, but also the selling price is assumed to be time-dependent. In addition, the demand rate is dependent on advertisement, which is modeled as the frequency of advertisement in each cycle. To characterize the optimal solution, some useful theoretical results are derived based on which an iterative and simple solution algorithm is developed.

The remainder of the paper is organized as follows: In section 2, the literature body of the problem is briefly reviewed and the related research gaps are distinguished. The assumptions and notations of the model are presented in section 3. In section 4, the mathematical model of the inventory system is formulated. Section 5 provides the theoretical results and the solution algorithm, which are applied to obtain the optimal solution. Numerical results are represented in section 6. Finally section 7, finishes the paper with conclusion and recommended future research directions.

2. Literature review

Dynamic pricing and inventory control of deteriorating items have attracted a great deal of academic attention. Thus, there exists a wide-spread literature in this area, which is briefly reviewed as follows.
In the context of the discussed problem, the first model was proposed by Cohen (1977). In this paper, demand was price-dependent and the deterioration rate is assumed to be constant. This paper was extended in Kang and Kim (1983) by considering a finite replenishment rate. Abad (2003) conducted one of the noble researches in the field of pricing. In the proposed economic production quantity model, demand was defined as a general function of selling price. This model is studied by Yang (2004) for economic order quantity problem. Teng et al. (2007) extended Abad's model by incorporating shortage cost and cost of lost reputation. For the first time, Wee (1997) proposed a time-dependent deterioration rate which followed Weibull distribution. The proposed model was extended in Wee (1999) by adding discount option for purchasing cost and modeling shortage in the form of partial backlog. Papachristos and Skouri (2003) extended the model of Wee (1999) by considering demand as a general, convex, and descending function of price, modeling shortage in form of time-dependent backlog, and modifying the revenue function. Mukhopadhyay et al. (2004, 2005) are other worth mentioning research instances with similar structures. Tsao and Sheen (2008) presented one of the rare studies on effect of advertisement on demand. Demand was considered as a linear function of price, exponential function of time and quadratic function of the cost of the advertisement. Shah et al. (2013) incorporated advertisement into the demand model as well. In their paper advertisement is modeled as the frequency of advertisement in each replenishment cycle.

Begum et al. (2009) developed an instantaneous replenishment policy for deteriorating items with price-dependent demand. They applied a three-parameter Weibull distribution to present time-dependency of inventory deterioration rate. Begum et al. (2012) conducted another study on the previous model by neglecting shortage and assuming demand as a non-linear function of price. Cai et al. (2013) proposed one of the very few researches on dynamic pricing which modeled price as a function of time. Optimal policy was obtained by considering the feedback of the price on demand per time unit. To validate the proposed structure, a numerical simulation was employed. Wang et al. (2013) also considered price as a function of time and modeled a non-instantaneous deterioration pattern. In this context, Geetha and Uthayakumar (2010), Valliathal and Uthayakumar (2011), Maihami and Abadi (2012), Ghoreishi et al. (2013) and Soni and Patel (2012, 2013) studied non-instantaneous deterioration as well. Dye et al. (2007) Considered an economic order quantity model where demand and deterioration rate were continuous functions of price and time, respectively. Shortage was possible and completely backlogged. In Dye (2007), some extensions have been made by assuming partial backlog rate as an exponential function and a reverse function of remained time. Abad (2008) studied a model with similar structure and considered partial backlog rate as a general descending function of the remained time to the next review of the inventory system. Soni and Joshi (2013) proposed a framework for pricing and inventory control of deteriorating items under bi-level trade credit in fuzzy settings.

Considering time-dependent deterioration rate as well as price and time dependent demand rate simultaneously, forms a more complicated variation of pricing problems, which has not been investigated sufficiently. Hsu et al. (2007) modeled the deterioration rate as a general function of time. A seasonal pattern was assumed for demand, which was time and price-dependent. Uncertain delivery time, backlogging and limited budget are some of the assumptions in the proposed model. Hsieh and Dye (2010) presented an inventory control model for deteriorating items under inflation. The model was solved by descending cash flow method. In addition, demand rate was dependent to time and price and shortage was partially backlogged. Shah and Raykandaliya (2010) considered demand as a decreasing function of price and time and assumed Weibull distribution for deterioration rate. Tripathy and Pradhan (2011) proposed a similar model but considered a three-parameter Weibull distribution for deterioration rate. Dye (2012) modeled demand as a general decreasing function of time and price. Purchasing price and product deterioration rate were defined as general functions of time. Finite time horizon and partial time-dependent backlog were considered in the structure of the proposed model. Delayed payment provided by both supplier and retailer is incorporated in the model as well. Avinadav et al. (2013) also provided a mathematical model with price and time-dependent demand for perishable items.

Coming up to a conclusion, there exist number of research gaps: Firstly in the area of dynamic pricing, there are few studies formulating price as a function of time by defining discount variable, while this study has formulated the selling price as a time-dependent function by defining discount fraction variable as well as initial price. Secondly, in spite of the high significance of advertisement in stimulating demand, there are few related research works linking demand to the advertisement factor. In this study, the advertisement factor is modeled as the frequency of advertisement in each replenishment cycle.

3. Notations and assumptions

Notations

Parameters

c	The constant purchasing cost per unit
h	The holding cost per unit per time unit
k	The ordering cost per order
G	Cost of each advertisement
$\theta(t)$	The deterioration rate

Variable

A	The frequency of advertisement in each cycle (decision variable)
$p(t)$	The dynamic price of product per unit at any time t (decision variable)
T	The replenishment cycle of the product (decision variable)
$I(t)$	The inventory level at time t
Q	The order quantity
OC	The total ordering cost
HC	The total inventory holding cost
PC	The total purchasing cost

AC	The total advertisement cost
DC	The total disposing cost
$TP(p(t), A, T)$	The total profit per time unit of the inventory system

Assumptions

1. The replenishment rate and planning horizon are infinite.
2. The lead time is zero and shortage is not allowed.
3. The inventory system involves single deteriorating item.
4. The dynamic price of the product at time t is modeled as: $p(t) = p_0 \exp(-\sigma t)$ where p_0 is the initial price and $\sigma \in \pi$ is discount variable for each time unit passing after the start of deterioration. In this paper we have considered $\pi = \{0.2, 0.5, 0.8\}$ naming low, medium and high discount rates.
5. The demand rate is a function of the selling price and frequency of advertisement, which is formulated as $D(p(t), A) = (MB - \omega p(t))(1 + A)^{\lambda}$ where MB is the potential demand where price is equal to zero, $\omega > 0$ is the price sensitivity factor and $0 \le \lambda < 1$ is the shape parameter of the advertisement.
6. The deterioration rate $\theta(t)$ at any time $t \ge 0$ follows Weibull distribution given by $\alpha \beta t^{\beta - 1}$. Where $0 < \alpha \le 1$ is the scale parameter and $\beta > 0$ is the shape parameter. In this paper we have assumed $\beta = 2$.

4. The mathematical formulation

The inventory system evolves as follows: Q units arrive at the inventory system at the beginning of each cycle. Then inventory level declines due to demand and deterioration during time interval $[0, T]$. Based on this description, during time interval $[0, T]$ the inventory status is represented by the following differential equation:

$$\frac{dI_2(t)}{dt} = -D(p(t), A) - \theta(t)\, I(t) \qquad 0 \le t \le T \tag{1}$$

With boundary condition $I(T) = 0$ solving Eq. (1) yields:

$$I(t) = e^{-\alpha t^2} \int_t^T D(p(u), A)\, e^{\alpha u^2}\, du \qquad 0 \le t \le T \tag{2}$$

Therefore the order quantity is equal to:

$$Q = I(0) = \int_0^T D(p(u), A)\, e^{\alpha u^2}\, du \tag{3}$$

The total profit of the inventory system involves the following components:

1. SR: The sales revenue

$$SR = \int_0^T p(t)\, D(p(t), A)\, dt \tag{4}$$

2. OC: The ordering cost

$$OC = k \tag{5}$$

3. HC: The inventory holding cost

$$HC = h\int_0^T I(t)dt = h\int_0^T e^{-\alpha t^2}\int_t^T D(p(u),A)\,e^{\alpha u^2}du\,dt \tag{6}$$

4. PC: The purchasing cost

$$PC = cQ = c(\int_0^T D(p(u),A)\,e^{\alpha u^2}du\ ) \tag{7}$$

5. AC: The advertisement cost

$$AC = G\,A \tag{8}$$

Therefore, the total profit per time unit ($TP(p(t),A,T)$) is given by:

$$TP(p(t),A,T) = \frac{1}{T}(SR - OC - HC - PC - AC) \tag{9}$$

5. Solution methodology

Since the formulated equations are complex, the concavity of the total profit per time unit cannot be proved by using Hessian matrix. Therefore, the problem is solved applying the following search procedure which is similar to ones used in Wu et al. (2009) and Shah et al. (2013) as well. We first prove that for given values of σ, p_0 and T there exist a unique optimal value of A. Then for fixed values of σ, p_0 and A ,a unique optimal value of T is obtained and finally, for fixed σ, A and T, a unique optimal value of p_0 is determined which maximizes the total profit per time unit. Since the discount fraction is defined as a discrete variable, the above mentioned procedure is applied for different values of σ and finally the optimal solution is obtained by comparing the results. Firstly, for fixed σ, p_0 and T, the second order derivative of $TP(p(t),A,T)$ is obtained as follows:

$$\frac{\partial^2 TP}{\partial A^2} = \frac{\lambda(\lambda-1)(1+A)^{\lambda}}{T}(Expression1) \tag{10}$$

where:

$$Expression1 = Expression\,a - Expression\,b - Expression\,c \tag{11}$$

$$Expression\,a = \int_0^T (MB - \omega p_0(e^{-\sigma t}))p_0(e^{-\sigma t})dt \tag{12}$$

$$Expression\,b = h\int_0^T (e^{-\alpha t^2})\int_t^T [(MB - \omega p_0(e^{-\sigma u}))(e^{\alpha u^2})du]dt \tag{13}$$

$$Expression\,c = c\int_0^T (MB - \omega p_0(e^{-\sigma t}))(e^{\alpha t^2})dt \tag{14}$$

Since $\lambda < 1$ it is obvious that $\frac{\partial^2 TP}{\partial A^2} < 0$, therefore $TP(p(t),A,T)$ is a concave function of A and the search to find the optimal frequency of advertisement is restricted to find a local optimum. Now some useful theoretical results are derived in order to find the optimal length of replenishment (T^*) and the optimal initial price (p_0^*) for two aforementioned possible cases.

Lemma 1. For fixed A and p_0 there exists a unique T^* which maximizes $TP(p(t),A,T)$.

Proof. The first order partial derivative of $TP(p(t),A,T)$ with respect to $T$is given by

$$\frac{\partial TP}{\partial T} = \frac{(SR' - HC' - PC')T - (SR - OC - HC - PC - AC)}{T^2} \tag{15}$$

where:

$$SR' = p_0(1+A)^{\lambda}(e^{-\sigma T})(MB - \omega p_0(e^{-\sigma T})) \tag{16}$$

$$HC' = h(1+A)^{\lambda}(MB - \omega p_0(e^{-\sigma T}))(e^{\alpha T^2})\int_0^T e^{-\alpha t^2}dt \tag{17}$$

$$PC' = c(1+A)^{\lambda}(MB - \omega p_0(e^{-\sigma T}))(e^{\alpha T^2}) \tag{18}$$

Motivated by Eq. (15) the auxiliary function $R(T)$ is defined as:

$$R(T) = (SR' - HC' - PC')T - (SR - OC - HC - PC - AC) \tag{19}$$

The first order derivative of $R(T)$ with respect to $T \in [t_d, \infty)$ gives:

$$\frac{dR(T)}{dT} = (SR'' - HC'' - PC'')T \tag{20}$$

where:

$$SR'' = -\sigma p_0(1+A)^{\lambda} e^{-\sigma T}(MB - 2\omega p_0 e^{-\sigma T}) \tag{21}$$

$$HC'' = h(1+A)^{\lambda}\left\{\begin{matrix}\left[2\alpha T e^{\alpha T^2}(MB - \omega p_0(e^{-\sigma T})) + \sigma\omega p_0(e^{-\sigma T})(e^{\alpha T^2})\right]\int_0^T (e^{-\alpha t^2})dt \\ (MB - \omega p_0(e^{-\sigma T}))\end{matrix}\right\} \tag{22}$$

$$PC'' = c(1+A)^{\lambda}\left[2\alpha T e^{\alpha T^2}(MB - \omega p_0(e^{-\sigma T})) + \sigma\omega p_0(e^{-\sigma T})(e^{\alpha T^2})\right] \tag{23}$$

It is shown that $R(T)$ is a strictly decreasing function, moreover from Eq. (19) it follows that:

$$R(0) = k + AG > 0 \tag{24}$$

$$\text{and } \lim_{T\to\infty} R(T) = -\infty \tag{25}$$

Applying the intermediate value theorem, there exist a unique value of T (say $T^* \in [0, \infty)$) where $R(T) = 0$ which means that T^* is the unique solution of $\frac{\partial TP(p(t), A, T)}{\partial T} = 0$.

From Eq. (15) and Eq. (19) we have

$$\frac{\partial TP(p(t), A, T)}{\partial T} = \frac{R(T)}{T^2} \tag{26}$$

Since $R(T)$ is strictly decreasing (i.e. $\frac{dR(T)}{dT} = (SR'' - HC'' - PC'')T < 0$) we have:

$$\frac{\partial^2 TP(p(t), A, T)}{\partial T^2}\Big|_{T=T^*} = \frac{(SR'' - HC'' - PC'')}{T} < 0 \tag{27}$$

Therefore, There exist a unique global maximum solution of $TP(p(t), A, T)$.

Lemma 2. For fixed A and T there exists a unique p_0^*, which maximizes $TP(p(t), A, T)$.

Proof. The first order derivative of $TP(p(t), A, T)$ with respect to p_0 gives

$$\frac{\partial TP}{\partial p_0} = (1+A)^{\lambda}\{MB\int_0^T e^{-\sigma t}dt - 2\omega p_0\int_0^T e^{-\sigma t}dt + h\omega\int_0^T (e^{-\alpha t^2})\int_t^{T_1}(e^{\alpha u^2})(e^{-\sigma u})du dt + c\omega\int_0^T (e^{\alpha t^2})(e^{-\sigma t})dt \tag{28}$$

By solving $\frac{\partial TP}{\partial p_0} = 0$, p_0^* yields

$$p_0^* = \{MB\int_0^{T^*} e^{-\sigma t}dt + h\omega\int_0^{T^*}(e^{-\alpha t^2})\int_t^{T^*}(e^{\alpha u^2})(e^{-\sigma u})dudt + c\omega\int_0^{T^*}(e^{\alpha t^2})(e^{-\sigma t})dt\} / -2\omega\int_0^{T^*} e^{-\sigma t}dt \tag{29}$$

At point $p_0 = p_0^*$

$$\frac{\partial^2 TP}{\partial p_0^{\,2}}\Big|_{p_0=p_0^*} = -2\omega(1+A)^{\lambda}\int_0^{T^*} e^{-\sigma t}dt + h\omega \tag{30}$$

$\frac{\partial^2 TP}{\partial p_0^{\,2}}\Big|_{p_0=p_0^*} < 0$. Thus p_0^* is the global optimum solution for fixed A and $T^* \in [0,\infty)$.

Remark. For fixed A and σ the unique optimal solution for (T, p_0) which maximizes $TP(p(t), A, T)$ exists. The optimal solution can be obtained through some iterative numerical procedure. Based on the concavity of the objective function, the following algorithm which is similar to the one proposed by Wu et al. (2009) is developed to identify global optimal solution for (A, p_0, σ, T).

Algorithm

Step 1: Set $j = 1$ and $\sigma = 0.2$.
Step 2: Set $A^j = 0$.
Step 3: Set $k = 1$ and initialize the value of $p_0^{k,j} = c$.
Step 4: Obtain the value of $T^{k,j}$ by solving $\frac{\partial TP(p(t), A, T)}{\partial T} = 0$. Substitute $T^{k,j}$ into Equation (27)in order to calculate $p_0^{k,j}$. Set $p_0^{k+1,j} = p_0^{k,j}$.
Step 5: If $\left|p_0^{k+1,j} - p_0^{k,j}\right| < Epsilon$ (*Epsilon* is considered to be a very small value), then set $(p_0^{j*}, T^{j*}) = (p_0^{(k+1,j)^*}, T^{(k,j)*})$ and go to Step 6. Otherwise, $k = k+1$ and go back to Step 4.
*Step 6:*Calculate $TP(p(t)^{j*}, A^j, T^{j*})$, then (p_0^{j*}, T^{j*}) is the optimal solution and $TP(p(t)^{j*}, A^j, T^{j*})$ is the maximum value of the objective function for fixed A^j and σ.
Step 7: Set $A'^j = A^j + 1$ and repeat Step 3 to Step 6 to obtain $TP(p(t)^{j*}, A'^j, T^{j*})$.
*Step 8:*If $TP(p(t)^{j*}, A'^j, T^{j*}) > TP(p(t)^{j*}, A^j, T^{j*})$, then $A^j = A'^j$ and go back to Step 7. Otherwise go to Step 9.
Step 9: Set $(p_0^{j*}, A^{j*}, T^{j*}) = (p_0^{j*}, A^j, T^{j*})$ which is the optimal solution for fixed σ.
Step 10: Set $j = j+1$ and $\sigma = \sigma + 0.3$. If $\sigma \le 0.8$ go back to Step 2; otherwise go to Step 11.
Step 11: Set $TP(p(t)^{p*}, A^p, T^{p*}) = \max_j\{TP(p(t)^{j*}, A^j, T^{j*})\}$ and $(p_0^*, \sigma^*, A^*, T^*) = (p_0^{p*}, \sigma^{p*}, A^{p*}, T^{p*})$.
Step 12: End

6. Experimental results

In this section, the developed algorithm is applied to solve the following problem in order to show validity of the proposed model and applicability of the developed algorithm. The values of the parameters of the problem are defined in Table 2. Identical parameters of model are taken from Shah et al. (2013) and adopted to our model. The expanded results of the executed procedure of the proposed algorithm are represented in Table 3. In order to show the effect of related deterioration and advertisement parameters, Table 4 provides the computational results for different values of α and λ.

Table 2
The value of the parameters of the problem

Parameter	MB	ω	k	G	h	c	α	λ
Value	300	12	250	80	0.4	3	0.75	0.04

Table 3
Illustration of solution procedure of the proposed algorithm

σ	A	p_0	T	Q	TP
0.2	0	4.2221	0.5199	197.5145	9719.384
	1	4.2485	0.5525	202.8615	10098.724
	2	4.2714	0.5705	208.7295	10217.574
	3	4.3108	0.6283	240.1519	10187.446
0.5	0	4.7874	0.5798	190.1285	10864.274
	1	4.8217	0.5979	212.0535	11359.493
	2	4.8522	0.6248	228.4295	11560.514
	3	4.8794	0.6305	242.5535	11636.628
	4	4.9294	0.6585	266.7615	11611.554
0.8	0	5.5495	0.5766	197.4795	12593.551
	1	5.5814	0.5865	216.4575	13233.156
	2	5.6103	0.5955	230.0738	13523.818
	3	5.6370	0.6039	241.3152	13666.859
	4	5.6620	0.6118	251.1525	13727.742
	5	5.6855	0.6392	260.0795	13736.595
	6	5.7078	0.6262	268.3591	13709.344

Table 4
Computational results for different values of α and λ

α	λ	σ	A	p_0	T	Q	TP
0.3	0.03	0.8	3	5.6011	0.6425	257.4321	13742.143
	0.04	0.8	5	5.9921	0.6642	270.3246	14417.142
	0.05	0.8	9	6.3348	0.7121	286.5532	14705.132
0.45	0.03	0.8	3	5.4771	0.6374	242.5512	13579.439
	0.04	0.8	5	5.8012	0.6513	264.8742	14227.169
	0.05	0.8	9	6.2189	0.7003	285.1327	14502.045
0.6	0.03	0.8	3	5.4581	0.6213	234.8765	13327.912
	0.04	0.8	5	5.6991	0.6426	252.4384	14013.762
	0.05	0.8	9	6,0872	0.6821	277.1132	14313.261
0.75	0.03	0.8	3	5.4327	0.6107	220.3471	13181.654
	0.04	0.8	5	5.6855	0.6392	260.0795	13736.595
	0.05	0.8	9	6.0431	0.6682	278.9554	14121.842

Based on the computational results, following managerial insights are obtained about deterioration and advertisement factors which are analogous to Shah et al. (2013):

1. For fixed α, increasing λ causes an increase in optimal replenishment cycle (T^*), optimal order quantity (Q^*), optimal advertisement frequency (A^*), optimal initial price (p_0^*) and total profit per time unit of the inventory system (TP). Increasing the shape of advertisement (λ) increase A^* and therefore, results in an increase in demand and the total profit of the system. The optimal discount fraction (σ^*) is not sensitive to changes in λ.
2. For fixed λ, with an increase in the value of α, optimal advertisement frequency (A^*) and optimal discount fraction (σ^*) remain unchanged, while increasing α results ina decrease in the value of optimal order quantity (Q^*) and total profit per time unit of the inventory system (TP). In comparison to the other variables, changes in the value of optimal initial price (p_0^*) is

imperceptible. It can be concluded that, implementing better inventory preservation technologies lowers the effect of deterioration and enhances the total profit of the inventory system.

7. Conclusion

Coordination of inventory management and marketing policies play an important role in maximizing profit of firms. In this paper, we proposed an integrated model for dynamic pricing and inventory control of non-instantaneous deteriorating items. The selling price was defined as a time-dependent function of the initial price and discount rate, which is one of the noble features of the proposed model. In addition to price, the demand rate was a function of advertisement as a powerful marketing tool. In spite of the high significance of advertisement in stimulating demand, there exist few related research works linking demand to the advertisement factor. In this study, the advertisement factor was modeled as the frequency of advertisement in each cycle. An iterative algorithm was developed based on derived theoretical results. We illustrated through the experimental results the way the optimal solution was obtained. Computational results indicated that increasing the impact of the deterioration decreases total profit of the inventory system, which was due to the imposed costs on the system. Therefore, implementing better inventory preservation technologies can efficiently enhance the profit of the system by lowering the negative effect of the deterioration. The proposed model can be extended by considering shortages, trade credit and time value of money. Considering non-instantaneous deterioration pattern is another future research direction.

References

Abad, P. L. (2003). Optimal pricing and lot-sizing under conditions of perishability, finite production and partial backordering and lost sale. *European Journal of Operational Research*, 144(3), 677–685.

Abad, P. L. (2008). Optimal price and order size under partial backordering incorporating shortage, backorder and lost sale costs. *International Journal of Production Economics*, *114*(1), 179–186.

Avinadav, T., Herbon, A., & Spiegel, U. (2013). Optimal Inventory Policy for a Perishable Item with Demand Function Sensitive to Price and Time. *International Journal of Production Economics*, *144*(2),497–506.

Begum, R., Sahoo, R. R., Sahu, S. K., & Mishra, M. (2009). An EOQ Model for Varying Items with Weibull Distribution Deterioration and Price-dependent Demand. *Journal of Scientific Research*, *2*(1), 24–36.

Begum, R., Sahoo, R. R., & Sahu, S. K. (2012). A replenishment policy for items with price-dependent demand, time-proportional deterioration and no shortages. *International Journal of Systems Science*, *43*(5), 903–910.

Cai, X., Feng, Y., Li, Y., & Shi, D. (2013). Optimal pricing policy for a deteriorating product by dynamic tracking control. *International Journal of Production Research*, 51(8), 2491–2504.

Cohen, M. A. (1977). Joint pricing and ordering policy for exponentially decaying inventory with known demand. *Naval Research Logistics Quarterly*, 24(2), 257–268.

Coy, P. (2000). The power of smart pricing. Business week, (April 10), 160.

Dye, C. Y. (2007). Joint pricing and ordering policy for a deteriorating inventory with partial backlogging. *Omega*, *35*(2), 184–189.

Dye, C. Y. (2012). A finite horizon deteriorating inventory model with two-phase pricing and time-varying demand and cost under trade credit financing using particle swarm optimization. *Swarm and Evolutionary Computation*, 5, 37–53.

Dye, C. Y., Hsieh, T. P., & Ouyang, L. Y. (2007). Determining optimal selling price and lot size with a varying rate of deterioration and exponential partial backlogging. *European Journal of Operational Research*, *181*(2), 668–678.

Geetha, K. V., & Uthayakumar, R. (2010). Economic design of an inventory policy for non-instantaneous deteriorating items under permissible delay in payments. *Journal of Computational and Applied Mathematics*, 233(10), 2492-2505.

Ghoreishi, M., Mirzazadeh, A., & Weber, G. W. (2013). Optimal pricing and ordering policy for non-instantaneous deteriorating items under inflation and customer returns. *Optimization*, (ahead-of-print), 1-20.

Hsieh, T. P., & Dye, C. Y. (2010). Pricing and lot-sizing policies for deteriorating items with partial backlogging under inflation. *Expert Systems with Applications*, 37(10), 7234-7242.

Hsu, P. H., Wee, H. M., & Teng, H. M. (2007). Optimal ordering decision for deteriorating items with expiration date and uncertain lead time. *Computers & Industrial Engineering*, 52(4), 448–458.

Kang, S., & Kim, I. T., (1983). A study on the price and production level of the deteriorating inventory system. *The International Journal of Production Research*, 21(6), 899–908.

Khanlarzade, N., Yegane, B. Y., Kamalabadi, I. N., & Farughi, H. (2014). Inventory control with deteriorating items: A state-of-the-art literature review. *International Journal of Industrial Engineering Computations*, 5(2), 179–198.

Maihami, R., &Abadi, I. N. K. (2012). Joint control of inventory and its pricing for non-instantaneously deteriorating items under permissible delay in payments and partial backlogging. *Mathematical and Computer Modelling*, 55(5), 1722–1733.

Mukhopadhyay, S., Mukherjee, R. N., & Chaudhuri, K. S. (2004). Joint pricing and ordering policy for a deteriorating inventory. *Computers & Industrial Engineering, 47*(4), 339–349.

Mukhopadhyay, S., Mukherjee, R. N., & Chaudhuri, K. S. (2005). An EOQ model with two-parameter Weibull distribution deterioration and price-dependent demand. *International Journal of Mathematical Education in Science and Technology*, *36*(1), 25–33.

Papachristos, S., & Skouri, K. (2003). An inventory model with deteriorating items, quantity discount, pricing and time-dependent partial backlogging. *International Journal of Production Economics*, 83(3), 247–256.

Shah, N. H., & Raykundaliya, N. (2010). Retailers' Pricing and Ordering Strategy for Weibull Distribution Deterioration under Trade Credit in Declining Market. *Applied Mathematical Sciences*, 4(21), 1011–1020.

Shah, N. H., Soni, H. N., & Patel, K. A. (2013). Optimizing inventory and marketing policy for non-instantaneous deteriorating items with generalized type deterioration and holding cost rates. *Omega,* 41(2), 421-430.

Soni, H. N., & Joshi, M. (2013). A fuzzy framework for coordinating pricing and inventory policies for deteriorating items under retailer partial trade credit financing. *Computers & Industrial Engineering*, 66, 865-878.

Soni, H. N., & Patel, K. A. (2012). Optimal pricing and inventory policies for non-instantaneous deteriorating items with permissible delay in payment: Fuzzy expected value model. *International Journal of Industrial Engineering Computations*, 3, 281-300.

Soni, H. N., & Patel, K. A. (2013). Joint pricing and replenishment policies for non-instantaneous deteriorating items with imprecise deterioration free time and credibility constraint. *Computers & Industrial Engineering*, 66, 944-951.

Teng, J. T., Ouyang, L. Y., & Chen, L. H. (2007). A comparison between two pricing and lot-sizing models with partial backlogging and deteriorated items. *International Journal of Production Economics*, 105(1), 190–203.

Tripathy, C. K., & Pradhan, L. M. (2011). Optimal Pricing & Ordering Policy for three parameter Weibull deterioration under trade credit. *International Journal of Mathematical Analysis*, 5(6), 275–284.

Tsao, Y. C., & Sheen, G. J. (2008). Dynamic pricing, promotion and replenishment policies for a deteriorating item under permissible delay in payments. *Computers & Operations Research*, 35(11), 3562–3580

Valliathal, M., &Uthayakumar, R. (2011). Optimal pricing and replenishment policies of an EOQ model for non-instantaneous deteriorating items with shortages. *International Journal of Advanced Manufacturing Technology*, 54(1-4), 361-371.

Wang, Y., Zhang, J., & Tang, W. (2013). Dynamic pricing for non-instantaneous deteriorating items. *Journal of Intelligent Manufacturing*, 1–12.

Wee, H. M. (1997). A replenishment policy for items with a price-dependent demand and a varying rate of deterioration. *Production Planning & Control*, 8(5), 494–499.

Wee, H. M. (1999). Deteriorating inventory model with quantity discount, pricing and partial backordering. *International Journal of Production Economics*, 59(1), 511–518.

Wu, K. S., Ouyang, L. Y., & Yang, C. T. (2009). Coordinating replenishment and pricing policies for non-instantaneous deteriorating items with price-sensitive demand. *International Journal of Systems Science*, 40(12), 1273–1281.

Yang, P. C. (2004). Pricing strategy for deteriorating items using quantity discount when demand is price sensitive. *European Journal of Operational Research*, 157(2), 389–397.

Yang, P. C., & Wee, H. M. (2003). An integrated multi-lot-size production inventory model for deteriorating item. *Computers & Operations Research*, 30(5), 671-682.

4

Selection of industrial robots using the Polygons area method

Mortaza Honarmande Azimi[a*], Houshang Taghizadeh[b], Nasser Fegh-hi Farahmand[b] and Jafar Pourmahmoud[c]

[a]*Department of Management, College of Management, Economics & Accounting, Tabriz Branch, Islamic Azad University, Tabriz, Iran*
[b]*Associate Professor, Department of Management, College of Management, Economics & Accounting, Tabriz Branch, Islamic Azad University, Tabriz, Iran*
[c]*Assistant Professor, Department of Applied Mathematics, Azarbaijan Shahid Madani University, Tabriz, Iran*

C H R O N I C L E

Keywords:
Robot Selection
Multi-attribute Decision Making (MADM)
Polygons Area Method (PAM)
SAW
TOPSIS
VIKOR
WPM

A B S T R A C T

Selection of robots from the several proposed alternatives is a very important and tedious task. Decision makers are not limited to one method and several methods have been proposed for solving this problem. This study presents Polygons Area Method (PAM) as a multi attribute decision making method for robot selection problem. In this method, the maximum polygons area obtained from the attributes of an alternative robot on the radar chart is introduced as a decision-making criterion. The results of this method are compared with other typical multiple attribute decision-making methods (SAW, WPM, TOPSIS, and VIKOR) by giving two examples. To find similarity in ranking given by different methods, Spearman's rank correlation coefficients are obtained for different pairs of MADM methods. It was observed that the introduced method is in good agreement with other well-known MADM methods in the robot selection problem.

1. Introduction

During the past few years, there have been tremendous efforts to use robots for different hard circumstances. Shivanand et al. (2006) define robots as "automatically controlled, reprogrammable, multi-purpose manipulators programmable in three or more axes". Indeed, robots are still novel concept in industry as a whole, and so it is not unusual for an industry to be a first-time robot purchaser (Rao, 2013). Robots with various capabilities and specifications are available for a wide range of applications. Hence, selection of the most appropriate robot has long been a difficult decision task. MCDM methods are the most common approach applied for robot selection (Tansel İç et al. 2013). Multiple Criteria Decision Making (MCDM) is associated with making decisions where there are different and usually conflicting criteria. These methods are divided into two general Multiple Objective Decision Making (MODM) and Multiple Attribute Decision Making (MADM) categories. MODM has decision variables in the form of a continuous variable or an integer. MODM can have one or multiple alternatives, which the selected alternative is the one, which fulfills the constraints of the

* Corresponding author.

problem and satisfaction of the decision-maker (Zanakis et al., 1998). Most MODM methods employ mathematical programming under various criteria. These methods are not easily compatible with problems of the manufacturing environment due to their specific mathematical difficulties. Although many applications and programs have been presented for these methods, lack of simplicity of understanding the theories placed behind them made them an unattractive option for users.

MADM methods are widely applied for real-world problems (Rao, 2007, 2013; Fendel & Spronk, 1983; Hwang & Yoon, 1981). Methods such as the Technique for Order Preference by Similarity to Ideal Solution (TOPSIS), Simple Additive Weighting (SAW) or weighted sum model (WSM), Weighted Product Method (WPM) or Multiplicative Exponent Weighting (MEW), Analytic Hierarchy Process (AHP), ELimination Et Choix Traduisant la REalité (ELECTRE), Vise Kriterijumska Optimizacija Kompromisno Resenje (VIKOR), and Linear Assignment are widely used methods. However, these methods have their own advantages and disadvantages (Rao, 2007). Their major deficiency of MADM is that in a single problem, various methods present different results for selection or ranking. By comparing the methods, the researchers found that in 40% of cases, each method presents a result, which is different from the other one (Voogd, 1983). To solve this problem, some methods have been introduced (Wang et al., 2005; Momeni, 2007; Rao, 2013) known as "aggregation methods". In this method, a problem with several MADM methods is ranked, and then, the final selection may be made on the basis of an aggregation of the results of those methods that have better Spearman's rank correlation coefficients (Rao, 2013). Another way of deciding is to choose an alternative suggested by the "majority" of the MADM methods (Rao, 2013). Based on MADM literature, the reason that researchers seek new methods for decision-making is to enhance selection confidence or ranking methods, which perform the selection or ranking of alternatives easily and reliably, in addition to previous methods.

This study presents the polygons area method for the robot selection problem. In this method, the maximum polygons area obtained from the attributes of an alternative robot on the radar chart is introduced as a decision-making criterion (Honarmande Azimi & Pourmahmoud, 2012). An advantage of the presented method is simplicity of the algorithm and primary calculations required for finding a decision criterion. The required calculations in this method are easily performed by four basic operations. In addition, the decision-making criterion is obtained based on a concrete and extremely precise method with proven experimental and mathematical formulas. Another advantage of this method is the sound agreement of the results from this method with other major MADM methods. The simplicity of learning and training is another advantage of this method. To show the applicability of this method in robot selection, two examples already studied by other researchers with other MADM methods have been proposed. To demonstrate the high correlation of the results from this method with main MADM methods, Spearman's correlation coefficient has been employed (Sheskin, 2004; Yurdakul & İç, 2009; Rao, 2013). Investigating the proposed examples indicates the high correlation of ranking results from this method with other MADM methods, and this method can be used along with other main MADM methods due to its simplicity and convenience for robot selection.

2. Literature Review

Industrial robot usage by manufactures has increased extensively in the past decades. There are over 75 attributes, which can be considered while selecting a robot (Bhangale et al., 2004). These attributes can be either objective or subjective. Objective attributes include cost, velocity, load capacity of the robot, load carrying capacity, repeatability, memory capacity, manipulator reach, and degree of freedom, which can be measured by numerical values. On the other hand, subjective criteria are qualitative in nature (e.g. vendor's service quality and robot's programming flexibility etc.) (Rao, 2007). A decision maker that can use robots is faced with different alternative robots for a given industrial application. Therefore, selection of a robot becomes an important problem, because improper selection of robots may adversely influences on the profitability of the company (Rao, 2013). There are many models for robot selection; different MADM methods can be effectively used to solve this problem (Athawale & Chakraborty, 2011; Tansel İç et al., 2013). Studies which use MADM methods are presented for robot selection in the Literature (Table 1). As seen in Table 1, MADM methods are the most common approach type applied for ranking robots.

Table 1
Robot selection literature

N0.	Name of the authors	Year	Method for robot selection
1	Agrawal et al.	1991	Employed the TOPSIS method
2	Wang et al.	1991	Decision support system applies a fuzzy set
3	Khouja	1995	DEA
4	Goh	1997	AHP
5	Bhangale et al.	2004	Using TOPSIS and graphical methods
6	Rao and Padmanabhan	2006	Digraph and matrix method
7	kahraman et al.	2007	Fuzzy TOPSIS
8	Karsak	2008	QFD and fuzzy linear regression
9	Chatterjee et al	2010	VIKOR & ELECTRE II
10	Singh & Rao	2011	Analytical hierarchy graph theory and matrix approach(AHGTAMA)
11	Rao and Patel	2011	subjective and objective integrated MADM method
12	Devi	2011	Fuzzy VIKOR
13	Athawale & Chakraborty	2012	Fuzzy VIKOR
14	Tao et al.	2012	DEA, AHP,TOPSIS,AFS(axiomatic fuzzy set)
15	Karsak	2012	Fuzzy regression-based decision-making approach
16	Vahdani et al.	2013	Fuzzy TOPSIS
17	Tansel İç et al.	2013	Fuzzy AHP,ROBSEL(ROBot SELection)
18	Mondal & Chakraborty	2013	DEA

The idea of using the polygons area on the radar chart or spider chart as a decision-making criterion has been addressed in industry and management literature in recent decades, most of which are in graphical and non-quantitative form. Results obtained from investigations indicate that few similar investigations have been conducted to find the area of polygons on the radar chart as a decision criterion. Using these methods, Bhangale et al. (2004) ranked robots by representing TOPSIS and graphical method. We can mention Waigaonkar et al. (2008) and Rao and Baral (2011) as later works conducted based on of Bhangale's method. The method proposed by Bhangale et al. (2004) was conducted regardless of placement order of the criteria on the chart and this method had insufficient stability. In other words, variations in area value will result in different decision-making criteria from one order to the other. Thus, since the decision criterion is not stable, therefore, the results will not be acceptable in practice. Honarmande Azimi and Pourmahmoud (2012) proposed an algorithm to form polygons on the radar chart, which on that basis, the maximum area formed on the chart is considered as a decision-making criterion by an alternative.

This study proposes a method for solving the robot selection problem based on the algorithm presented by Honarmande Azimi and Pourmahmoud (2012). In previous MADM methods presented by researchers, a limited number of studies have been conducted in terms of graphical method, and no method has been presented based on the proposed algorithm. Among the many MADM methods available, in our research, we have chosen SAW, WPM, TOPSIS, and VIKOR methods for comparison. The rationale for selection was that most of these methods are among the more popular and widely used methods (Rao, 2007, 2013; Behzadian et al., 2012). SAW and WPM assume additive and multiplicative weighted preferences in an interval scale (Zanakis et al., 1998); and SAW was selected as the basis for which to compare the other methods, because its simplicity makes it often used by practitioners. Even some researchers argue that SAW should be the standard for comparisons (Triantaphyllou & Mann, 1989). The MADM methods VIKOR and TOPSIS are both based on an aggregating function representing ''closeness to the ideal'' (Zanakis et al., 1998). The VIKOR method uses linear normalization; the TOPSIS method uses vector normalization.

The basic principle of the TOPSIS method is that the solution should have the shortest distance from the ideal solution, and the farthest distance from the negative-ideal solution; however, it does not consider the relative importance of these points (Hwang and Yoon, 1981). VIKOR is a helpful tool in a situation where the decision maker is not able, or does not know how to express their preference at the beginning of system design (Opricovic & Tzeng, 2004).

3. MADM Methods

In this paper, we concentrate on single decision maker deterministic MADM methods. Nowadays, SAW, WPM, TOPSIS, VIKOR, and AHP methods are more widely used in practice. These methods are logical decision-making approaches and have a higher potential to solve decision-making problems in manufacturing environments (Rao, 2007).

3.1 SAW Method

This is also called Weighted Sum Method (WSM). SAW is the simplest and the most commonly used approach (Triantaphyllo et al., 1998). Each attribute is given a weight, and the sum of all weights must be one. Each alternative is assessed with respect to every criterion. Then, the performance score of the alternative is obtained using Eq. (1).

$$S_i = \sum_{j=1}^{n} w_j x_{ij} \qquad for \quad i = 1,2,\ldots,m \tag{1}$$

where m is the number of alternatives, and n is the number of decision criteria, x_{ij} is the actual value of the ith alternative in terms of the jth criteria, w_j is the weight of importance of the jth criterion. This method should be used when all the units employ identical units of measure. If the units of the decision matrix are normalized, then this method can be used for any type of attribute. The overall performance score (S_i) of the ith alternative is calculated as follows,

$$S_i = \sum_{j=1}^{n} w_j x'_{ij} \qquad for \quad i = 1,2,\ldots,m \tag{2}$$

where x'_{ij} is the normalized value of x_{ij}. The alternative having the highest S_i value is the best choice (Rao, 2007).

3.2 WPM Method

This method is similar to the SAW method. The main difference is that instead of addition, there is multiplication in this method. The overall performance score (P_i) for the ith alternative is computed as (Rao, 2007):

$$p_i = \prod_{j=1}^{n} [\ x'_{ij}\ ]^{w_j} \qquad for \quad i = 1,2,\ldots,m \tag{3}$$

Here, the normalized value of the ith alternative on the jth criterion is raised to the power of the relative weight of the corresponding criterion. The best alternative is the one having the highest Pi value.

3.3 TOPSIS method

Hwang and Yoon (1981) developed TOPSIS. This method is widely used because of its simple and programmable nature. The basic concept of this method is that the proper alternative has the shortest

Euclidean distance from the ideal solution, and the longest Euclidean distance from the negative-ideal solution. TOPSIS ranks the alternatives according to these two distance measures. More information can be found in Yoon and Hwang (1995). The main steps of TOPSIS are as follows (Rao, 2007; Triantaphyllou et al., 1998):

Step 1: Prepare decision matrix ($\boldsymbol{D} = [x_{ij}]_{m\times n}$)

Where m is the number of alternatives and n is the number of criteria, x_{ij} is the actual value of the ith alternative in terms of the jth criteria.

Step 2: Convert **D** matrix to normalized decision matrix ($\mathbf{D}'=[x'_{ij}]_{m\times n}$) using the following equation:

$$x'_{ij} = \frac{x_{ij}}{\sqrt{x_{i1}^2 + \ldots + x_{in}^2}} \quad for \quad i = 1,2,\ldots,m \quad j = 1,2,\ldots,n \tag{4}$$

Step 3: Obtain the weighted normalized matrix (**V**).

$$V_{ij} = w_j \times x'_{ij} \tag{5}$$

Where x'_{ij} is the normalized value of x_{ij} and w_j is the weight of importance j^{th} criterion.

Step 4: Determine the ideal (best) and the negative-ideal (worst) solutions as follows:

$$V^+ = \left\{\left[\max_i V_{ij} \middle| j \in j\right], \left[\min_i V_{ij} \middle| j \in j'\right] i = 1,\cdots,m\right\} = \left(V_1^+,\cdots,V_n^+\right) \tag{6}$$

$$V^- = \left\{\left[\min_i V_{ij} \middle| j \in j\right], \left[\max_i V_{ij} \middle| j \in j'\right] i = 1,\cdots,m\right\} = \left(V_1^-,\cdots,V_n^-\right) \tag{7}$$

where V_j^+ indicates the most preferable alternative and V_j^- indicates the least preferable alternative. J=(j=1,2,...,n) / j is associated with beneficial criteria and J′=(j=1,2,...,n) / j′ is associated with non-beneficial criteria.

Step 5: Calculate the separation measures of each alternative from the ideal and the negative ideal solutions using the following equations:

$$S_i^+ = \sqrt{\sum_{j=1}^{n} (v_{ij} - v_j^+)^2}, i = 1,\ldots, m$$
$$S_i^- = \sqrt{\sum_{j=1}^{n} (v_{ij} - v_j^-)^2}, i = 1,\ldots, m \tag{8}$$

where Si+ is the separation of each alternative from the ideal solution and S_i^- is the separation of each alternative from the negative-ideal solution.

Step 6: Calculate the relative closeness to ideal Solution.

The relative closeness of an alternative to the ideal solution is defined as below:

$$C_i = \frac{S_i^-}{(S_i^+ + S_i^-)} \tag{10}$$

where C_i is the priority criterion; the alternatives are ranked in descending order.

Step 7: Rank the preference order

Based on the relative closeness measures, the alternatives can be ranked in descending order. Therefore, the best alternative is the one that has the shortest distance to the ideal solution and the farthest distance from the negative-ideal solution.

3.4 VIKOR method

The VIKOR method was established by Yu (1973) and Zeleny (2002). The Serbian name is 'Vlse Kriterijumska Optimizacija Kompromisno Resenje' which means multiple criteria optimization (MCO) and compromise solution,. Later, this method was introduced as one applicable technique to implement within MADM (Opricovic, 1998). Opricovic and Tzeng (2004, 2005) developed the VIKOR method as an MADM method. This method focuses on ranking and selecting from a set of alternatives in the presence of conflicting criteria. The VIKOR method began with the form of L_P- metric (Opricovic & Tzeng, 2004):

$$L_{p,i} = \left\{ \sum_{j=1}^{n} [w_j \, (x_j^* - x_{ij})/(x_j^* - x_j^-)]^p \right\}^{\frac{1}{p}} \qquad (11)$$

$$1 \leq p \leq \infty \qquad i = 1, 2, \ldots, m$$

If we assume the jth criteria represents as a benefit

$$x_j^* = max_{\ i} \ x_{ij} \qquad \text{and} \qquad x_j^- = min_{\ i} \ x_{ij}$$

The alternatives are denoted as $a_1, a_2, a_3, \ldots, a_m$ and w_j is the weight of the jth criterion, where j=1, 2, …, n and x_{ij} is the performance score of the jth criterion for alternative a_i. The procedural steps for the VIKOR method are as follows:

Step 1: Prepare the decision matrix.

Step 2:

a) Determine the best x_j^* and the worst x_j^- values of all criterion functions, j=1, 2,…, n

b) Calculate S_i and R_i values,

$$S_i = L_{1,i} = \textstyle\sum_{j=1}^{n}(w_j[(x_j^* - x_{ij})/(x_j^* - x_j^-)]) \qquad i = 1, \ldots, m \qquad (12)$$

$$R_i = L_{\infty,i} = max^m \ of \ \{w_j[(x_j^* - x_{ij})/(x_j^* - x_j^-)]\} \quad j=1, \ldots, n \qquad (13)$$

For non-beneficial criteria , x_j^*, x_j^- can be rewritten as:

$$x_j^* = min_{\ i} \ x_{ij} \qquad \text{and} \qquad x_j^- = max_{\ i} \ x_{ij}$$

c) Calculate Q_i value

$$Q_i = v[(S_i - S^*)/(S^- - S^*)] + (1 - v)[(R_i - R^*)\ /(R^- - R^*) \qquad (14)$$

where

$$S^* = min_{\ i} \ S_i \qquad S^- = max_{\ i} \ S_i \qquad R^* = min_{\ i} \ R_i \qquad R^- = max_{\ i} \ R_i$$

v is introduced as weight of the strategy of 'the majority of attributes' (or 'the maximum group utility'). The value of v lies in the range of 0 to 1. Normally, its value is taken as 0.5. However, any value from 0 to 1 can be taken.

d) Arrange the alternatives in ascending order, according to the values of Qi, Ri, and Si values. The best alternative is the one having the minimum Q_i value.

Condition 1: Acceptable advantage:

$$Q(a'')-Q(a')\geq DQ \tag{15}$$

where a'' is the alternative with second place in the ranking list by Q; $DQ=1/(m-1)$. m is the number of alternatives.

Condition 2: Acceptable stability in decision-making: The alternative a' should also be the best ranked by S or/and R. If one of the conditions is not satisfied, then a set of compromise solutions is proposed, which consists of:

- Alternatives a' and a'' if only the C_2 is not satisfied,
- Or alternatives a', a'',…, a^m if the C_1 is not satisfied; and a^m is determined by the below relation for maximum m.

$$Q(a^m)-Q(a')<DQ \tag{16}$$

The alternative having the minimum value of Q is the best alternative. The core ranking result is the compromise-ranking list of alternatives and the compromise solution with the "advantage rate" (Opricovic & Tzeng, 2004).

3.5 Aggregate method

Decision makers are not limited to one method of MADM in critical situations, because various possible MADM methods attain different results. In order to overcome this problem, various methods called aggregate methods have been introduced. These methods include Rank Average method, Borda method, and Copeland method (Momeni, 2007).

The Average Method ranks alternatives based on the average of calculated ranks from different MADM methods. Each MADM method ranks all of the alternatives. If there are m alternatives, each alternative receives $(m-1)$ points for the first choice, $(m-2)$ points for the second choice, and so on. The alternative with the most points is declared winner in the Borda method.
The Copeland Method starts at the end of the Borda method. It calculates the number of losses for all alternatives. Subtracting the number of loses from numbers of wins determines the prominence of any alternatives (Wang et al., 2005). By considering ranking strategies (i.e. rank average, Borda and Copeland methods), and through creating one Partially Ordered Set (POSET), it will reach consensus" (Azar & Radjab zadeh, 2010).

3.6 Proposed method: Polygons Area Method(PAM)

One of the widely used charts in management problems and practical situations is the radar (Spider) chart. This chart graphically shows areas of relative strengths and relative weaknesses, as well as depicting general overall performance. Performance of each alternative in MADM problems can be shown on the radar chart as a polygon. In this paper, we calculate the maximum polygons area obtained as a decision criterion on the radar chart.

The stepwise procedure of the proposed method is given as follows (Honarmande Azimi and Pourmahmoud, 2012):

Step 1: Prepare decision matrix (**D**).

For an MADM problem when there are 'm' alternatives and 'n' attributes, the general form of decision matrix *(**D**)* is given as follows:

$$D = \begin{bmatrix} x_{11} & \cdots & x_{1j} & \cdots & x_{1n} \\ \cdots & \cdots & \cdots & \cdots & \cdots \\ x_{i1} & \cdots & x_{ij} & \cdots & x_{in} \\ \cdots & \cdots & \cdots & \cdots & \cdots \\ x_{m1} & \cdots & x_{mi} & \cdots & x_{mn} \end{bmatrix}_{m \times n} \quad (17)$$

The attributes may be objective, subjective or a combination of both.

Step 2: Normalization/Normalized decision matrix (**D'**)

There are many normalization procedures available in decision-making literature. In the proposed method, the linear normalization procedure is adopted (Rao, 2007). Let x'_{ij} for attribute j compared with alternative i as follows:

$$x'_{ij} = \frac{x_{ij}}{\max(x_{ij})}, \quad \text{if attribute is beneficial} \quad (18)$$

$$x'_{ij} = \frac{\min\left(x_{ij}\right)}{x_{ij}}, \quad \text{if attribute is non-beneficial} \quad (19)$$

Using linear normalization for this method, the decision matrix will transform into the following (Rao, 2007; Singh & Rao, 2011).

$$D' = \left[x'_{ij}\right]_{m \times n} \quad (20)$$

Step 3: Relative importance of attributes (***W***$_j$)

In MADM methods, it is also required to determine the priority weight (w_j) of each criterion such that the sum of weights for all criteria equals one. These priority weights can be determined using AHP or entropy methods (Rao, 2007). A pair wise comparison matrix is constructed using a scale of relative importance. Judgments are entered using the fundamental scale of the analytical hierarchy process (Saaty, 1980).

Step 4: Formation of weighted normalized decision matrix (**V**)

$$V_{ij} = w_j \times x'_{ij} \quad (21)$$

where w_j is the weight of jth criteria.

Step 5: Calculating maximum area of polygon on the radar chart

The attributes have been considered to form the radar (spider) chart. The attributes are plotted on the axes, and each alternative will form a polygon. From a different set of attributes, different polygons can be obtained, which seems to be unacceptable in reality. Hence, the polygon with maximum area is considered as a decision criterion. The polygon with maximum area is obtained using the following algorithm if $n \geq 4$:

a) In weighted decision matrix (V), for alternative *i*, attributes are ranked from the largest to the smallest in order.
b) An attribute with first raking is introduced over the base axis of the chart.
c) Attributes with even rankings, each having v_{ij} in length and $\theta = 2\pi / n$ angle of its previous attribute, are placed clockwise on the chart in a descending order.
d) Attributes with odd rankings are also placed counter clockwise similar to the previous step.
e) Connecting the points obtained from the two previous steps; draw the irregular polygon on the radar chart.
f) Represent the apexes of irregular polygon by beginning from the first ranking and moving counter clockwise by v'_{ij}.
g) The area of triangle (S_Δ) given SAS (Side, Angle, Side) for example (a, θ, b) is calculated as follows (Murray, 1968):

$$S_\Delta = 1/2ab \sin \theta \qquad (22)$$

Divide polygon into n triangle(s) and calculate the area of each one as follows:

$$S_\Delta = \frac{1}{2} v'_{ij} \times v'_{i(j+1)} Sin \frac{2\pi}{n} \qquad (23)$$

h) The area of an irregular polygon is obtained from the sum of the areas of these triangles, and is calculated as follows (Murray, 1968):

$$S_{pi} = \frac{1}{2} Sin \frac{2\pi}{n} (V'_{i1}V'_{i2} + V'_{i2}V'_{i3} + \ldots V'_{i(n-1)}V'_{in} + V'_{in} \times V'_{i1}) \qquad (24)$$

where S_{pi} is the area of polygon for the ith alternative on the chart.

Step 6: Selecting the best alternative

The purpose of this step is selecting the best alternative, which is performed with the help of calculated areas of the irregular polygon of S_{pi} s. In this formula, all S_{pi} s is a common (1/2sin2π/n) value, which can be relinquished from all comparisons.

$$C_i = S_{pi} / \frac{1}{2} \sin \frac{2\pi}{n} \qquad (25)$$

Shifting the priority criterion from S_{pi} to C_i s, in which C_i s are calculated as follows:

$$C_i = \sum_{j=1}^{n} (V'_{ij} \times V'_{i(j+1)}) + V'_{in} \times V'_{i1} \qquad i = 1, \ldots, m \qquad (26)$$

b) If the number of attributes N=3, C_i is calculated from the V matrix as follows:

$$C_i = v_{i1} \times v_{i2} + v_{i2} \times v_{i3} + v_{i3} \times v_{i1} \quad i = 1, \ldots, m \qquad (27)$$

The best alternative is the one having the highest C_i value.

Fig. 1 shows the framework of decision making in this paper.

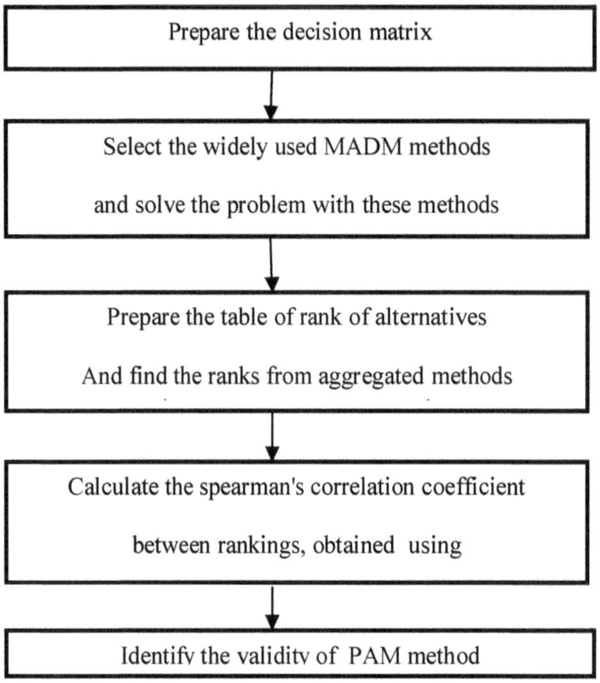

Fig. 1. Framework of decision making

4 Examples & Findings

4.1 Robot Selection for supporting NC machines

In this example, a jointed-arm robot is to be selected for supporting Numerical Control (NC) machines. The actual data was provided by the manufactures (Rao, 2007, 2013). Karasak and Ahiska (2005), Rao (2007, 2013), Wang and Chin (2009), Singh and Rao (2011), and Mondala and Chakraborty (2013) solved this example using MCDM methods. The decision matrix with twelve alternative robots and five attributes is given in Table 2. The attributes are **P**urchasing **C**ost (PC) in US$, Handling **C**oefficient (HC), **L**oad **C**apacity (HC) in kg, 1/ **RE**epeatability (RE) in 1/mm, and **VE**locity (VE) (m/s), among which HC, LC, RE, and VE are beneficial attributes (where higher values are desirable); whereas, PC is non-beneficial type (where lower values are preferable).

Table 2
Decision Matrix (D)

Robots	*PC*	*HC*	*LC*	*RE*	*VE*
R1	100000	0.995	85	1.7	3
R2	75000	0.933	45	2.5	3.6
R3	56250	0.875	18	5	2.2
R4	28125	0.409	16	1.7	1.5
R5	46875	0.818	20	5	1.1
R6	78125	0.664	60	2.5	1.35
R7	87500	0.88	90	2	1.4
R8	56250	0.633	10	8	2.5
R9	56250	0.653	25	4	2.5
R10	87500	0.747	100	2	2.5
R11	68750	0.88	100	4	1.5
R12	43750	0.633	70	5	3

The normalization of attribute data is carried out using Eq. (18), (19) and the normalized decision matrix (D') is given in Table 3.

Table 3

Normalized Matrix (D')

Robots	*PC*	*HC*	*LC*	*RE*	*VE*
R1	0.281	1	0.85	0.213	0.833
R2	0.375	0.938	0.45	0.313	1
R3	0.5	0.879	0.18	0.625	0.611
R4	1	0.411	0.16	0.213	0.417
R5	0.6	0.822	0.2	0.625	0.306
R6	0.36	0.667	0.6	0.313	0.375
R7	0.321	0.884	0.9	0.25	0.389
R8	0.5	0.636	0.1	1	0.694
R9	0.5	0.656	0.25	0.5	0.694
R10	0.321	0.751	1	0.25	0.694
R11	0.409	0.884	1	0.5	0.417
R12	0.643	0.636	0.7	0.625	0.833

The judgments made by decision makers or relative importance of attributes are given in Table 4. The normalized weight of each attribute is obtained from Super Decision software, which are w_{PC}=0.385, w_{HC}=0.092, w_{LC}=0.042, w_{RE}=0.219, w_{VE}=0.262 and consistency ratio (CR) is 0.08, which is much less than the allowed CR value of 0.1.

Table 4

The results of judgment

Attributes	PC	HC	LC	RE	VE
PC	1	5	5	1	3
HC	1/5	1	3	1/2	1/3
LC	1/5	1/3	1	1/6	1/7
RE	1	2	6	1	1/2
VE	1/3	3	7	2	1

Multiplying the (D') by weight matrix, the weighted normalized matrix (V) is given in Table 5.

Table 5

Weighted Normalized Matrix (V)

Robots	*PC*	*HC*	*LC*	*RE*	*VE*
R1	0.1082	0.0920	0.0357	0.0466	0.2182
R2	0.1444	0.0863	0.0189	0.0685	0.2620
R3	0.1925	0.0809	0.0076	0.1369	0.1601
R4	0.3850	0.0378	0.0067	0.0466	0.1093
R5	0.2310	0.0756	0.0084	0.1369	0.0802
R6	0.1386	0.0614	0.0252	0.0685	0.0983
R7	0.1236	0.0813	0.0378	0.0548	0.1019
R8	0.1925	0.0585	0.0042	0.2190	0.1818
R9	0.1925	0.0604	0.0105	0.1095	0.1818
R10	0.1236	0.0691	0.0420	0.0548	0.1818
R11	0.1575	0.0813	0.0420	0.1095	0.1093
R12	0.2476	0.0585	0.0294	0.1369	0.2182

Now the proposed PAM method, SAW, WPM, TOPSIS and VIKOR methods are applied to this problem as explained before. Comparison of rankings obtained by these methods is given in Table 7. The value of V'_{ij} for PAM is calculated using the introduced algorithm in step 5. For example, the C_1 for robot 1 is calculated using Eq. (26) as:

V'_{11}= 0.2182; V'_{12}=0.092; V'_{13}=0.0357; V'_{14}=0.0466; V'_{15}=0.1082

$$C_1= V'_{11}\times V'_{12}+V'_{12}\times V'_{13}+V'_{13}\times V'_{14}+V'_{14}\times V'_{15}+V'_{15}\times V'_{11}$$

$$C_1=0.2182\times0.092+0.092\times0.03257+0.0357\times0.0466+0.0466\times0.1082+0.1082\times0.2182=0.05368$$

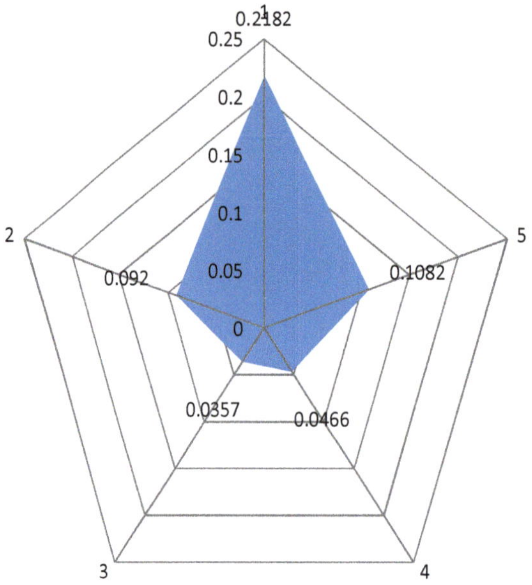

Fig. 2. Polygon area for R1

Table 6
Performance scores using MADM methods

Robots	*SAW(S_i)*	*WPM(P_i)*	*TOPSIS(C_i)*	*VIKOR(Q_i)*	*PAM(C_i)*
R1	0.50077	0.41392	0.3031	0.88431	0.05368
R2	0.58012	0.51098	0.4498	0.43169	0.07325
R3	0.57788	0.55836	0.5413	0.23216	0.07175
R4	0.58543	0.48355	0.4892	0.47751	0.06471
R5	0.53207	0.49883	0.5025	0.51023	0.06179
R6	0.39196	0.38159	0.2252	0.77553	0.03242
R7	0.39938	0.36633	0.1736	0.88074	0.03336
R8	0.65604	0.60598	0.6769	0.12958	0.09425
R9	0.55468	0.54260	0.5139	0.27025	0.06883
R10	0.47125	0.42184	0.2811	0.77264	0.04700
R11	0.49954	0.47878	0.3729	0.50426	0.05129
R12	0.69058	0.68562	0.6777	-3.7E-07	0.10642

Table 7
Robots Rank

Robots	*SAW*	*WPM*	*TOPSIS*	*VIKOR*	*PAM*	*Average**	*Borda***	*Copeland*
R1	8	10	9	12	8	10	9	9
R2	4	5	7	5	3	5	5	5
R3	5	3	3	3	4	3	3	3
R4	3	7	6	6	6	6	6	6
R5	7	6	5	8	7	7	6	6
R6	12	11	11	10	12	11	11	11
R7	11	12	12	11	11	12	12	12
R8	2	2	2	2	2	2	2	2
R9	6	4	4	4	5	4	4	4
R10	10	9	10	9	10	9	9	9
R11	9	8	8	7	9	8	8	8
R12	1	1	1	1	1	1	1	1

*Average of SAW; WPM; TOPSIS and VIKOR methods ranks
**Borda & Copeland methods for SAW; WPM; TOPSIS and VIKOR methods ranks
Poset: R12>R8>R3>R9>R2> (R4,R5)>R11>R9>R10>R6>R7

Karsak and Ahiska (2005) using common weight multi-criteria decision making (MCDM), Wang and Chin (2009) using DEA approach with double frontiers, Singh and Rao (2011) using AHGTMA method , Mondal and Chakraborty (2013) using DEA method identified R12 Robot as the best choice. The best alternative obtained using PAM method is R12, which is the same as the one suggested by previous researchers. This validates the PAM method for decision-making situations of manufacturing environment.

To find similarity in rankings given by different methods, Spearman's rank correlation coefficients (Sheskin, 2004; Myers & Well, 2003; Yurdakul & İç, 2009; Rao, 2013) were obtained for different pairs of the five MADM methods and aggregate methods using SPSS software. Table 8 shows the Spearman's rank correlation coefficients.

Table 8
The results of Spearman's rank correlation coefficients

Method	*SAW*	*WPM*	*TOPSIS*	*VIKOR*	*PAM*
SAW	1	0.881*	0.881*	0.846**	0.958*
WPM		1	0.972*	0.958*	0.944*
TOPSIS			1	0.909*	0.909*
VIKOR				1	0.888*
PAM					1
Average					.951*
Borda					.958*
Copeland					.958*

*sig. = 0.000 Correlation is significant at the 0.01 level (2 tailed)
**sig. =0.001 Correlation is significant at the 0.01 level (2 tailed)

The values of Spearman's rank correlation coefficients show that there is perfect agreement between the introduced method (PAM) and other typical MADM methods.

4.2 Robot Selection Example 2

Mondal and Chakraborty (2013) applied Data Envelopment Analysis (DEA) to solve a robot selection problem in a given industrial scenario while considering seven competitive robot alternatives and five criteria (i.e. repeatability (RE in mm), load capacity (LC in Kg), maximum tip speed (MTS in mm/sec), memory capacity (MC in points or steps) and manipulator reach (MR in mm)) as predominant robot selection attributes. Among these, LC, MTS, MC, and MR are beneficial in nature, and RE is the only non-beneficial attribute. Athawale and Chakraborty (2011) estimated criteria weights as w_{LC}=0.1574, w_{RE}=0.1825, w_{MTS}=0.2668 w_{MC}=0.2430, and w_{MR}=0.2043. The decision matrix with seven robot alternatives and five selection criteria is shown in Table 9.

Table 9
Decision Matrix (D)

Robots	*LC*	*RE*	*MTS*	*MC*	*MR*
R1	60	0.4	2540	500	990
R2	6.35	0.15	1016	3000	1041
R3	6.8	0.1	1727.2	1500	1676
R4	10	0.2	1000	2000	965
R5	2.5	0.1	560	500	915
R6	4.5	0.08	1016	350	508
R7	3	0.1	177	1000	920

Then, the proposed PAM method and SAW, WPM, TOPSIS, and VIKOR methods were applied to this problem as explained before. Comparison of rankings obtained by these methods for this robot selection example is given in Table 11.

Table 10
Performance scores using MADM methods

Robots	*SAW(S_i)*	*WPM(P_i)*	*TOPSIS(C_i)*	*VIKOR(Q_i)*	*PAM(C_i)*
R1	0.58912	0.45349	0.54464	0.5182	0.077713
R2	0.55349	0.45695	0.50262	0.1668	0.064012
R3	0.63891	0.53518	0.52372	0.0000	0.085088
R4	0.45552	0.41819	0.41568	0.3335	0.043548
R5	0.35284	0.24333	0.32016	0.7909	0.02856
R6	0.37695	0.26293	0.34807	0.8853	0.032155
R7	0.35502	0.22145	0.32295	0.9656	0.029483

Table 11

The results of average of SAW; WPM; TOPSIS and VIKOR methods ranks

Robots	*SAW*	*WPM*	*TOPSIS*	*VIKOR*	*PAM*	*Average*[*]	*Borda*	*Copeland*
R1	2	3	1	4	2	2	2	2
R2	3	2	3	2	3	3	3	3
R3	1	1	2	1	1	1	1	1
R4	4	4	4	3	4	4	4	4
R5	7	6	7	5	7	7	7	7
R6	5	5	5	6	5	5	5	5
R7	6	7	6	7	6	6	6	6

Bhangale et al. (2004), Rao (2007), Chatterjee et al. (2010), Mondal and Chakraborty (2013) also identified R3 Robot as the best choice. While comparing the obtained rankings with those of the past researchers, excellent consistency is observed. To find similarity in rankings given by different methods, Spearman's rank correlation coefficients were obtained for different pairs of the five MADM methods and aggregate method using SPSS software. Table 12 shows the Spearman's rank correlation coefficients.

Table 12

The results of Spearman's rank correlation coefficients

Method	*SAW*	*WPM*	*TOPSIS*	*VIKOR*	*PAM*
SAW	1	0.929**	0.964**	0.786*	1**
Sig.		0.03	0.00	.036	
WPM		1	0.854*	0.922**	0.929**
Sig.			0.014	0.003	0.003
TOPSIS			1	0.679	0.964**
Sig.				0.094	0.00
VIKOR				1	0.786*
Sig.					0.036
PAM					1
Sig.					
Average					0.964**
Sig.					0.00
Borda					1
Sig.					
Copeland					1
Sig.					

*Correlation is significant at the 0.05 level (2 tailed)
** Correlation is significant at the 0.01 level (2 tailed)

The values of Spearman's rank correlation coefficients show that there is perfect agreement between the introduced method (PAM) and other typical MADM methods, especially with SAW method.

5 Conclusion

Selecting the best robot is an important problem in the industrial environment considering various multiple performance attributes. This paper presents the PAM method using polygons area for solving the robot selection problem. Two numerical robot selection examples were presented to indicate the validity and advantages of the proposed method in comparison with the four well-known MADM methods. Spearman's rank correlation coefficient was obtained for different pairs of MADM methods. Observations showed that the proposed method is in good agreement with the other methods. The two robot selection examples show that the PAM method can derive quite acceptable ranking results to assist decision makers in devising appropriate decisions. This validates the proposed method for decision-making situations of manufacturing environments, and can be extended to other decision-making situations. The PAM method can also be used for any type of decision-making problem. PAM is logical and simple to implement when compared to other MADM methods. It is suggested that further researches consider other MADM methods and analyze them comparatively. To rank fuzzy data, fuzzy model of the proposed method can be investigated by researchers in future studies.

Acknowledgement

The authors would like to thank the anonymous referees for their comments on this paper.

References

Agrawal, V. P., Kohli, V., & Gupta, S. (1991). Computer aided robot selection: the 'multiple attribute decision making approach. *The International Journal of Production Research*, *29*(8), 1629-1644.

Athawale, V. M., & Chakraborty, S. (2011). A comparative study on the ranking performance of some multi-criteria decision-making methods for industrial robot selection. *International journal of industrial engineering computations*, *2*(4), 831-850.

Azar, A., & Radjab Zadeh, A. (2010). Application Decision MADM Approach. Danesh Publication, 4th Ed.

Behzadian, M., Khanmohammadi Otaghsara, S., Yazdani, M., & Ignatius, J. (2012). A state-of the-art survey of TOPSIS applications. *Expert Systems with Applications*, *39*(17), 13051-13069.

Bhangale, P. P., Agrawal, V. P., & Saha, S. K. (2004). Attribute based specification, comparison and selection of a robot. *Mechanism and Machine Theory*, *39*(12), 1345-1366.

Chatterjee, P., Manikrao Athawale, V., & Chakraborty, S. (2010). Selection of industrial robots using compromise ranking and outranking methods. *Robotics and Computer-Integrated Manufacturing*, *26*(5), 483-489.

Devi, K. (2011). Extension of VIKOR method in intuitionistic fuzzy environment for robot selection. *Expert Systems with Applications*, *38*(11), 14163-14168.

Fendel, G., & Spronk, J. (1983). *Multiple Criteria Decision Methods and Applications*. New York: Spring-Verlag.

Goh, C. H. (1997). Analytic hierarchy process for robot selection. *Journal of Manufacturing Systems*, 16, 381–386.

Honarmande Azimi, M., & Pourmahmoud, J. (2012). A new multiple attribute decision making method. The 5th International Conference of Iranian Operations Research Society, Tabriz, Iran, 65-67.

Hwang, C. L., & Yoon, K. (1981). *Multiple attributes decision making methods and applications*. New York: Springer.

Tansel İç, Y., Yurdakul, M., & Dengiz, B. (2013). Development of a decision support system for robot selection. *Robotics and Computer-Integrated Manufacturing*, *29*(4), 142-157.

Kahraman, C., Çevik, S., Ates, N. Y., & Gülbay, M. (2007). Fuzzy multi-criteria evaluation of industrial robotic systems. *Computers & Industrial Engineering*,*52*(4), 414-433.

Karsak, E. E. (2008). 'Robot selection using an integrated approach based on quality function deployment and fuzzy regression'. International Journal of Production Research, 46, 723-738.

Karsak, E.E., & Ahiska, S.S. (2005). Practical common weight multi-criteria decision-making approach with an improved discriminating power for technology selection. International Journal of Production Research, 43, 1537-1554.

Karsak, E. E., Sener, Z., & Dursun, M. (2012). Robot selection using a fuzzy regression-based decision-making approach. *International Journal of Production Research*, *50*(23), 6826-6834.

Khouja, M. (1995). The use of data envelopment analysis for technology selection. *Computers & Industrial Engineering*, *28*(1), 123-132.

Momeni, M. (2007). *New topics in operations research*. Tehran: University of Tehran publication, 2nd Ed.

Mondal, S. & Chakraborty, S. (2013).A solution to robot selection problems using data envelopment analysis. *International Journal of Industrial Engineering Computations*, 4(3), 355-372.

Murray, R. S., & Liu, J. (1968). Mathematical handbook of formulas and tables. *Spiegel-Schaum's outline series*.

Myers, J. L., & Well, A. D. (2003). Research Design and Statistical Analysis, 2nd ed., Lawrence Erlbaum.

Opricovic, S. (1998). Multicriteria Optimization of Civil Engineering Systems. Faculty of Civil Engineering, Belgrade.

Opricovic, S., & Tzeng, G. H. (2004). Compromise Solution by MCDM Methods: A Comparative Analysis of VIKOR and TOPSIS. *European Journal of Operational Research*, 156, 445-455.

Opricovic, S., & Tzeng, G. H. (2007). Extended VIKOR Method in Comparison with Outranking Methods. *European Journal of Operational Research*, 178, 514-529.

Rao, P. V., & Baral, S. S. (2011). Attribute based specification, comparison and selection of feed stock for anaerobic digestion using MADM approach. *Journal of Hazardous Materials*, 186, 2009–2016.

Rao, R. V., & Padmanabhan, K. K. (2006). Selection, identification and comparison of industrial robots using digraph and matrix methods. *Robotics and Computer-Integrated Manufacturing*, *22*(4), 373-383.

Rao, R.V. (2007). *Decision making in the manufacturing environment using graph theory and fuzzy multiple attribute decision making methods*. London: Springer-Verlag.

Rao R.V., & Patel B.K. (2010). Decision making in the manufacturing environment using an improved PROMETHEE method. *International Journal of Production Research*, 48(16), 4665-4682.

Rao, R. V., Patel, B. K., & Parnichkun, M. (2011). Industrial robot selection using a novel decision making method considering objective and subjective preferences. *Robotics and Autonomous Systems*, *59*(6), 367-375.

Rao, R.V. (2013). *Decision making in the manufacturing environment using graph theory and fuzzy multiple attribute decision making methods*, 2nd ed., London: Springer-Verlag.

Saaty, T.L. (1980). *The analytic hierarchy process*. New York: McGraw-Hill.

Sheskin, D. J. (2004). *Handbook of parametric and nonparametric statistical procedures*. Chapman and Hall /CRC.

Shivanand, H. K. (2006). *Flexible manufacturing system*. New Age International.

Singh, D., & Rao, R. (2011). A hybrid multiple attribute decision making method for solving problems of industrial environment. *International Journal of Industrial Engineering Computations*, *2*(3), 631-644.

Tao, L., Chen, Y., Liu, X., & Wang, X. (2012). An integrated multiple criteria decision making model applying axiomatic fuzzy set theory. *Applied Mathematical Modelling*, *36*(10), 5046-5058.

Triantaphyllou, E., & Mann, S. H. (1989). An examination of the effectiveness of multi-dimensional decision-making methods: A decision-making paradox. *Decision Support Systems*, 5, 303-312.

Triantaphyllou, E. (1998). Multi-criteria decision making methods: An operation research approach. *Encyclopedia of Electrical and Electronic*, 15,175-186.

Vahdani, B., Tavakkoli-Moghaddam, R., Mousavi, S. M., & Ghodratnama, A. (2013). Soft computing based on new interval-valued fuzzy modified multi-criteria decision-making method. *Applied Soft Computing*, *13*(1), 165-172.

Voogd, H. (1983). *Multicriteria Evaluation for Urban and Regional Planning*. Pion, London.

Waigaonkar, S., Babu, B. J. C., & Prabhakaran, R. D. (2008). A new approach for resin selection in rotational molding. *Journal of Reinforced Plastics and Composites*, 27, 1021-1037.

Wang, M. J. J., Singh, H. P., & Huang, W. V. (1991). A decision support system for robot selection. *Decision Support Systems*, *7*(3), 273-283.

Wang, Y. M., Yang, J. B., & Xu, D. L. (2005). A preference aggregation method through the estimation of utility intervals. *Computers & Operations Research*,*32*(8), 2027-2049.

Yoon, K. P., & Hwang, C. L. (1995). *Multiple attribute decision making: an introduction* (Vol. 104). Sage Publications.

Yu, P.L. (1973). A class of solutions for group decision problems. *Management Science*, 19 (8), 936–946.

Yurdakul, M., & Ic, Y. T. (2009). Application of correlation test to criteria selection for multi criteria decision making (MCDM) models. *The International Journal of Advanced Manufacturing Technology*, *40*(3-4), 403-412.

Zanakis, S. H., Solomon, A., Wishart, N., & Dublish, S. (1998). Multi-attribute decision making: A simulation comparison of select methods. *European journal of operational research*, *107*(3), 507-529.

Zeleny, M. (2002). *Multiple Criteria Decision Making*. New York: McGraw Hill.

Experimental scrutiny to induce the ramification of cutting parameters in CNC turning of AISI H21 steel employing response surface methodology

Rajan Jindal[a] and Deepak Choudhary[b*]

[a]*M.Tech. Scholar, Department of Mechanical Engineering, Yamuna Institute of Engineering & Technology, Gadholi, Distt. Yamuna Nagar, 133103, India*
[b]*Assistant Professor, Department of Mechanical Engineering, Yamuna Institute of Engineering & Technology, Gadholi, Distt. Yamuna Nagar, 133103, India*

CHRONICLE

Keywords:
Analysis of variance
Face centered central composite design
Response surface methodology
Surface roughness
Turning process

ABSTRACT

This paper demonstrates an experimental scrutiny into turning process of hot work tool steel AISI H21 under dry machining plight. In this paper, face centered central composite design concealed by response surface methodology is practiced and analysis of variance is implemented to analyze the eloquent benefaction of machining parameters on responses. To access accommodate between the surface roughness and the MRR, an approach for concurrent optimization of multi-objective characteristics based on comprehensive desirability function is employed. The multi objective optimization concludes a spindle speed of 1599.568 rpm, feed rate of 0.262 mm/rev and depth of cut of 2 mm.

1. Introduction

In any machining operation, along with accomplishing the factual dimensions, increase metal removal rate and a good surface trait are also important. Quality influences the degree of amusement of the customers. At the same time, higher MRR is coveted by the industry to cope up with mass production product in shorter time without enduring the product trait. Higher MRR is accomplished by increasing the process parameters like depth of cut, feed and cutting speed. However, very high cutting speed craves the larger power which may eclipse the power accessible in the machine tool. Also at the same time, the cutting temperature increases with the increase in the process parameters. This influences both the tool as well as the product as it causes dimensional inaccuracies by built-up-edge formation, thermal deformation and amends the keenness of the tool and results in reverberation of the machine tool. So, excerption of pertinent process parameter plays a very vital aspect in the efficiency, effectiveness and comprehensive economy of manufacturing to accomplish the targets higher MRR and higher product trait.

This leads optimization problem which shots to access best parametric combination for the said manufacturing process. Optimization of input variables is one of the most important characteristics in any process planning of materials to lessen the cost and time for machining. However, optimization of multi-objective problems is a great commitment of today's producers to yield the precision parts at little costs. In order to advance and optimize a surface roughness and material removal rate model, it is indispensable to perceive the current status of work in this area. A number of researchers have been focused on an appropriate method to evaluate the optimal value of the process parameters to predict the surface roughness and material removal rate. Jiang et al. (1997) examined the effect of austenite grain size on tool life & chip deformation in turning of AISI 304L austenitic stainless steel bar and showed that inhomogeneous distribution of grain size up to a depth of 15 mm of the bar, resulted in tool edge breakage & lower tool life when turning hot-forged bar as compared with quenched bars. Noordin et al. (2004) described the performance of a multi-layer WC tool using RSM when turning AISI 1045 steel. The experimental results indicated that feed was the most important parameter that influenced the tangential force & the surface roughness.

Gaitonde et al. (2008) determined the optimum amount of MQL and the most appropriate cutting speed and feed rate during turning of brass using K10 carbide tool. The optimization results indicated that MQL of 200 ml/h, cutting speed of 200 m/min and a feed rate of 0.05 mm/rev were essential to simultaneously minimize surface roughness and specific cutting force. Aggarwal et al. (2008) presented an experimental investigation into the effect of feed rate, depth of cut, cutting speed, cutting environment and nose radius in CNC turning of AISI P-20 tool steel and revealed that cryogenic environment was the most prominent factor in minimizing power consumption followed by depth of cut and cutting speed & also concluded that although both techniques predicted approximately similar result, RSM technique, however, seemed to an edge over the Taguchi's technique. Kaladhar et al. (2010) optimized the process parameters in turning of AISI 202 austenitic stainless steel using CVD coated cemented carbide tools. From the analysis, it was observed that the feed was the most prominent factor that affected the surface roughness followed by nose radius. Mahdavinejad and Saeedy (2011) optimized turning parameters of AISI 304 stainless steel. It was showed that cutting speed and feed rate had the main effect on the flank wear & surface roughness respectively and the use of cutting fluid resulted in greater tool life and better surface finish.

Rodríguez et al. (2011) conducted experiments on AISI 316L, AISI 304 and AISI 420 steels during a turning process and observed that the cutting temp. increased when feed, cutting speed, depth of cut and material maximum strength increased and cutting temperature decreased with the increased of material's thermal conductivity. Asilturk et al. (2011) focused on optimizing turning parameters based on the Taguchi method to minimize surface roughness (Ra and Rz). Dry turning tests were carried out on AISI 4140 (51 HRC) with coated carbide cutting tools. Results indicated that the feed rate had the most significant effect on Ra and Rz. Sivaraman et al. (2012a) turned the multiphase (ferrite-bainite-martensite) micro alloyed steel to study the effect of machining parameters such as feed, cutting speed and depth of cut on cutting forces. The result showed that feed and depth of cut influenced more on cutting force than cutting speed. Kumar et al. (2012) examined the effect of process parameters in turning of carbon alloy steels in a CNC lathe. They used SAE8620, EN8, EN19, EN24 and EN47 carbon alloy steels for turning. It was observed that the surface roughness increased with increased feed rate and was higher at lower speeds and vice versa for all feed rates. Sivaraman et al. (2012b) carried out the machining of multiphase (ferrite-bainite-martensite) microalloyed steel in a high speed lathe to assess the machinability. The result showed that the feed rate and depth of cut influenced more on cutting force and for surface roughness the only influencing parameter was feed rate. Khamel et al. (2012) investigated the effect of process variables (depth of cut, feed rate & cutting speed) on performance characteristics such as surface roughness, cutting forces and tool life in hard turning of AISI 52100 bearing steel with CBN tool. The results showed that feed rate and cutting speed greatly affected the tool life and surface roughness. However, depth of cut revealed maximum influenced on cutting forces.

Barik and Mandal (2012) presented an experimental study of roughness characteristics of surface roughness generated in CNC turning of EN 31 alloy steel. It was seen that the surface roughness parameter decreased with increased in spindle speed and depth of cut but increased with increased in feed rate. Kumbhar and Waghmare (2013) used Taguchi approach to find optimum process parameters for turning hardened EN31 alloy steel. The conclusion revealed that the feed rate was the most effective parameter on surface roughness & tool life. Ahmed et al. (2013) investigated the effect of tool overhang in the turning process on surface quality of the work piece& tool wear. They observed that the effect of depth of cut on the surface roughness was negligible and deflection of the cutting tool increased with increased in tool overhang.

This is winded up from literature review that the Taguchi design of experiments & response surface methodology techniques are being broadly employed in the current & past research works on turning process. Despite the techniques RSM and Taguchi predicted near similar results, however, RSM technique sounds to an edge over the Taguchi's technique. It has also been noted that during turning, the cutting parameters which has prominent consequence on performance characteristics are speed, feed and depth of cut. Therefore, these are the parameters which are preferred to perform the experimental work on AISI H21 steel.

2. Design of Experiments (DOE)

The most widely employed techniques for surface roughness and material removal rate prediction in terms of machining parameters is the RSM. Therefore, face centered central composite design concealed by Response surface methodology is employed for the experiment plan in this work.

3. Experimental Campaign

In the pageant work, a set of experiments are run on the work piece AISI H21 hot work tool steel (as illustrated in Fig. 1) to appraise the consequence of machining parameters such as feed rate, spindle speed & depth of cut on material removal rate and surface roughness. The cutting insert which is employed for the experiment is Taegu Tech make TT8135 grade CNMG 120412 MP TiN coated carbide insert as depicted in Fig. 2. It is clenched onto a tool holder, ISO designation DCLNR 20 20 K 12. The total length of the work piece is seized as 750 mm which is cut into 7 pieces in the cylindrical pattern of steel bars with diameter of 50 mm and length of 90 mm by employing Power Hacksaw. Then, 30mm length of each bar is retained in the chuck and 60 mm is turned in dry plight to perform 3 experiments in a single piece.

Fig. 1. Hot work tool steel (H21) rod

Fig. 2. Turning insert

AISI H21 steel is employed for high stressed hot work tools such as mandrels, dies and containers for metal tube and rod extrusion, screws, rivets, hot extrusion tools, tools for manufacture of hollows, die casting tools, die inserts, extrusion dies for brass, bronze and steel, hot-press dies, drawing and hot-swaging dies etc. The Design Expert_ software (Stat-Ease Inc., USA) version 9.0.3.1 is employed to advance the experimental design matrix for RSM and to interpret the data possessed from experimentation. The range of each parameter is associated at three different levels, namely low, medium, and high based on tool manufacturer recommendation. The process parameters, their designated symbols and ranges are demonstrated in Table 1.

Table 1
Levels of Independent Control Parameters

Sr. No.	Cutting Parameters	Symbol	Level of Parameters			Unit
			-1	0	1	
1	Spindle speed	N	400	1000	1600	RPM
2	Feed rate	f	0.15	0.25	0.35	mm/rev
3	Depth of cut	a	1.5	1.75	2	mm

3.1 Composition testing

Composition Testing employs the EDAX analysis which exemplifies Energy Dispersive X-ray spectroscopy. Composition of AISI H21 steel is approved on Polyvac 181 TJM Spectrometer. The chemical composition of AISI H21 hot work tool steel is exhibited in Table 2.

Table 2
Percentage of Elements in H21 hot work tool steel

Elements	%age
Carbon, C	0.321
Silicon, Si	0.242
Manganese, Mn	0.335
Chromium, Cr	3.300
Phosphorus, P	0.023
Sulphur, S	0.020
Vanadium, V	0.393
Tungsten, W	9.120
Molybdenum, Mo	0.350
Cobalt, Co	0.150
Tin, Sn	0.015
Iron, Fe	85.630

3.2 Properties of the material

The various physical and mechanical properties of AISI H21steel are shown in Tables 3 and Table 4.

Table 3
Physical Properties

Physical Properties	Metric
Specific gravity g/cc	8.19
Density (kg/m^3)	8.28 x 1000

Table 4
Mechanical Properties

Mechanical Properties	Metric	Conditions T (°C)
Poisson's ratio	0.27 - 0.30	25
Elastic modulus (GPa)	190 - 210	25

3.3 Equipment employed

A HMT CNC turning center STALLION 100HS is employed for experimentation as presented in Fig. 3. The lathe equipped with variable spindle speed from 100 rpm to 3000 rpm, and a 5.5 kW motor drive is employed for the tests.

Fig. 3. A HMT CNC turning center STALLION 100HS

Fig. 4. Mitutoyo surftest-4 surface roughness tester

3.4 Roughness measurement

Surface roughness is consistent employing stylus type Mitutoyo surftest-4 on a turned length of 20 mm as exposed in Fig. 4. Three measurements are run along the length of cut on each work piece and the average Ra value is listed.

Table 5
Experimental Design matrix with uncoded values and observed responses

Stadard order	Spindle speed, r.p.m.	Feed rate mm/rev	Depth of cut, mm	Material removal rate, mm^3/sec	Ra 1	Ra 2	Ra 3	Mean surface roughness, Ra in µm
1	400	0.15	1.50	226.08	2.63	2.18	2.75	2.520
2	1600	0.15	1.50	904.32	2.29	2.00	2.10	2.130
3	400	0.35	1.50	527.52	2.83	3.15	2.49	2.823
4	1600	0.35	1.50	2110.08	3.02	3.12	3.02	3.053
5	400	0.15	2.00	301.44	2.75	2.78	2.77	2.767
6	1600	0.15	2.00	1205.76	2.47	2.18	2.36	2.337
7	400	0.35	2.00	703.36	2.81	3.11	3.35	3.090
8	1600	0.35	2.00	2813.44	2.86	3.00	2.82	2.893
9	400	0.25	1.75	439.6	3.04	3.00	2.47	2.837
10	1600	0.25	1.75	1758.4	2.05	1.95	2.30	2.100
11	1000	0.15	1.75	659.4	2.59	2.98	2.55	2.707
12	1000	0.35	1.75	1538.6	2.94	2.86	2.77	2.857
13	1000	0.25	1.50	942.00	2.20	2.67	2.64	2.503
14	1000	0.25	2.00	1256.00	2.50	3.16	2.76	2.807
15	1000	0.25	1.75	1099.00	2.06	2.48	3.20	2.580
16	1000	0.25	1.75	1099.00	2.27	3.42	2.80	2.830
17	1000	0.25	1.75	1099.00	2.02	2.64	2.81	2.490
18	1000	0.25	1.75	1099.00	2.53	2.81	3.15	2.830
19	1000	0.25	1.75	1099.00	2.72	2.7	2.40	2.607
20	1000	0.25	1.75	1099.00	2.74	2.57	2.42	2.577

In Table 5 material removal rate is computed by the product of cutting speed (Vc), feed rate (f) and depth of cut (a) and revealed in mm^3/sec as:

$$\mathrm{MRR} = 1000 \times \mathrm{Vc} \times \mathrm{f} \times \mathrm{a} \ \ (\mathrm{mm^3/min})$$

And cutting speed is calculated as,

$$Vc = \frac{\pi \times D \times N}{1000}$$

Where Vc = cutting speed in m/min; D = Diameter of work piece in mm; N = Spindle Speed in r.p.m.

4. Results and Discussion

4.1 Development of empirical models

Employing the experimental data, analytical model for surface roughness and material removal rate is developed using multiple linear regression (MLR) analysis. The dependent variable surface roughness and MRR is conceived as a linear consolidation of the independent variables namely feed rate, spindle speed & depth of cut.

Since, there are large numbers of variables governing the process, so empirical models are imperative to represent the process. However, these models are advanced using only the momentous factors.

4.2 Final equation in terms of actual factors for MRR and Ra

$$MRR = 1099 - 1.099N - 4396f - 628a + 4.396\,Nf + 0.628Na + 2512fa \quad (1)$$

$$Ra = 2.362 - (2.550 \times 10^{-4}N) + 2.240f \quad (2)$$

These equations are in terms of actual factors which can be employed to build predictions about the responses MRR and surface roughness (Ra) for given levels of each factor.

4.3 Analysis of variance (ANOVA)

In order to develop empirical models, statistical analysis of the experimental results is indispensable by employing analysis of variance. ANOVA is a computational technique that empowers the estimation of the relative contributions of each of the control factors to the comprehensive deliberated response.

Table 6

Analysis of variance table for MRR after backward elimination

Source	Sum of Squares	df	Mean Square	F Value	p-value Prob> F	Contribution %age	
Model	7.19×10^6	6	1.20×10^6	1370.81	< 0.0001		significant
A-Spindle speed	4.35×10^6	1	4.35×10^6	4976.56	< 0.0001	60.4056	significant
B-Feed rate	1.93×10^6	1	1.93×10^6	2211.81	< 0.0001	26.8407	significant
C-Depth of cut	2.47×10^5	1	2.47×10^5	282.12	< 0.0001	3.4245	significant
AB	5.57×10^5	1	5.57×10^5	637	< 0.0001	7.7327	significant
AC	70989.12	1	70989.12	81.25	< 0.0001	0.986	significant
BC	31550.72	1	31550.72	36.11	< 0.0001	0.438	significant
Residual	11358.26	13	873.71	....	...	0.157	
Lack of fit	11358.26	8	1419.78	....	....		
Pure error	0	5	0	....	....		
Cor Total	7.20×10^6	19	...	...	....		

Table 6 reveals that model is significant and there is only a 0.01% incidental that an F-value of model can be large due to noise. If the p- value probability > F is less than 0.05 then, it depicts model terms are significant. In this case A (spindle speed), B (feed rate), C (depth of cut), AB, AC, BC are significant model terms.

4.3.1 ANOVA for response surface linear model i.e. for Ra

Result of ANOVA for the Ra model is delineated in Table 7. It represents that model is significant and there is only a 0.19% contingent that an F-value of model can be large due to noise. In this case A (spindle speed) and B (feed rate) are significant model terms. The values > 0.100 manifests that the model terms are not significant.

Table 7
Analysis of variance table for Ra

Source	Sum of Squares	df	Mean Square	F Value	p-value Prob> F	
Model	0.81	3	0.27	7.82	0.0019	significant
A-Spindle speed	0.23	1	0.23	6.75	0.0194	significant
B-Feed Rate	0.5	1	0.5	14.47	0.0016	significant
C-Depth of cut	0.077	1	0.077	2.23	0.1545	not significant
Residual	0.55	16	0.035	....	...	....
Lack of fit	0.45	11	0.041	2.02	0.2255	not significant
Pure error	0.1	5	0.02	....	....	....
Cor Total	1.37	19	...	....	....	...

4.3.2 ANOVA for response surface reduced linear model i.e. for Ra

The ANOVA table for the reduced linear model for Ra is laid out in Table 1.8. The F-value of lack of fit i.e. 2.17 implies that lack of fit is insignificant relative to the pure error. There is a 20.16 % incidental that a lack of fit, F-value can be large due to noise.

Table 8
Analysis of variance table for Ra after backward elimination

Source	Sum of Squares	df	Mean Square	F Value	p-value Prob> F	Contribution %age	
Model	0.74	2	0.37	9.89	0.0014		significant
A-Spindle speed	0.23	1	0.23	6.3	0.0225	16.788	significant
B-Feed rate	0.5	1	0.5	13.49	0.0019	36.496	significant
Residual	0.63	17	0.037	....	....		....
Lack of fit	0.53	12	0.044	2.17	0.2016		not significant
Pure error	0.1	5	0.02	...	....		....
Cor Total	1.37	19	....	...	...		....

Table 1.9 represents that the value of predicted R^2 i.e. 0.9791 is in reasonable agreement with the value of adjusted R^2 i.e. 0.9977 for MRR since the difference is less than 0.2. The value of adequate precision > 4 is desirable which manifests an adequate signal and summons that model can be employed to navigate the design space.

Table 9
Various R^2 statistics for MRR

Standard deviation	29.56	R^2 (Coefficient of determination)	0.9984
Mean	1099	Adjusted R^2	0.9977
	2.69	Predicted R^2	0.9791
	1.50 x 10^5	Adequate Precision	143.649

Table 10 illustrates that the value of predicted R^2 is in reasonable agreement with the value of adjusted R^2 for the Ra.

Table 10
Various R^2 statistics for Ra

Standard deviation	0.19	R^2	0.5379
Mean	2.67	Adjusted R^2	0.4835
	7.23	Predicted R^2	0.3229
	0.93	Adequate Precision	10.096

4.4 Influence of cutting parameters on MRR & Ra

The influence of process parameters on output responses i.e. MRR and Ra are presented in Figures below.

4.4.1 Residuals vs. Run plot

The plots below illustrate a random pattern of residuals on both sides of 0.00 and do not expose any recognizable patterns. Thus, it implies that there is nothing awesome about the residuals in Fig. 5 and Fig. 6.

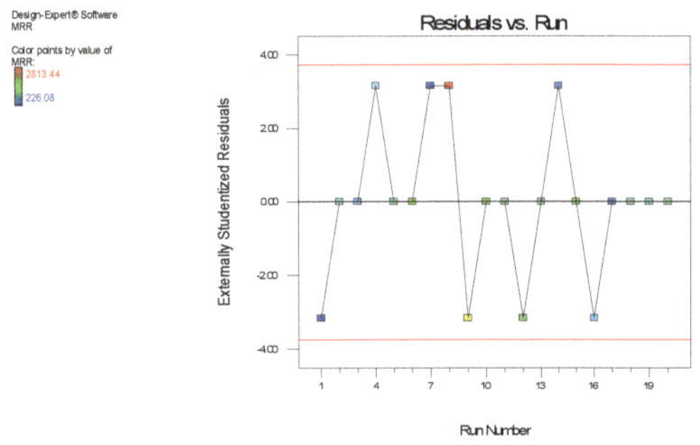

Fig. 5. Residuals vs Run plot for MRR

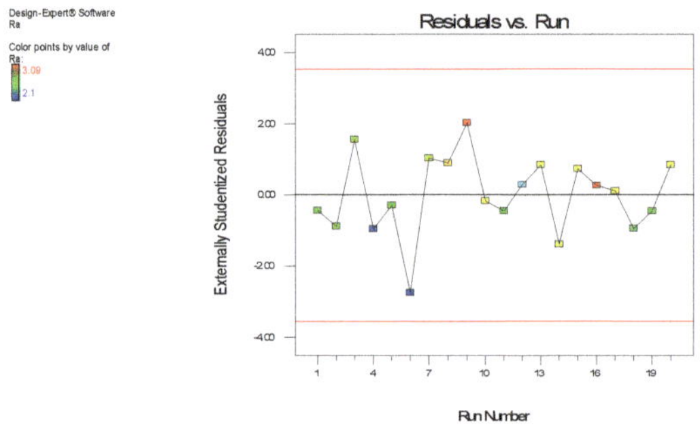

Fig. 6. Residuals vs Run plot for Ra

4.4.2 Interaction plot

An interaction occurs when the response is disparate, anticipating on the settings of two factors. When the lines are parallel, interaction influences are zero. The more distinctive the slopes, the more influence the interaction repercussion on the results. The interaction plots for MRR vs spindle speed and feed rate delineate that MRR increases with increase in spindle speed, however, the influence of spindle speed is large when feed rate is at 0.35 mm/rev as shown in Fig. 7. Similarly, the influence of feed rate on MRR is more, when depth of cut is 2 mm as demonstrate in Fig. 8. Thus, the maximum value of MRR is achieved at the highest range of the input parameters in all the interaction plots.

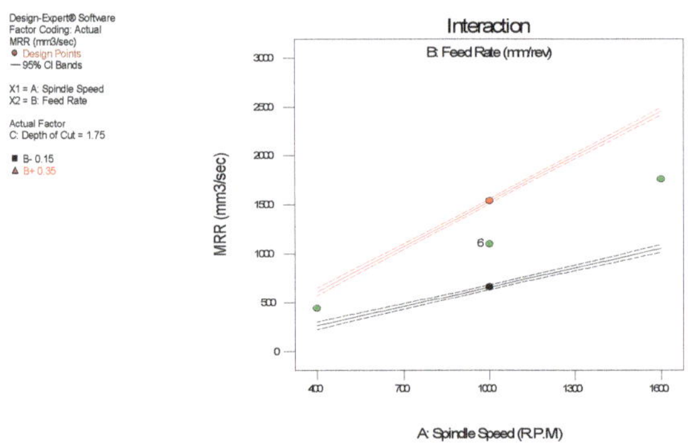

Fig. 7. Interaction plot for MRR vs Spindle speed and feed rate

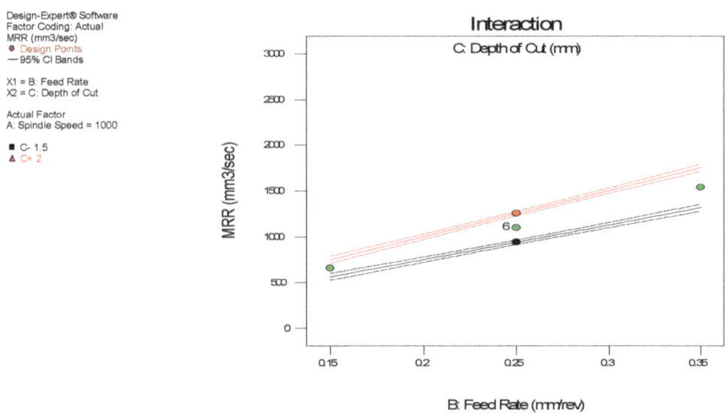

Fig. 8. Interaction plot for MRR vs Feed rate and depth of cut

Fig. 9 reveals interaction plot for surface roughness vs feed rate and spindle speed. This plot represents that Ra is minimum when spindle speed is at 1600 r.p.m.

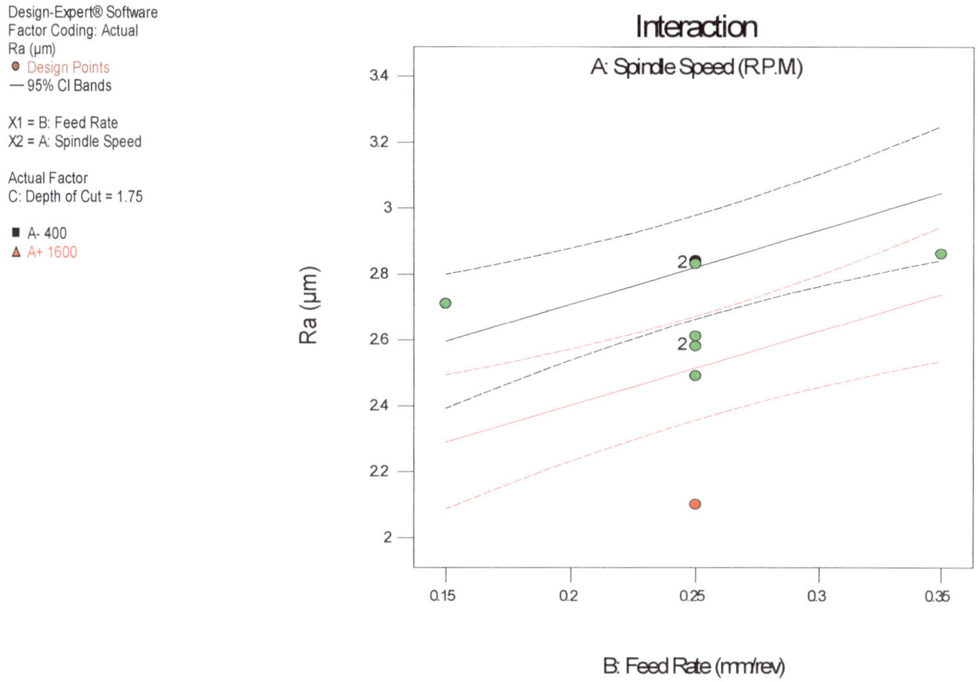

Fig. 9. Interaction plot for Ra vs Feed rate and spindle speed

4.4.3 3-D surface plots

It is contemplated that increase in spindle speed and feed rate lean to increase the MRR as exhibit in Fig. 10. It is noted from Fig. 11 that the increase in depth of cut causes the MRR marginally increase. Thus, increasing the feed rate, spindle speed & depth of cut expedite an increase in the extent of material removal rate.

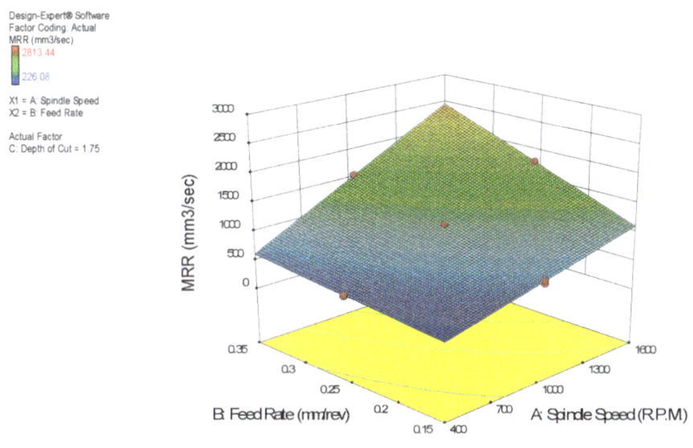

Fig. 10. Influence of Feed rate & Spindle speed on MRR

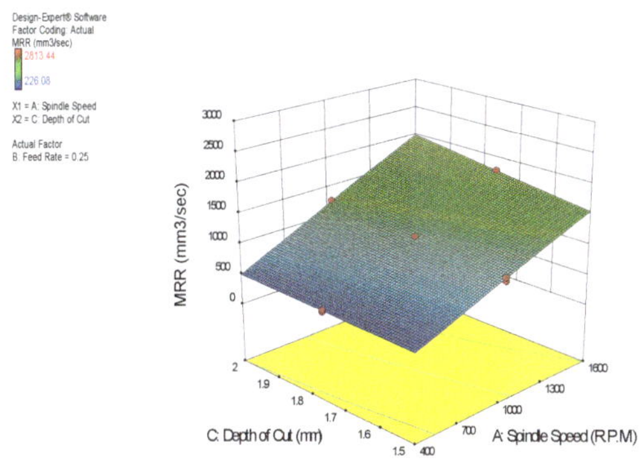

Fig. 11. Influence of Depth of cut & Spindle speed on MRR

The consequence of process parameters on output response, surface roughness is shown in Fig. 12. From this Fig, it is ascertained that as the feed rate increases, Ra also increases but as the spindle speed increases then surface roughness decreases.

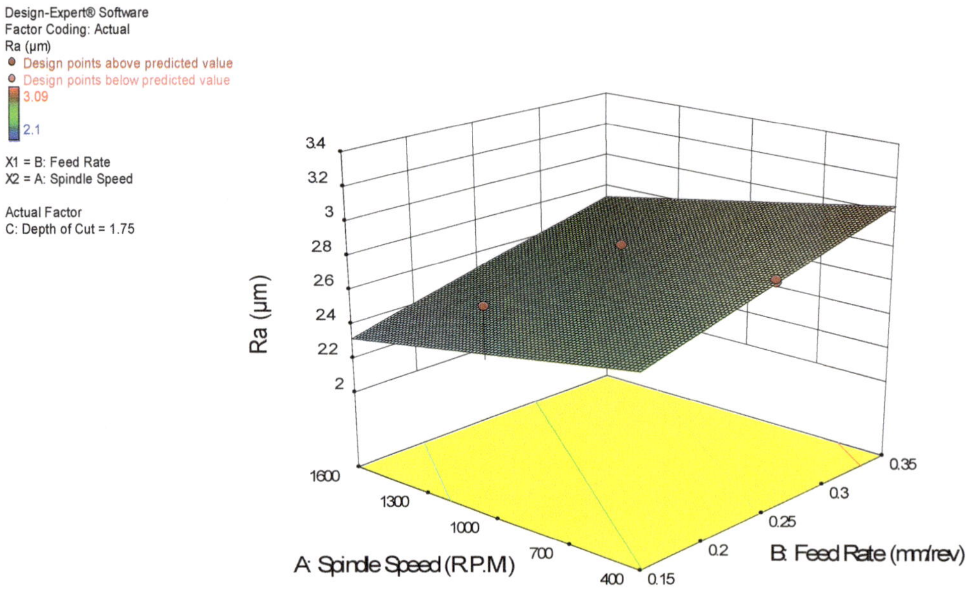

Fig. 12. Influence of Spindle speed & Feed rate on Ra

5. Optimization of the problem

Desirability is quietly a mathematical method to access the optimum. By default, the input factors are set "in range", thus preventing extrapolation as laid out in Table 11.

Table 11
Constraints for combined MRR and Ra

Name	Goal	Lower Limit	Upper Limit	Lower Weight	Upper Weight	Importance
A: Spindle speed	is in range	400	1600	1	1	3
B: Feed rate	is in range	0.15	0.35	1	1	3
C: Depth of cut	is in range	1.5	2	1	1	3
MRR	maximize	226.08	2813.44	1	1	3
Ra	minimize	2.1	3.09	1	1	3

Three solutions are attained. They are presented in Table 12. Solution 1, which is having maximum value of desirability i.e. 0.634, is tabbed. The optimum values of spindle speed, feed rate and depth of cut to maximize the MRR (2097.3 mm^3/sec) & minimize the Ra (2.54 µm) are 1599.568 r.p.m., 0.262 mm/rev and 2 mm respectively.

Table 12
Optimization solutions for combined MRR and Ra

Number	Spindle speed	Feed rate	Depth of Cut	MRR	Ra	Desirability	
1	1599.568	0.262	2	2097.3	2.54	0.634	Selected
2	1600	0.272	2	2176.9	2.563	0.633	.....
3	1600	0.222	2	1796.2	2.452	0.625	.....

5.1 Numerical optimization Ramps

Ramps view reveals the desirability for each factor and each response. The ramp function graph for overall desirability for MRR and Ra is illustrated in Fig. 13. In this figure, red mark on curves of spindle speed, feed rate and depth of cut are delineating the optimum values. The corresponding optimum value of response i.e. MRR and Ra is also exposed by blue dot on curves of these responses. Fig. 13 also depicts the individual desirability value of these multi-objective characteristics.

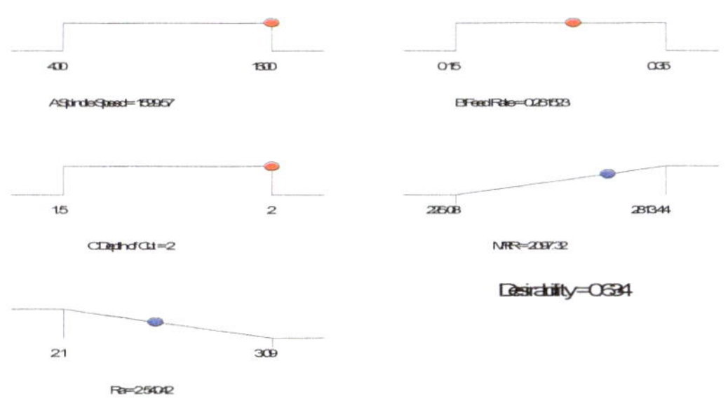

Fig. 13. Ramp function plot for combined for MRR & Ra

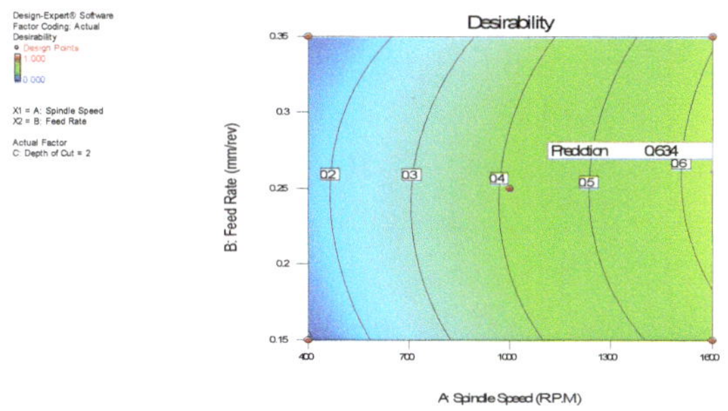

Fig. 14. Contour plot at maximum desirability value for combined MRR & Ra

5.2 Contour plot at maximum desirability value of responses

Contour graph (Fig. 14) at maximum desirability value (0.634) presents optimum values of spindle speed and feed rate. This plot manifests that increase in spindle speed and feed rate result in increase in desirability value of MRR & Ra.

6. Conclusion

It can be winded up from above analysis that response surface method can be successfully employed to induce optimal values of cutting parameters for multi-objective problem. Surface roughness & MRR parameters greatly rely on work piece material. Material removal rate increases with the increase in feed rate, spindle speed & depth of cut. The ramification of the depth of cut on the surface roughness is negligible. Surface roughness parameter decreases with increase in spindle speed but increases with increase in feed rate. The values of cutting parameters: spindle speed of 1599.568 rpm, feed rate of 0.262 mm/rev and depth of cut of 2 mm are foreseen to counter with a minimum surface roughness and maximum MRR.

Acknowledgement

We would like to acknowledge Research & Development Centre for Bicycle & Sewing Machine, Ludhiana, India for providing CNC machining facility and the surface roughness tester to carry out this experimental work.

References

Aggarwal, A., Singh, H., Kumar, P., & Singh, M. (2008). Optimizing power consumption for CNC turned parts using response surface methodology and Taguchi's technique—a comparative analysis. *Journal of Materials Processing Technology*, *200*(1), 373-384.

Ahmed, G. S., Ahmed, H., & Samad, S. S. (2013). Experimental investigation of effect of tool length on surface roughness during turning operation and its optimization. *IOSR Journal of Mechanical and Civil Engineering*, *7*(2), 73-80.

Asiltürk, I., & Akkuş, H. (2011). Determining the effect of cutting parameters on surface roughness in hard turning using the Taguchi method. *Measurement*, *44*(9), 1697-1704.

Barik, C. R., & Mandel, N. K. (2010). Parametric effect and optimization of surface roughness of EN 31 in CNC dry turning. *International Journal of Lean Thinking*, *3*(2), 1-13.

Gaitonde, V. N., Karnik, S. R., & Davim, J. P. (2008). Selection of optimal MQL and cutting conditions for enhancing machinability in turning of brass. *Journal of Materials Processing Technology*, *204*(1), 459-464.

Jiang, L., Roos, Å., & Liu, P. (1997). The influence of austenite grain size and its distribution on chip deformation and tool life during machining of AISI 304L.*Metallurgical and Materials Transactions A*, *28*(11), 2415-2422.

Kaladhar, M., Subbaiah, K. V., Rao, C. S., & Rao, K. N. (2010). Optimization of process parameters in turning of AISI202 austenitic stainless steel. *ARPN Journal of Engineering and Applied Sciences*, *5*(9), 79-87.

Khamel, S., Ouelaa, N., & Bouacha, K. (2012). Analysis and prediction of tool wear, surface roughness and cutting forces in hard turning with CBN tool. *Journal of mechanical science and technology*, *26*(11), 3605-3616.

Kumar, N. S., Shetty, A., Shetty, A., Ananth, K., & Shetty, H. (2012). Effect of spindle speed and feed rate on surface roughness of Carbon Steels in CNC turning. *Procedia Engineering*, *38*, 691-697.

Kumbhar, Y. B., & Waghmare, C. A. (2013). Tool Life And Surface Roughness Optimization Of PVD TiAlN/TiN Multilayer Coated Carbide Inserts In Semi Hard Turning Of Hardened EN31 Alloy Steel Under Dry Cutting Conditions. *International Journal of Advanced Engineering Resources Studies*, *2*(4), 22-27.

Mahdavinejad, R. A., & Saeedy, S. (2011). Investigation of the influential parameters of machining of AISI 304 stainless steel. *Sadhana*, *36*(6), 963-970.

Noordin, M. Y., Venkatesh, V. C., Sharif, S., Elting, S., & Abdullah, A. (2004). Application of response surface methodology in describing the performance of coated carbide tools when turning AISI 1045 steel. *Journal of Materials Processing Technology*, *145*(1), 46-58.

Rodríguez, J., Muñoz-Escalona, P., & Cassier, Z. (2011). Influence of cutting parameters and material properties on cutting temperature when turning stainless steel. *Revista de la Facultad de Ingeniería Universidad Central de Venezuela*, *26*(1), 71-80.

Sivaraman, V., Sankaran, S., & Vijayaraghavan, L. (2012a). The Effect of Cutting Parameters on Cutting Force During Turning Multiphase Microalloyed Steel. *Procedia CIRP*, *4*, 157-160.

Sivaraman, V., Sankaran, S., & Vijayaraghavan, L. (2012b). Machinability of multiphase microalloyed steel. *Procedia CIRP*, *2*, 55-59.

Scheduling for production teams

Yuri Mauergauz[*]

Sophus Group, Moscow 119296, Russian Federation

CHRONICLE

Keywords:
Scheduling
Production intensity
Order utility

ABSTRACT

This paper presents a method of calendar (weekly) scheduling for production teams, when the average orders utility function is used as the quality criterion. The method is based on the concept of "production intensity", which is a dynamic parameter of production process. Applied software package allows scheduling for medium quantity of jobs. The result of software application is the team load on the planning horizon. The computed schedule may be corrected and recalculated in interactive mode. Current load of every team is taken into account at each recalculation. The method may be used for any combination of complex and specialized teams.

1. Introduction

Contemporary methods of production planning consider each department of an enterprise, for which the schedule is developed, as the "working center". This paper relates to scheduling for working centers, which are production teams. The production teams are organized for assembly of compound machines, equipment repair or transfer to a new place and so on. Each production team consists of regular or temporary staff, completing a joint job (order) and having joint responsibility for their results. Team members may have the same or different professions. Accordingly, there are specialized or complex teams. Each job may comprise some tasks (operations) of various types that require personnel of appropriate professions. Depending on compliance between task types and professions of team workers, a complex team may perform all tasks of a job or its part only. A specialized team may only perform a task corresponding to its profile.

The main advantage of using teams as working centers is associated with great flexibility inherent in such systems. Two types of flexibility are possible in scheduling. Routing flexibility is possibility to choose among two or more working centers to carry out a given operation. According to the classification (Blackburn & Millen, 1986) this type of flexibility is associated with hardware flexibility. The other type of flexibility named as sequencing flexibility is associated with software flexibility. This type of flexibility makes it possible to change the sequence of operations within the job. The possibility to appoint one of available complex teams when scheduling for any job relates to routing flexibility of scheduling. At the same time, the sequence of operations needed for a job may be changed by agreement of team members, depending on such factors as worker's load, availability of facilities, etc.

Scheduling for teams is a difficult problem as it is often difficult for a manager to determine the load level of each worker and expected completion dates of the team tasks. The only possible version of production planning here is the combination of tasks scheduling for all teams within a planning period (a week, a month) and the daily plan, which is elaborated by the team itself. Since for completion of a specific planned job several teams of various specialization may be engaged, sequence of their work may only be directly determined by team leaders and may vary depending on the situation.

If production teams are considered as "machines", then according to the classification of planning problems (Graham et al., 1979), the set of production teams with non-determined sequence of their use may be determined as Open Shop. At the same time, possibility to several teams for a given operation provides the flexibility of such machine set. Therefore, we may suggest that scheduling for the set of production teams is associated with the Flexible Open Shop problem.

Apparently, the better is the preliminary schedule the less is the necessity of its revision and daily amendment. The schedule quality depends on the data set available to a planner at the moment of scheduling, and on the planning algorithm. In this article the criterion of average orders utility is used for scheduling estimation. The remainder of this paper is organized as follows. Section 2 provides a review of the relevant literature. Section 3 determines the main notions and the function of current order utility. In Section 4 the problem is formulated and the algorithm for its solving is provided. The example of scheduling for a given set of jobs is provided in Section 5. Section 6 contains some concluding remarks.

2. Literature review

As far as the author knows, there is only one article, which is dedicated to scheduling for technical structure of Flexible Open Shop type. In the paper by Witkowski et al. (2011) the problem of this type is considered, when all jobs at the moment of planning are available. In this paper the makespan $C_{\max}$ is considered as a criterion. The criterion $C_{\max}$ has been used in most other studies on Open Shop Scheduling problem. Gonzalez and Sahni (1976) elaborated the exact algorithm for this problem, when preemption is possible. Bai and Tang (2013) considered the task with the given release dates. In the book (Gupta et al., 2013) several algorithms were studied for scheduling at two operation stages with the criterion $C_{\max}$. Shabtai and Kaspi (2006) researched the Open Shop Scheduling problem with the criterion $C_{\max}$ when job duration may vary. In some papers other criteria are used. Brasel et al. (2008) designed some heuristic algorithms for the Open Shop Scheduling problem with the criterion of mean flow time $\bar{F}$. Liaw (2005) studied this problem for the case of summary tardiness minimization $\sum T_i$, Naderi et al. (2011) used the criterion of summary duration minimization $\sum C_i$. Doulabi (2010) elaborated the algorithm for the Open Shop Scheduling problem based on the method of mixed integer linear programming. They used the criterion of earliness–tardiness penalties minimization.

The important version of the Open Shop Scheduling problem relates to a task of Concurrent Open Shop Scheduling. The latter may be considered as a variant of the classical Open Shop model, in which operations belonging to the same job may be processed concurrently on several machines. Ng et al. (2003) studied this problem for the criterion of weighted tardy jobs number $\sum w_i U_i$. Leung et al. (2005) used definition of the task as PD Open Shop Scheduling, if a certain machine is assigned to each operation. For example, Mastrolilli et al. (2010) for this problem applied the linear programming method with the criterion of minimal summary weighted duration $\sum w_i C_i$; Ma and Jin (2013) also used linear programming with the criterion $C_{\max}$. All criteria above do not take into account the process time of planned jobs, which is a substantial disadvantage. For the Open Shop Scheduling problem Blazewicz et al. (2004) developed "non-classic" quality criterion, which is equal to the summary processing time of tardy jobs $Y = \sum Y_i$. In this criterion for each job $Y_i = \min(T_i, p_i)$, i.e. this value is equal to a smaller value of job tardiness T_i and processing time p_i.

3. Main problem definitions and utility functions

Let us assume that it is necessary to perform n various jobs at the facility of scheduling (enterprise, vessel, building, etc.) for a certain period of time after the planned start. The planning horizon of the working enterprise is usually equal to a certain reporting period, for example, a month or a week. When a vessel or a building is constructed, the horizon may be determined for the reporting period, or to the entire period of construction. Let us assume each job may be performed by one or several production teams, and their number is equal to M. Each job i comprises several tasks (operations) and has to be completed on due date d_i.

Assumptions

a) Within this scheduling the sequence of tasks (operations) that belong to the same job to be performed was not taken into account. In general, these tasks may be executed in any sequence.
b) The priority coefficient can be determined for each job.
c) Release date is known for every job.
d) Each job can be executed by any number of production teams simultaneously.
e) Process time of a job as a whole and process time of every task belonging to a job is known and deterministic.
f) The summary duration of job completion is assigned normatively.
g) Each job has one task of a certain type as the main one.
h) In the beginning, for every production team it is known what a job is being performed, and when this work will be completed.

Notation

Indices

$i = 1, 2 \ldots n$ Index of order (job)
$l = 1, 2 \ldots J$ Index of operation execution tree level
$m = 1, 2 \ldots M$ Index of specific production team in team list
$j = 1, 2 \ldots S$ Index of task (operation) type
$z = 1, 2 \ldots Z_l$ Index of decision tree node on level l

Parameters

d_i Due date of job i in calendar days
G Duration of plan period in calendar days
E Quantity of hours per work day
α "Psychological" coefficient
h Horizon of planning in work days
p_{ij} Processing time of operation j for job i in hours
r_i Release moment for job i in calendar days

w_i	Weight coefficient for job i
p_i	Normative duration for job i in hours
N_i	Main operation type for job i

Variables

g_i	Necessary release moment of job i in calendar days
S_i	Pool of teams, which are able to perform the task of main type for job i
C_{lz}	Operation completion moment for tree node z at level l in hours
C_m	Last job completion moment for team m in hours
t	Current time
$t_{l+1,k}$	Start moment for operation k at level $l+1$ in hours
H_i	Current production intensity of job i
V_i	Current utility of job i
V	Current utility of all remaining operations
$\bar{V}_l$	Average utility of all jobs at level l
n_m	Number of jobs for team m
$\bar{V}_{l+1,k}$	Average utility of all jobs at level $l+1$ for operation k

Criterion

$\bar{V}$ Average utility of all jobs on the planning horizon

Let us assume that the average job duration in calendar days is 1.4 more than the corresponding duration in working days. Accordingly, the necessary release date for job i is

$$g_i = d_i - 1.4\, p_i / E. \tag{1}$$

The customer service level may be assessed with the help of the current orders utility function V. From the manufacturer's point of view, the order value increases proportionately to work amount p_i, since staff engagement increases. Besides, the more the time reserve is for completing an order, the more attractive is the order, since there is an opportunity to prepare for order execution. Eventually the order time reserve is decreasing, and the order value is diminishing. After all, if due date has expired, the order value becomes negative.

The manufacturer's attitude to the order changes with time, and the appropriate function is named *production intensity* (Mauergaz, 2012).

$$H_i = \frac{w_i p_i}{EG} \frac{1}{(d_i - t)/\alpha G + 1} \text{ at } d_i - t \geq 0$$
$$H_i = \frac{w_i p_i}{EG}[(t - d_i)/\alpha G + 1] \text{ at } d_i - t \leq 0 \tag{2}$$

The diagram of this function is shown in Fig. 1. X-axis in this figure represents time reserve. The reserve is equal to subtraction between due date and current time. In the positive part of the diagram ($d_i > t$) the values of intensity decrease in hyperbolic mode as available time reserve is growing. When the time reserve is negative ($d_i < t$) and there is delay of order completion, the production intensity linearly increases. Since production intensity is dimensionless, it has no physical sense, but it has psychological sense. Indeed, when order delay is augmenting, nervousness about order execution is increasing. Two curves in Fig. 1 differ in the psychological coefficient value. The higher is the α coefficient, the more placid is the attitude to delays and the lower is the intensity.

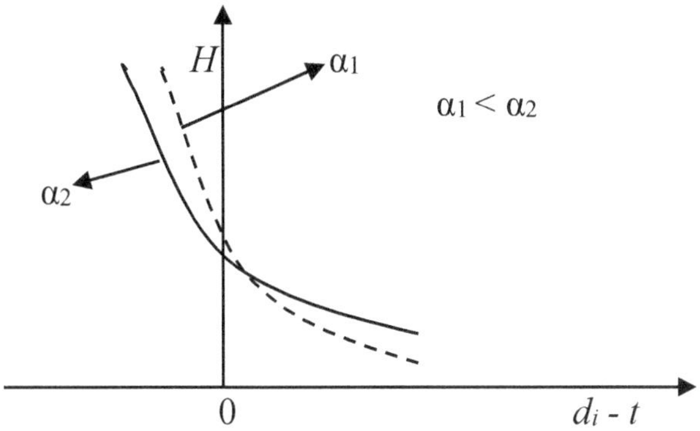

Fig. 1 Production intensity dependence on time reserve

The production intensity concept may be used for determination of the current order utility function V. The current utility (Fig. 2) for an order i is (Mauergauz, 2012):

$$V_i = \frac{w_i p_i}{EG} - H_i. \tag{3}$$

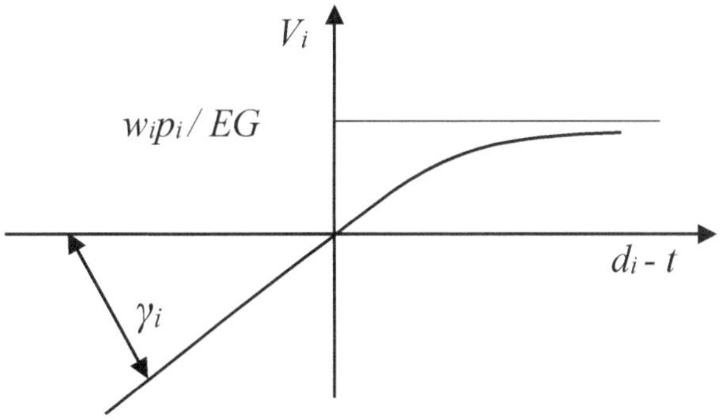

Fig. 2. Order utility function

The curve in Fig. 2 for the positive value $d_i - t \geq 0$ tends to the horizontal asymptote

$$V_i = w_i p_i / EG \tag{4}$$

In the negative part $d_i - t \leq 0$ the curve turns into the inclined straight line with

$$tg\gamma_i = \frac{w_i p_i}{\alpha EG^2}. \tag{5}$$

Assume the order quantity on the planning horizon equals n. Then their total utility V amounts to the sum of all order utilities, because orders, as a rule, are independent, i.e.

$$V = \sum_{i=1}^{N} V_i = \frac{1}{EG}\sum_{i=1}^{N} w_i p_i - \sum_{i=1}^{N} H_i. \tag{6}$$

The value of the function V changes in time, since the time reserve to the moment of scheduled execution changes. Besides, some orders are completed and new orders appear.

This paper suggests building a multi-level tree of works performance for making a schedule of parallel operating teams, while consistently considering feasibility of adding new tree nodes. Besides, these nodes determine both new jobs and teams that perform these jobs. New branches of the tree outgoing from each built node, may correspond both to the parent node team or other teams. Let us assume that a certain job corresponding to the tree node at the level l is completed at the moment of time C_{lz}. At the next level one of a set of uncompleted jobs has to be performed. Let us also assume that the job k starts at the moment t_{lk}, so this job has to be completed at the moment $t_k + p_k$. It was shown in (Mauergauz, 2013), the whole utility value at each level of the job fulfillment tree can be calculated with one of the recurrent formulas, which are used if utility value is known at the previous level:

$$\overline{V}_{l+1,k} = \frac{1}{t_k + p_k}(\overline{V}_l \times C_{zl} + \int_{C_{zl}}^{t_k+p_k} V_k dt), \qquad \text{if } t_k + p_k \geq C_{zl} \tag{7}$$

$$\overline{V}_{l+1,k} = \overline{V}_l + \frac{1}{C_{zl}} \int_{C_{jq}}^{t_k+p_k} V_k dt, \qquad \text{if } t_k + p_k < C_{zl}. \tag{8}$$

In Eqs. (7-8) the variable value $\overline{V}_l$ is equal to average utility of the whole set of jobs, which are planned from the start $t = 0$ until completion C_{zl} of the last planned job; j is a team number, which is selected for a job J_k in a new tree branch; C_{jq} is a completion moment of the last job, which was performed by the team j at the level q; V_k is the current utility of the whole set of all non-completed jobs; $\overline{V}_{l+1,k}$ is the average utility of the whole set of all planned jobs between the start moment $t = 0$ and the completion moment $t_k + p_k$ of the next planned job.

The upper integral limits in Eq. (7) and Eq. (8) are the same and equal to the completion moment of a job, which is planned at the next tree level construction. The bottom integral limits in Eq. (7) and Eq. (8) are different since they are equal to work completion moments of different teams. In the Eq. (7) this is the team z, which corresponds to the last built level; in the Eq. (8) this is the team j, which must perform a job at the next level. Apparently, this team may be used for tree construction at the previous levels, so the last level of its participation in scheduling $q < l$. The rules of integral calculations in Eqs. (7-8) are determined in (Mauergauz, 2014).

4. Problem formulation and its solution algorithm

In accordance with the well-known three-part scheduling classification, the problem may be recorded as:

$$FO \mid d_i, r_i \mid \overline{V}, \tag{9}$$

where FO is the designation of Flexible Open Shop problem, in which any sequence of parallel working teams is possible.

To solve the problem (9), we should apply a greedy algorithm, in which at each step a solution is selected with the best corresponding criterion. In problems with the average utility criterion $\overline{V}$ there is an additional condition of start due date g_i (Mauergauz, 2014). Since the number of solution-selecting conditions, which are possible at each step, is equal to 2, the number of such solutions (tree nodes) may be more than 1. Accordingly, the solution tree comprises all branches that grow from each node. At each step some possible solution versions that do not dominate each other are selected, and the algorithm below is used.

Step 1. *(Initial computation of utility functions)*

Let us assume that the level number is l=0; the initial orders utility function V_0 may be computed by the formula (6); number of nodes Z_0=1.

External cycle

Step 2. *(Determination of possible operations at next levels)*

For each node z of the built tree at the level l all possible jobs k at the level $l+1$ are determined, and values g_k are computed by the formula (1).

Intermediate cycle

Step 3. *(Determination of necessary main teams at next levels)*

For each possible job the pool S_i of teams, which are able to perform the main task of this job, is determined.

Step 4. *(Determination of necessary subsidiary teams at next levels)*

For each possible job the least charged teams, which are not the main for this job, are determined.

Internal cycle

Step 5. *(Utility function computation at next levels)*

For each team m related to the pool S_i values $\overline{V}_{l+1,k,m}$ are computed using the formulas (7,8).

End of internal cycle

End of intermediate cycle

Step 6. *(Determination of dominated tree nodes)*

If the level $l+1$ is not last, then for domination on the level $l+1$ of the tree node y with a job i over the tree node x it is sufficient to comply with the following equations

$$\overline{V}_{l+1,y} \geq \overline{V}_{l+1,x} \text{ and } g_{l+1,y} < g_{l+1,x}, \tag{10}$$

Otherwise: on the last level $l+1$ domination is possible, if

$$\overline{V}_{l+1,y} \geq \overline{V}_{l+1,x} \tag{11}$$

Step 7. *(Transition to the next level or stopping)*

If the level is more than the last (all operations are completed), then STOP.

Otherwise: level number increment $l = l+1$ and go to Step 2.

End of external cycle

5. Example of teams load scheduling with given set of jobs

Let us assume it is necessary to perform 40 various jobs at the facility of scheduling (enterprise, vessel, building etc.) during some period after planned start. For example, assume there is maintenance of various equipment going on, and each job comprises tasks, every of which relates to one of five possible types (Table 1).

Table 1
Types of tasks (operations)

Type No.	Type name
1	Mechanical
2	Electrical
3	Instrument
4	Building
5	Transport

Assume in this case there are five teams, which may be drawn in. Besides, some of them are specialized and can perform tasks of a single type, and other teams are complex. In Table 2 the example of highest throughput for each task type is shown.

Table 2
Highest team throughput in hours/day

Task type No.	Team No.				
	1	2	3	4	5
1	24	16	16	0	0
2	8	16	8	0	0
3	0	8	8	0	0
4	0	0	0	16	0
5	0	0	0	0	16

Each planned job is described by a whole set of parameters. In Table 3 the fragment of the main set of data on jobs is shown.

Table 3
Fragment of main data on jobs

Job No.	Calendar due date	First possible day	Whole process time in hours	Normative duration in working days	Main task type	Weight coefficient	Completion mark
1	-2	0	50	4	1	5	0.5
2	3	0	200	10	1	1	0.5
3	5	0	50	4	5	1	0.5
4	5	0	150	10	1	1	0.5
5	7	0	100	7	2	1	0.5
6	9	0	50	4	5	1	0
7	9	0	150	10	3	1	0
8	12	8	100	7	1	1	0
9	12	8	50	4	2	1	0
10	14	8	150	10	2	1	0

From Table 3 it follows that jobs are sorted by increment of due date, and this date on a day of scheduling may be exceeded. The job due date, the possible start date, the whole process time and the job duration are defined by given norms for scheduled maintenance. Any job has priority, which is determined by the weight coefficient. If a job has already been started at the moment of scheduling, the value 0.5 is input into the column "Completion mark". Since, as it was mentioned above, a job may consist of five tasks, Table 3 specifies the main task for each job, i.e. the one that determines possibility to start the entire job. For scheduling it is necessary to have initial data for started jobs at the planning moment (Table 4). If a task has to be completed but its execution has not started, the number -1 is input into the appropriate column.

Table 4
Fulfillment of started jobs at planning moment in %

Job No.	Task type	Team No.				
		1	2	3	4	5
1	1	0	80	0	0	0
1	2	0	-1	0	0	0
2	1	20	0	0	0	0
2	2	20	0	0	0	0
2	3	0	-1	0	0	0
3	4	0	0	0	-1	0
3	5	0	0	0	0	80
4	1	0	0	70	0	0
4	2	0	0	100	0	0
4	3	0	0	-1	0	0
4	4	0	0	0	100	0
5	1	0	30	0	0	0
5	2	0	70	0	0	0
5	5	0	0	0	0	60

As it follows from Table 4, at the moment of scheduling the team 1 performs only the job 2, and the team 3 performs the job 4. The team 2 at this moment started execution of the jobs 1 and 5; besides, it must take part in the job 2. The team 4 completed its part of the job 4, and must take part in the job 3. The team 5 performs the jobs 3 and 5 in parallel. In scheduling process it is necessary to take into account that the team throughput during following weeks may be below high values in Table 2, because some workers may be absent for any reasons. A special table for expected throughput on the planning horizon of 6 weeks is made. All tables are recorded in the MS Excel sheet.

To search for solutions on the basis of the criterion $\overline{V}$, the program using the VBA language for MS Excel has been designed. In Fig. 3 the results of scheduling computation are shown, when parameters have the following values: psychological coefficient α=0.1; duration of plan period G = 5 days; quantity of hours per work day E=8. For each job calendar days of the start and the completion are recorded. Numbers in brackets correspond to the teams, which take part in the job; the team, which executes the main job component, is the first.

	A	B
1	Job 1: 0 – 2 (2)	
2	Job 2: 0 – 12 (1, 2)	
3	Job 3: 0 – 2 (5, 4)	
4	Job 4: 0 – 5 (3)	
5	Job 5: 0 – 5 (2, 5)	
6	Job 6: 0 – 6 (5, 3)	
7	Job 7: 0 – 14 (2, 4)	
8	Job 8: 0 – 10 (1, 3)	
9	Job 9: 8 – 14 (2)	
10	Job 10: 8 – 22 (1)	
11	Job 11: 8 – 18 (2, 5)	
12	Job 12: 8 – 18 (2, 4, 5)	

Fig. 3. Fragment of jobs distribution

Planned teams load for the following seven weeks is shown in Fig. 4. In this case we can see that during the first and the second weeks the team load is not more than 100%; in the third week all teams are overloaded, in following weeks there are load oscillations. The computed plan solution generally ensures timeliness of jobs completion, but it is not optimal, and may be substantially improved. For this purpose, it is necessary to analyze the list of jobs, which are planned for the teams on the planning horizon. In Fig. 5 the fragment of the team timetable for jobs execution in the following 3 weeks is shown. Information in Fig. 4 shows that the team 3 is the most overloaded team, which on the third week has the load of 161%. Studying the list of jobs for the team 3 in the third week, we can suggest that this load could be less, if execution of the job 22 was postponed to the next week.

	A	B	C	D	E	F	G	H
13								
14			Weekly teams load in % of throughput					
15	Weeks	1	2	3	4	5	6	7
16	Team 1:	72	83	112	62	146	136	28
17	Team 2:	76	86	100	112	121	49	0
18	Team 3:	96	76	161	65	125	76	9
19	Team 4:	52	95	128	58	142	85	16
20	Team 5:	50	28	112	78	138	98	10

Fig. 4. Planned teams load

	A	B	C	D	E	F	G	H
21								
22			Weekly jobs list					
23	Weeks	1		2			3	
24	Team 1:	2, 8		2, 8, 10			10, 26, 20	
25	Team 2:	1, 2, 7, 5		2, 7, 11, 12, 9			11, 12, 20, 24, 30	
26	Team 3:	4, 8, 6		8, 14, 19, 17, 13, 15			14, 19, 17, 13, 15, 22	
27	Team 4:	7, 3		7, 12, 19, 13, 15			12, 19, 13, 15, 18, 24	
28	Team 5:	3, 5, 6		11, 12, 14			11, 12, 14, 26, 18, 30, 22	

Fig. 5. Fragment of jobs list scheduled for teams

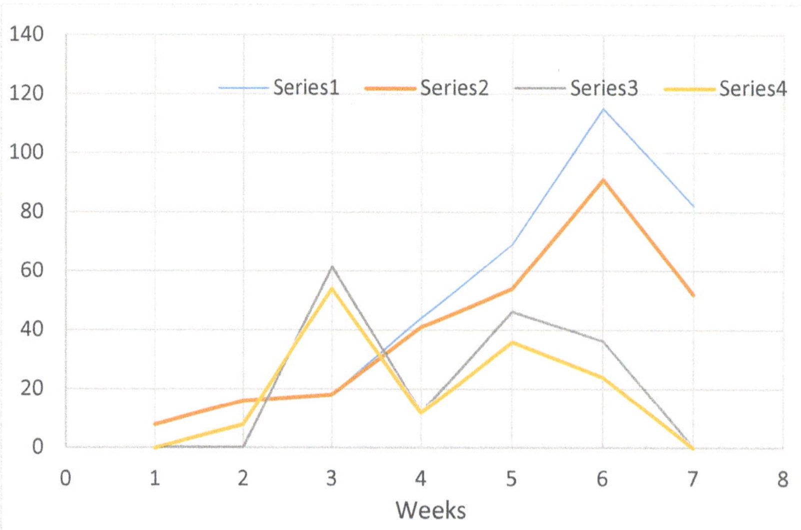

Fig. 6. Jobs tardiness and highest team overload

The plan improvement achieved by correction is shown on the diagrams in Fig. 6. The diagrams show the summary tardiness for all planned jobs in hours, and the overload of the most charged teams as percentage on the planning horizon. Thin lines refer to the initial planning solution, thick lines relate to the corrected solution. As it follows from Fig. 6, the correction diminishes both the jobs tardiness and the most team overload, appreciably. However, it is impossible to eliminate unevenness completely using only the scheduling process. Therefore, for timely jobs completion we have to change the team structure or duration of working time.

1) Summary weekly jobs tardiness according initial schedule in hours.
2) Summary weekly jobs tardiness according corrected schedule in hours.
3) Overload of the charged team according to initial schedule as percentage.
4) Overload of the charged team according to corrected schedule as percentage.

6. Conclusion

Results above confirm that the approach to production teams scheduling, which is based on application of order utility functions, produces a satisfactory schedule. The problem solution, which uses the "greedy" algorithm, is able to construct the search tree starting from the initial system state. This algorithm determines the domination bound for tree nodes at each level according to the criterion $\overline{V}$, and in addition according to the value of necessary job release date g. The program in VBA language for MS Excel was designed, and the example of its application for scheduling was made. The calculated schedule may be improved in interactive mode, if a certain job is postponed to the next week.

Scheduling is a regular process that repeats with certain, but not always constant frequency. For this purpose it is convenient to use new MS Excel sheets, where information from previous sheets may be contained. By changing or inserting new data, the user may correct the previous plan or design a new one. Analyzing a set of calculated schedules, one may detect systematic deviations of teams load and optimize professional team structures. In practice various additional constraints may be necessary for scheduling. For example, often it is needed to take into account impossibility of job execution in a certain time interval. In the nearest future it is planned to elaborate some solutions for such problems.

References

Bai, D., & Tang, L. (2013). Open shop scheduling problem to minimize makespan with release dates. *Applied Mathematical Modeling, 37*, 2008-2015.

Blackburn, J., & Millen, R. (1986). Perspectives on flexibility in manufacturing: hardware versus software. *Modeling and DeMgn of Flexible Manufacturing Systems*, *2*(2), 116-117.

Błażewicz, J., Pesch, E., Sterna, M., & Werner, F. (2004). Open shop scheduling problems with late work criteria. *Discrete Applied Mathematics,134*(1), 1-24.

Bräsel, H., Herms, A., Mörig, M., Tautenhahn, T., Tusch, J., & Werner, F. (2008). Heuristic constructive algorithms for open shop scheduling to minimize mean flow time. *European Journal of Operational Research, 189*(3), 856-870.

Doulabi, S. H. H. (2010). A mixed integer linear formulation for the open shop earliness-tardiness scheduling problem. *Appl Math Sci*, *4*(35), 1703-1710.

Gonzalez, T., & Sahni, S. (1976). Open shop scheduling to minimize finish time. *Journal of the ACM (JACM)*, *23*(4), 665-679.

Graham, R. L., Lawler, E. L., Lenstra, J. K., & Kan, A. R. (1979). Optimization and approximation in deterministic sequencing and scheduling: a survey.*Annals of discrete mathematics*, *5*, 287-326.

Gupta, D., Renuka, M., & Singla, P. (2013). *Open Shop Scheduling: Minimizing Makespan*. Saarbrucken, LAP Lambert.

Leung, J. Y. T., Li, H., & Pinedo, M. (2005). Order scheduling in an environment with dedicated resources in parallel. *Journal of Scheduling*, *8*(5), 355-386.

Liaw, C. F. (2005). Scheduling preemptive open shops to minimize total tardiness. *European Journal of Operational Research*, *162*(1), 173-183.

Mastrolilli, M., Queyranne, M., Schulz, A. S., Svensson, O., & Uhan, N. A. (2010). Minimizing the sum of weighted completion times in a concurrent open shop. *Operations Research Letters, 38*(5), 390-395.

YanMin Ma, Y. M., & Di Jin, D. J. (2013). Concurrent Open-shop Scheduling Accurate Algorithm Research. *International Journal of u-and e-Service, Science and Technology*, *6*(5), 1-16.

Mauergauz, Y. (2012). Objectives and constraints in advanced planning problems with regard to scale of production output and plan hierarchical level.*International Journal of Industrial and Systems Engineering*, *12*(4), 369-393.

Mauergauz, Y. (2013). Cost-efficiency method for production scheduling. In*Proceedings of the world congress on engineering* (Vol. 1, pp. 587-593).

Mauergauz, Y. (2014). Dynamic Pareto-optimal group scheduling for single machine. *International Journal of Industrial and Systems Engineering, 16,* 537-559.

Naderi, B., Ghomi, S. F., Aminnayeri, M., & Zandieh, M. (2011). Scheduling open shops with parallel machines to minimize total completion time. *Journal of Computational and Applied Mathematics*, *235*(5), 1275-1287.

Ng, C. T., Cheng, T. C. E., & Yuan, J. J. (2003). Concurrent open shop scheduling to minimize the weighted number of tardy jobs. *Journal of Scheduling*, *6*(4), 405-412.

Shabtay, D., & Kaspi, M. (2006). Minimizing the makespan in open-shop scheduling problems with a convex resource consumption function. *Naval Research Logistics (NRL)*, *53*(3), 204-216.

Witkowski, T., Antczak, P., & Antczak, A. (2012, February). Hybrid method for solving flexible open shop scheduling problem with simulated annealing algorithm and multi-agent approach. In *Advanced Materials Research* (Vol. 383, pp. 4612-4619).

7

Dealing with customers enquiries simultaneously under contingent situation

Sujan Piya*

Department of Mechanical and Industrial Engineering, College of Engineering, Sultan Qaboos University, AL-Khod 123, Muscat, Sultanate of Oman

CHRONICLE

Keywords:
Quotation
Contingent order
Negotiation
Make-to-Order

ABSTRACT

This paper proposes a method to quote the due date and the price of incoming orders to multiple customers simultaneously when the contingent orders exist. The proposed method utilizes probabilistic information on contingent orders and incorporates some negotiation theories to generate quotations. Rather than improving the acceptance probability of quotation for single customer, the method improves the overall acceptance probability of quotations being submitted to the multiple customers. This helps increase the total expected contribution of company and acceptance probability of entire new orders rather than increasing these measures only for a single customer. Numerical analysis is conducted to demonstrate the working mechanism of proposed method and its effectiveness in contrast to sequential method of quotation.

1. Introduction

Companies that operate in Make-to-Order (MTO) environment must respond to enquiries by preparing quotations that are attractive to their customers and feasible for them to fill. Dealing properly with customers enquiries has a huge impact on orders being confirmed by the customer. Lots of researchers have contributed to the area of quotation. However, from the relevance of industrial domain, due date and price quotation are the most researched areas. With an objective of minimizing expected aggregate cost per job, Seidman and Smith (1981) have proposed a method to identify due date for incoming order. On the other hand, Bertrand (1983) and Baker (1984) have considered the objective of minimizing the average weighted due date quoted to the customers. Ragatz and Mabert (1984), Shantikumar and Sumita (1988), Wein (1991) and Chand and Chhajed (1992) have advocated that due date based on shop-floor congestion information achieves better delivery performance. Most of these researchers have dealt with due date and/or pricing decisions under a single-product and single-stage production system. Enns (1995) and Van and Bertrand (2001) have investigated the setting of cost optimal internal due dates for determining the priorities on the shop floor, and in determining the external due dates that can be quoted to the customer. Lawrence (1995) has developed a model for due date estimation using empirical distribution of forecasted error in production settings with machine breakdowns. Elhafse (2000) has proposed exact and heuristic algorithms for determining the lead-time and price to be quoted to a single order by dividing total orders into regular and rushed orders. Proposing two different models, Keskinocak et al. (2001) has shown that different strategies for quotation are needed for different categories of customers. Charnsirisakskul et al. (2006) has presented an optimization model for due date and price sensitive customers under deterministic demand function. Liu et al. (2007) has considered pricing and lead-time decisions from a supplier-retailer perspective where demand is deterministic and sensitive to price and lead-time decisions. Similarly, Kaminsky and Kaya (2008) have developed due date quotation model in the supply chain environment where the model is affected by the performance of supplier. Based on the result of empirical analysis, Zorzini et al. (2008) has proposed a method of quoting due date that took into account the known average lead time. Slotnick (2011) has presented stochastic dynamic programming model for lead time quotation for a production system in which bottleneck process requires a minimum batch size. Chaharsooghi et al. (2011) has extended the model of Charnsirisakskul et al. (2006) to include multi class customers with stochastic demand. Recently, Xianfei et al. (2013) has studied production scheduling problem with an objective of maximizing profit when customers are sensitive to quoted delivery time.

When an enquiry about new order arrives, the company may already have several orders either in the form of confirmed orders or contingent orders. The time between customer enquiry and order acceptance (OA) decision can be extremely long and exhaustive enquiry may, unfortunately, not be converted into confirmed order. As a matter of fact, the completion time and production costs required to fill new order that arrive, when contingent orders exist, cannot be anticipated with any degree of precision. This complicates matters for company that have to prepare quote for new order. Easton and Moodie (1999) is the first paper to explicitly coin the effect of contingent orders on quotation. The paper dealt with this problem in a single resource and single job production environment. The same problem was later expanded into a multiple job environment with different routines for each job by Cakravastia and Takahashi (2003). To prepare quotation for different class of customer under contingency environment EDD and FCFS method of scheduling was utilized in Watanapa and Techanitisawad (2005a). Further, Watanapa and Techanitisawad (2005b) have proposed GA based quotation method where limited number of tardy jobs is allowed. In the same line of research, Corti et al. (2006) has developed analytical tool to compare capacity requirement of order pool, including contingent orders, with respect to actual capacity of shop floor. All these papers have defined contingent orders as potential orders that were awaiting customer confirmation on whether to accept or reject quotation submitted by company. These papers utilized S-shaped logit model to calculate the probability of quotation being accepted by new customer. The model was constructed without utilizing any probabilistic information on contingent orders. Also, as contingent orders are a source of uncertainty in capacity, these papers fail to propose appropriate strategy to hedge against this uncertainty in case contingent order cannot be converted into confirmed order.

Quotation represents initial phase in OA decisions (Sujan et al., 2009). OA decisions, in MTO systems, are often the consequence of negotiations between the customer and the company over contested issues (Moodie and Bobrowski, 1999). Sujan et al. (2009) has proposed a method that can simultaneously quote the due date and the price by implementing the concept of negotiation margin. Further, defining contingent order as a customer engaged in negotiation with a company to reach an agreement, Piya et al. (2009b) has proposed a probabilistic method with a negotiation structure to counter the uncertainty created by contingent orders and prepare quotes for a new order. All these mentioned papers with contingent orders have considered generating quotation only for single customer at a time. However, once new customer makes an inquiry, there will be some time lapses before the company prepares quotation and submit it. Within the lapsed time, other new customers may arrive in the systems for an inquiry. In such situation, the company has to prepare quotation for the customer under the influence of uncertainties created by both the contingent orders and the other new orders. To cope with such circumstances, this paper proposes a method that can prepare quotations simultaneously for multiple new customers. The method aims to increase the probability of quotations being accepted by their customers. To the best of author's knowledge, this is the first paper to consider the research on quotations for multiple customers simultaneously under contingency effect. The method will first tackle the effect of contingent orders and then generate quotations for multiple customers simultaneously.

The rest of the paper is structured as follows. Section 2 describes the problem that will be addressed in this research. Section 3 explains the proposed method in detail. Section 4 discusses and presents the results of the numerical analysis. Finally, concluding remarks and future research directions are highlighted in section 5.

2. Problem Description

This paper reflects the method of quotation as a function of negotiation between the company and the customers of contingent orders. In MTO systems, once customer has arrived with some technical specifications, the company is asked to give quote, which basically will be on the due date and the price. After submitting quotation, the customer and the company may engage in negotiations to reach an agreement, if the quoted value is not acceptable to the customer. Negotiations begin with a customer counter-offering another due date and/or price against the quotation submitted by the company. Such orders represent contingent orders in this paper. The negotiations may continue for several rounds during which the company proposes new offer on due date and/or price and the customer opposes it by counter-offering another due date and/or price within their limit levels. In an attempt to reach an agreement, they move in the opposite direction which will reduce the distance between them on the contested issues. While undergoing negotiation with contingent orders, if a new order arrives, it is difficult for the company to quote the due date and the price for new order. This is because the company cannot be sure at this stage whether agreement can be reached on orders involving ongoing negotiations. Also, while preparing quote for new order, more orders may arrive in the system within small time window for quotations. This will further complicate situation for the company on assigning the due dates and the prices for all the new orders. As it is believed that customer prefers early delivery of an order at a cheaper price, in such situation, the company must try to minimize the average due dates and prices to be quoted to the customers by taking into account the possible outcome of negotiations with the contingent orders. But no quotation should be so ludicrous that they will send the company bankrupt or tarnish its reputation merely for the sake of appeasing customers.

This paper assumes above problem in a flow shop environment consisting of various work centers. An order consists of batch of one product type that will enter from the initial work center and exit through the last work center. Each order involves operations at all the work centers. The paper aims at providing company with a capability to manage contingent orders while generating quotes for multiple customers simultaneously, without compromising on profits they expect to achieve. For the purpose, following assumptions are considered.

i) There would be at least one offer from the company and one counter-offer from all the contingent orders when the new orders arrive.

ii) There are many factors that affect the length of negotiation (Bac, 2001). Before negotiation begins, the negotiating parties may basically fix a deadline for negotiation or the number of negotiation rounds. Based on this, it is assumed that the maximum rounds of negotiation (R) with contingent orders would be defined before starting negotiation.

iii) All the orders accumulated within period $\{t_z, (t+1)_z\}$ will be processed at period $(t+1)$.

3. Proposed Method

The proposed method can be explained by two steps; steps 1 and 2. These steps will be utilized to manage the contingent orders and prepare quotes for the new orders respectively.

Step 1: Managing contingent orders

To offset the effect of contingent orders on quotation, Piya et al. (2009b) proposed a method to classify these orders into different sets, namely negotiation set and rejection set. Proposed method is based on the statistical data on acceptance probability of orders that arrived in the past. The major shortcoming of this method is the subjective nature of standard deviation. If its value is very high or low, the method will either classify all the contingent orders into negotiation set or rejection set. To overcome above shortcoming, this paper proposes new method that helps select best contingent orders in terms of expected contribution and acceptance probability. As shown in equation (1), the objective here is to find the combination and processing sequence of confirmed and contingent orders such that their expected contribution and acceptance probability lies at the minimum distance from the maximum expected contribution and acceptance probability.

$$\min D = \left\{(Z_{\max} - Z^s)^2 + (A_{\max} - A^s)^2\right\}^{\frac{1}{2}} \tag{1}$$

To define the maximum expected contribution Z_{max}, at first, expected contribution Z^s for each combination and processing sequence of confirmed and contingent orders is calculated by Eq. (2). The expected contribution consists of selling price, production cost and acceptance probability.

Nomenclature

Index

o: Order (o = 1, 2,……, O); c, k, n, l
- c: Confirmed order
- k: Contingent order,
- n: New order
- l: Latest accepted order

t: Time/ period ($t = t_e, t_s, t_z, t_o$)
- t_e: End of the period t
- t_s: Start of period t
- t_z: Any time within period t

j: Issues $\in$ due date (d), price (p)

r: Rounds of negotiation (r= 1,….., R)

i : Operation (i = 1,……., I)

f: Unit in the order (f= 1, 2,….., F)

m: Machine (m=1,…., M)

g: Processing sequence of new orders (g=1, 2, ….., G)

s: Processing sequence of confirmed and contingent orders (s=1, 2,….,S)

Variable

D: Distance

Z^s: Expected contribution by sequence s

A^s: Acceptance probability of sequence s

Z_{max}: Maximum expected contribution

$\lim_{jo}$: Limit level on issue j of order o

Q_{jo}: Quoted value on issue j of order o

w_j^c : Expected weight of customer on issue j

$Prod_o$: Total cost of producing order o

$prod_o$: Total production time of order o

v_j : 0 when Y_{jor}= $\lim_{jo}$; 1 otherwise

$(A_{or})^s$: Acceptance probability for r negotiation round of order o with sequence s

$(X_{jor})^s$: Offer on issue j for r negotiation round of order o with sequence s

Y_{jor}: Counter-offer on issue j for r negotiation round of order o

h_o^s : Inventory holding cost of order with sequence s ($h_o^s \in h_o^{IP}$; In process, h_o^{FGI} ; Finished goods)

ER_{oif} / CT_{oif}: Earliest release date/ Completion time for the operation i and unit f of order o

Parameter

q_o: Total units in the order o

RM_o: Raw material cost for order o

cap_m: Capacity of machine m

N: Total new orders

α: Smoothing constant

$Wl(t)$:Workload at time t

g_{first}: Fist order in a sequence g

s_{last}: Last order in a sequence s

i_{max}: Operation with maximum processing time

Cd_{jo}: Cumulative difference on issue j of order o

A_{max}: Maximum acceptance probability

SU_{oi}: Set up cost for i operation of order o

p_{oi}: Processing time for i operation of order o

P_{oi}: Cost of processing operation i of order o

w_j: Weight assigned to issue j by the company

λ_o: Profit margin coefficient of order o

NM_{jo}: Negotiation margin on issue j of order o

a_{jo}: agreed value on issue j of order o

p_o: Price of order o

δ_{or}: Aspiration level of order o at r negotiation round

$$Z^S = \sum_{o \in c,k} \left\{ \left\{ q_o (X_{por})^s - \left[\left(\sum_{i=1}^{I} SU_{oi} + q_o RM_o + q_o \sum_{i=1}^{I} p_{oi} P_{oi} \right) + h_o^s \right] \right\} (A_{or})^s \right\} \tag{2}$$

i) *Selling price* $(X_{por})^s$: The selling price of confirmed order, for any sequence s, will be fixed and equal to the agreed price (p_o).

However, the selling price for the contingent order will be equal to the price that can be offered to the customer for the next round of negotiation. Basically, it depends on the strategy of negotiation used by the manufacturer. As proposed by Piya et al. (2009b), selling price will be calculated by considering a linear trade-off relationship between the due date and the price at a fixed aspiration level. Aspiration level here represents a level of benefit sought by the company for a particular round of negotiation (Cakravastia & Nakamura, 2002).

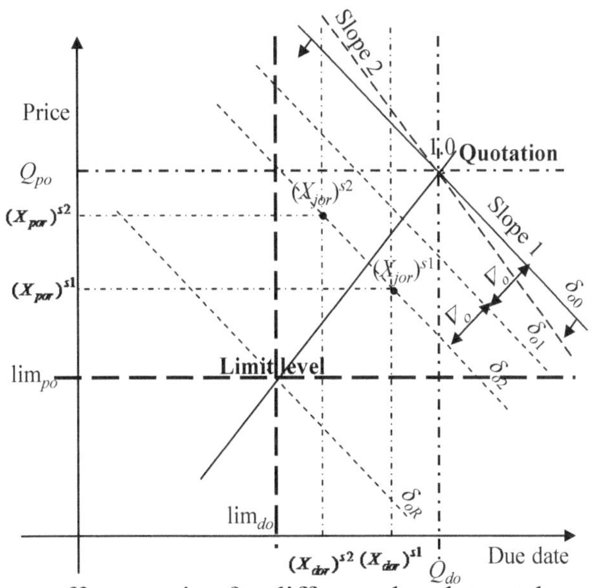

Fig. 1. Different new offer on price for different due date at the same aspiration level

Let the point (Q_{do}, Q_{po}) in Fig. 1 be the point of quotation for contingent order o and let the aspiration level at this point be the maximum aspiration level, δ_{o0}, of company. Normalizing this point and the point of limit level ($\lim_{do}$, $\lim_{po}$) within (1,0) and reducing aspiration level with the progress of negotiation by a fixed step size to approach nearer to the counter-offer of customer, aspiration level at any round can be calculate by Eq. (3).

$$\delta_{or} = \delta_{o(r-1)} - \frac{(\delta_{o0} - \delta_{oR})}{R} \tag{3}$$

As the completion time of contingent order will be different for the different combination and processing sequence s of confirmed and contingent orders, the price X_{por} in equation (2) for contingent order depends on X_{dor}. Suppose, by sequence $s1$, if the due date of contingent order is $(X_{dor})^{s1}$ in Fig. 1, then $(X_{por})^{s1}$ will be the price that will be offered if the aspiration level for round r is δ_{o2}. Based on this concept, the selling price for contingent order is given by the following equation.

$$(X_{por})^{s} = \frac{1}{w_p}\left[\delta_{or}(Q_{po} - \lim_{po}) - \frac{w_d\{(X_{dor})^{s} - \lim_{do}\}(Q_{po} - \lim_{po})}{(Q_{do} - \lim_{do})}\right] + \lim_{po} \quad ,\forall o \in k \tag{4}$$

From above equation, it can be noted that the price $(X_{por})^{s}$ also depends on the weight assigned by the company (w_j) on the given issue. The weight affects the slope of aspiration level (Piya et al., 2010).

The details on the method for calculating the limit level ($\lim_{jo}$) and quotation (Q_{jo}) on the two issues will be discussed in step 2 of proposed method.

ii) *Production cost*: Production cost consists of set up cost, raw material cost, cost of processing the order and inventory holding cost. Except inventory holding cost, all the other costs are fixed.

$$h_o^{s} = \sum_{i=1}^{I}\sum_{f=1}^{F}\left(ER_{o(i+1)f} - CT_{oif}\right)h_o^{IP} + \sum_{f=1}^{F}\left\{(X_{dor})^{s} - CT_{olf}\right\}h_o^{FGI} \tag{5}$$

As shown in Eq. (5), both in process (IP) and finished good (FGI) inventory holding cost makes up the total cost of carrying inventory. For contingent orders, FGI cost will not be affected by the processing sequence as it is assumed that, during the negotiation process with contingent order, the order will be delivered to the customer as soon as its processing is completed. On the other hand, for confirmed orders both IP cost and FGI cost depends on the order processing sequence s.

The earliest release date and completion time in Eq. (5) can be calculated by using iterative method that considers precedence relationship between the operations of order and the shop floor information.

iii) *Acceptance probability* $(A_{or})^{s}$: To calculate the probability of customer accepting a quotation submitted by the company, Easton and Moodie (1999) and Cakravastia and Takahashi (2003) have used an S-shaped logit model. This model was constructed without utilizing any information on contingent order. In the negotiation process, it is possible to obtain information on negotiating issues from the customer of contingent order. Use of such information will increase the authenticity of calculated acceptance probability. Proposed method utilizes information such as counter-offer on due date and price received from the customer, along with the due date and price that can be offered by the company, to calculate the acceptance probability of contingent orders. Therefore, acceptance probability here indicates the probability that the new offer of company will be accepted by the customer of contingent order.

$$(A_{or})^{s} = \left\{\exp\sum_{j\in d,p}\left[w_j^{c}\left\{v_j\frac{(X_{jor})^{s}(\lim_{jo} - Y_{jor})}{\lim_{jo}(Q_{jo})}\right\}^{\left(1-\frac{r-1}{R}\right)}\right]\right\}^{-1} \quad ,\forall o \in k \tag{6}$$

If the counter-offer, Y_{jor}, on any issue is more than or equal to the limit level, $\lim_{jo}$, the binary variable v_j in Eq. (6) will be 0, thus resulting in an acceptance probability equal to 1. This is because the limit level in this research indicates the level at which the company is willing to accept the order. From the equation note that the acceptance probability will be different for different rounds of negotiation r. It depends on the latest counter-offer of customer, the offer that will be submitted next by the company, the expected weight of customer on different issues and the remaining rounds of negotiation. It should be noted that the acceptance probability for confirmed order will be equal to 1. For the method to calculate expected weight of customer in Eq. (6), refer to Piya et al. (2010). After calculating expected contribution for all the possible combination and sequence of confirmed and contingent orders, the expected contribution that has the highest value will be selected as the maximum expected contribution Z_{max} in Eq. (1).

$$Z_{\max} = \max(Z^s) \tag{7}$$

Next, the acceptance probability of sequence s in Eq. (1) is calculated by taking the product of acceptance probability of all the orders that is included in the given sequence.

$$A^s = \prod_{o=1}^{O} (A_{or})^s, \qquad \forall o \in c, k \tag{8}$$

The maximum acceptance probability (A_{max}) in equation (1) can be directly assigned the value of 1.0. This is because acceptance probability for the combination of only confirmed orders with any sequence s will be 1.0.

As discussed before, the expected contribution and acceptance probability will be calculated for each combination and processing sequence of confirmed and contingent orders. The number of combinations will be equal to 2^k and for each combination the possible sequence will be equal to $(k+c)!$. It means that even for small problem instance, the possible combination and sequence will be very large. To avoid generation of such large combinations, two important constraints are employed.

$$\lim_{do} \le CT_{oIF} \le d_o, \qquad \forall o \in c \tag{9}$$

$$X^s_{dor} \ge \max(\lim_{do}, CT_{oIF}), \qquad \forall o \in k \tag{10}$$

Constraint in Eq. (9) indicates that the completion time of confirmed order should be greater than or equal to its limit level on due date and less than or equal to agreed due date. Similarly, constraint in Eq. (10) indicates that the new offer on due date should be more than or equal to the maximum value between its limit level on due date or completion time. Other constraints include the precedence constraint, capacity constraint and constraint on total weight as indicated by Eq. (11), Eq. (12) and Eq. (13), respectively.

$$ER_{o(i+1)f} \ge CT_{oif}, \qquad \forall i, \forall f, \forall o \in c, k \tag{11}$$

$$\sum_{o \in c,k} \sum_{i=1}^{I} (p_{oi})_{\{t_z,(t+1)_z\}} \le \sum_{m=1}^{M} (cap_m)_{(t_s,t_e)} \tag{12}$$

$$\sum_{j \in d,p} w_j = 1 \tag{13}$$

Step 2: Generating quotes

Step 1 helps identify the set of confirmed and contingent orders, with their processing sequence, which gives the maximum expected contribution and acceptance probability. The result of step1 is then used in step 2. From here onwards, sequence s means the sequence obtained for confirmed and contingent orders in step 1 and k represents only those contingent orders that are included in the sequence s. The objective of step 2 is to minimize the average of due dates and prices to be quoted to the multiple new customers. As shown in Eq. (14), the average of quoted values on two issues i.e., the due dates and the prices are integrated by vector normalization method (Van & Nijkamp, 1977).

$$\min V = \sum_{j\in d,p} \frac{Avg(Q_{jo})^g}{\left[\sum_{g=1}^{G}\left\{Avg(Q_{jo})^g\right\}^2\right]^{\frac{1}{2}}}, \forall o \in n \tag{14}$$

$$Avg(Q_{jo})^g = \frac{\sum_{o=1}^{O}(Q_{jo})^g}{N}, \qquad \forall o \in n \tag{15}$$

Sequence g in Eqs. (14) means the sequence of new orders. The starting time of sequence g will be equal to the completion time of sequence s. It means that the model considers all the possible sequence g of new orders, at fixed sequence s of confirmed and contingent orders, to find the minimum value of V. As shown in Eq. (16), the quoted value on issue j consists of limit level and negotiation margin on the given issue. It will be different for the different processing sequence g of new orders.

$$(Q_{jn})^g = (\lim_{jn})^g + (NM_{jn})^g \tag{16}$$

i) *Limit Level* ($\lim_{jn}$): Limit level represents the level below which company will not negotiate with customers on either issues i.e., due date and price.

(a) *Due date*: The limit level on due date, $\lim_{dn}$, is the date below which it is not possible to complete the processing of an order. It is calculated by subtracting the completion time of new order at given sequence g with the product of probability of rejection and the total production time of contingent orders.

$$(\lim_{dn})^g = (CT_{niF})^g - [1 - A^s]\sum_{o\in k} prod_o \tag{17}$$

Completion time of new order in Eq. (17) depends on its position in the sequence g. As shown in Eq. (18), if the position of new order lies in the beginning of sequence g, the completion time will be the summation of its total production time and the completion time of first operation of order o that lies at the end of sequence s. Otherwise, it will be the summation of its total production time and the completion time of first operation of other new order whose position lies before order n in the sequence g.

$$(CT_{niF})^{g_{first}} = (CT_{oif})^{s_{last}} + prod_n, \forall o \in c,k \tag{18}$$

$$(CT_{niF})^{g_{nth}} = (CT_{nif})^{g_{nth-1}} + prod_n \tag{19}$$

The total production time in the Eq. (17), Eq. (18), and Eq. (19) can be calculated based on the operation of order with the maximum processing time. As shown in Eq. (20), the first and second portions at the right hand side of equation are used to calculate the processing time without including service delays. However, the third and fourth portions add on the respective service delays before and after operation with maximum processing time. Service delays incurs if machine on which operation of an order is to be performed is occupied by the operation of any other order.

$$prod_o = (q_o - 1)\max_i p_{oi} + \sum_{i=1}^{I} p_{oi} + \sum_{i=1}^{i_{\max}-1}\left(ER_{o(i+1)1} - ER_{oi1} - p_{oi}\right) + \sum_{i=1}^{i_{\max}+1}(CT_{oiF} - CT_{o(i-1)F} - p_{oi}) \tag{20}$$

b) *Price*: The limit level on price, $\lim_{pn}$, indicates the price at which company wishes to reach an agreement after negotiation. As shown in Eq. (21), it consists of production cost ($Prod_n$), and profit margin coefficient (λ_n).

$$(\lim_{pn})^g = (\Pr od_n)^g(1+\lambda_n) \text{ where,} \tag{21}$$

$$(\Pr od_n)^g = \sum_{i=1}^{I} SU_{ni} + q_n RM_n + q_n \sum_{i=1}^{I} p_{ni} P_{ni} + (h_n^{IP})^g \sum_{i=1}^{I}\sum_{f=1}^{F}(ER_{n(i+1)f} - CT_{nif}) + (h_n^{FGI})^g \sum_{f=1}^{F}\{(Q_{dn})^g - CT_{nif}\} \tag{22}$$

The production cost of new order includes set up cost, raw material cost, cost of processing the order and the inventory holding costs. Here also, except in process and finished goods inventory holding cost all the other costs are fixed and is unaffected by the sequence g. Profit margin coefficient in Eq. (21), λ_n, adds some profit margin to the production cost. In the proposed research, it is assumed that the profit margin of company will be the function of expected workload in the system and the expected number of competitors. Equations for calculating profit margin coefficient are similar to Piya et al. (2009b) except for the calculation of expected workload. Eq. (23) indicates the summation, within a time frame of (t_s, t_e), of workload of all the confirmed orders, expected workload of contingent orders and the total processing time of new orders to calculate the expected workload.

$$Wl(t) = \left[\sum_{o \in c} \sum_{i=1}^{I} (CT_{oiF} - ER_{oi1}) + A^s \sum_{o \in k} \sum_{i=1}^{I} (CT_{oiF} - ER_{oi1}) + \sum_{o \in n} \left(q_o \sum_{i=1}^{I} p_{oi} \right) \right] \quad (23)$$

ii) *Negotiation margin* (NM_{jn},): The negotiation margin provides company with an allowance to negotiate on the contested issues with the customer. Therefore, if the customer requested to reduce the value on the quoted due date and/or price, the company can do it without risking the chance of order being tardy and without reducing their desired profit margin. The concept of cumulative difference is utilized to calculate negotiation margin. As shown in Eq. (24), the cumulative difference along with the limit level on the given issue at sequence g is used to calculate the negotiation margin for the given issue.

$$(NM_{jn})^g = \left[\frac{cd_{jn}}{(1 - cd_{jn})} \right] (\lim_{jn})^g \quad (24)$$

$$cd_{jn} = \alpha \left(\frac{Q_{jl} - a_{jl}}{Q_{jl}} \right) + (1 - \alpha) cd_{jl} \quad (25)$$

Cumulative difference is basically the cumulative information on the difference between the quoted and the agreed values on given issues related to the past orders. In equation (25), smoothing constant (α) represents the value company gives to latest information as compared to past information on agreement.

4. Numerical Analysis

The aim of numerical analysis is to show the working mechanism and effectiveness of proposed method.

Working mechanism: The analysis is carried out by considering 5 work centers with each center capable of performing specific task. Nine orders are generated with orders A, B and C as confirmed orders; orders D, E and F are contingent orders and orders G, H and I are new orders for which quotations are to be prepared. The agreed due dates and prices of confirmed orders and the information on contingent orders necessary to calculate its acceptance probability are selected randomly as shown in Table 1. Processing time, total units, set up cost, raw material cost, cost of processing the order and inventory holding costs are uniformly distributed within the value as shown in Table 2. The total number of negotiation rounds, current rounds of negotiation for entire contingent orders and the weights of manufacturer and customer for each issue are fixed at 5, 3 and 0.5 respectively.

Table 1

Status of confirmed and contingent orders

Order status	Order	d_o	p_o	Y_{dor}	Y_{por}	$\lim_{do}$	$\lim_{po}$	Q_{do}	Q_{po}
Confirmed order	A	24	15000	-	-	-	-	-	-
	B	60	12000	-	-	-	-	-	-
	C	80	13800	-	-	-	-	-	-
Contingent order	D	-	-	38	13000	45	18000	80	26000
	E	-	-	34	10000	38	12500	74	21000
	F	-	-	40	12000	42	14000	88	23000

Table 2

Value of other parameters considered in the analysis

p_{oi} (Units)	q_o (Time unit)	SU_{oi} ($)	RM_o ($)	P_{oi} ($)	h_o^{IP} ($)	h_o^{FGI} ($)
U(2, 5)	U(2, 4)	U(1000, 1750)	U(2500, 3200)	U(100, 125)	U(50, 80)	U(80, 120)

Step 1: *Managing contingent orders*

The analysis will first show the method of selecting contingent orders and its sequence with confirmed orders such that the objective of step 1 is accomplished. To manage the contingent orders, at first, proposed method elaborates all the possible combinations and sequences of confirmed and contingent orders for the given problem. Satisfying constraints in Eq. (9) and Eq. (10) fourteen different sequences are possible which are as shown in Table 3. Next, the due date and the price that can be submitted as a new offer in the next round i.e., (X_{dor}, X_{por}), is calculated for each contingent orders available in the given sequence. For example, for order D in the sequence A-B-C-D, the completion time will be equal to 51. The price for this completion time can be calculated by using Eq. (4).

$$X_{por} = \frac{1}{0.5}\left[0.6(26000-18000) - \frac{0.5(51-45)(26000-18000)}{(80-45)}\right] + 18000 = \$23{,}028$$

Aspiration level for round 3 in the above equation i.e., 0.6 is obtained by using Eq. (3). The acceptance probability of this offer is then calculated by Eq. (6).

$$A_{or} = \left\{\exp\left[0.5\left\{1.0\frac{51(45-38)}{45(80)}\right\}^{\left(1-\frac{3-1}{5}\right)} + 0.5\left\{1.0\frac{23028(18000-13000)}{18000(26000}\right\}^{\left(1-\frac{3-1}{5}\right)}\right]\right\}^{-1} = 0.725\%$$

Next, identifying fixed processing cost and inventory holding cost based on the given sequence, the expected contribution of an order is computed by Eq. (2). Then, summing the expected contribution of all the orders in the sequence will gives value of Z^s as shown in Table 3. From the table, it is seen that the maximum expected contribution, Z_{max}, is obtained for the sequence A-B-C-E-D-F. The acceptance probability A^s in Table 3 is calculated by taking the product of acceptance probability of all the orders in the sequence. For the example A-B-C-D, the acceptance probability of this sequence is calculated as shown below.

A^s= (1x1x1x0.725) =0.725%.

As the unit of Z^s and A^s is different, the value of Z^s is first normalized by equating Z_{max} with 1.0 and minimum Z^s with 0. Then, by considering normalized value of Z_{max}, Z^s, A_{max} and A^s, the distance D is calculated by Eq. (1). From Table 3 it is evident that the minimum distance is obtained for the sequence A-B-C-F-D.

Table 3

Results obtained from step 1

Sequence	Z_s ($)	Z_{max}	A_s (%)	A_{max}	Norm(Zs)	Distance D
A-B-C	115769		1.0		0	0.762
A-B-C-D	146620		0.725		0.328	0.725
A-B-C-D-E	169094		0.554		0.567	0.621
A-B-C-D-E-F	206573		0.466		0.965	0.535
A-B-C-E	121167		0.756		0.157	0.973
A-B-C-E-D	172344	209808	0.545	1.0	0.601	0.604
A-B-C-E-D-F	209808		0.46		1.000	0.540
A-B-C-E-F-D	209490		0.46		0.996	0.541
A-B-C-F	146932		0.85		0.331	0.685
A-B-C-F-E	170568		0.65		0.582	0.544
A-B-C-F-E-D	207669		0.46		0.977	0.540
A-B-C-D-F	190005		0.61		0.789	0.443
A-B-C-F-D	190217		0.63		0.792	0.424
A-B-C-E-F	172223		0.64		0.600	0.537

Step 2: *Generating quotes*

To generate quotes on due dates and prices, at first, the possible sequence of new orders with sequence A-B-C-F-D obtained in step 1 is elaborated. For new orders G, H and I there will be total of six sequences as shown in Table 4.

Table 4
Results obtained for different sequences

Sequence	*Avg* Q_{do}	*Avg* Q_{po}	V	*Min* V
A-B-C-F-D-G-I-H	153	28086	0.824	
A-B-C-F-D-G-H-I	159	28136	0.829	
A-B-C-F-D-I-G-H	133	27745	0.793	0.793
A-B-C-F-D-I-H-G	148	27986	0.822	
A-B-C-F-D-H-G-I	156	28038	0.826	
A-B-C-F-D-H-I-G	151	28149	0.831	

Then, the limit level on due date of each new order in the sequence is calculated. For example, for order G in the sequence A-B-C-F-D-G-I-H, completion time is 101 time units. Therefore,
$lim_{dG}= 101- \{(1-0.63) \times 56\} = 81$

In the above equation, 0.63 is the acceptance probability of sequence A-B-C-F-D and 56 is the total processing time of contingent orders D and F.

Next, the negotiation margin is calculated by Eq. (24). For the purpose, the cumulative difference is assumed to be 0.35.

$NM_{dG}= \{0.35/ (1-0.35)\}\ 81= 38$

Then, by summing the negotiation margin with the limit level, the due date that can be quoted for order G in this sequence will be 124 time units. Similarly, calculating the due dates for orders I and H and taking their average will gives 153 as shown in Table 4.

Next, calculate the limit level on prices that can be quoted to the new orders. Assuming λ_n as 0.30,

$lim_{pG}=14167(1+0.30) = \$18417$

In the above equation, 14167 represent the cost of producing order G with the sequence A-B-C-F-D-G-I-H. Then, using the concept similar to calculating negotiation margin on due date, the negotiation margin on price for order G is $9916. Summing the negotiation margin with the limit level, the price that can be quoted for order G in this sequence will be $28333. Similarly, calculating the prices for orders I and H and taking their average will gives $28086 as shown in Table 4.

After obtaining the average values on Q_{jo} for all the sequences g, the average on due dates and prices of each sequence is then integrated by Eq. (14) which are as shown by the fourth column of Table 4. From the table, it can be noticed that the sequence A-B-C-F-D-I-G-H gives the minimum value of V. Therefore, the due dates and prices corresponding to this sequence of new orders will be quoted to the customers G, H and I.

Comparing proposed method with sequential quotation: The proposed method is then compared with sequential method of quotation in terms of production cost, average quoted due date and price, and computational time to solve the problem. In the sequential method, quotation is generated for each order one by one, independent of other new orders. For this method also, step 1 is utilized to manage contingent orders. However, the major difference will be in a way quotation is generated in step 2. For the comparison several test problems are generated by taking various combinations of new orders with confirmed and contingent orders as shown in Table 5. In Table 5, at column one, the first and second elements (3×3) representing confirmed and contingent orders respectively for all test instances are same. The reason is that these two elements will not have any effect on the result between sequential and simultaneous methods of quotations because for both the methods step 1 i.e., managing contingent orders is same. All the information's related to these orders are similar to Table 1 and Table 2. The third element in test instances column of Table 5 represents new order for which quotations need to be prepared. Processing time, number of units and all the necessary costs for new orders are also generated by using information's in Table 2. For each test problem five replications are generated with the set up as mentioned above and conduct the analysis. The results obtained from these replications are then aggregated to obtain the average result. All test codes are executed using Matlab program of version 7.9.0 R2009b on an Intel® Core™ 2 Quad CPU @ 3.00 GHz with 4.00 GB RAM.

Table 5

Sequential verses simultaneous method of quotation

Test instances	Sequential Quotation				Simultaneous Quotation			
	Production cost ($)	Avg. Q_{do} ($)	Avg. Q_{po} ($)	CPU time (mins)	Production cost ($)	Avg. Q_{do} ($)	Avg. Q_{po} ($)	CPU time (mins)
3×3×1	21150	45	23500.00	00:08:54	21150	45	23500.00	00:08:54
3×3×2	39015	72	21675.00	00:08:54	36835.2	64	20464.00	00:08:56
3×3×3	52515	93	19450.00	00:09:03	49749.3	85	18425.67	00:09:04
3×3×4	69345	112	19262.50	00:09:08	64333.8	100	17870.50	00:09:28
3×3×5	77422.5	131	17205.00	00:09:12	70642.8	117	15698.40	00:09:29
3×3×6	96826.5	149	17930.83	00:09:10	86882.4	132	16089.33	00:19:48
3×3×7	119191.5	170	18919.29	00:09:18	104231.7	142	16544.71	00:36:38
3×3×8	133980.3	192	18608.38	00:09:24	116686.8	158	16206.50	01:12:08
3×3×9	154122.3	215	19027.44	00:09:32	132109.2	181	16309.78	01:56:12
3×3×10	178872.3	237	19874.70	00:09:34	153078.3	195	17008.70	02:52:23

From Table 5, it can be observed that simultaneous method outperform sequential method of quotation in terms of all the measure of performance except computational timing. The main reason for the result is the ability of simultaneous method to coordinate the processing sequence among new orders. It helps to complete processing of orders early. Also, due to this, the cost of carrying inventory will be less which in turn will reduce the production cost. Since the average of due dates and prices to be quoted to the customers is less, the average probability of quotations getting accepted will be more by simultaneous method as compared to sequential method. Fig. 1 shows the relationships between the differences on quoted values by sequential and simultaneous methods with respect to number of new orders. From the figure it is evident that as the number of new orders increases, the difference on the quoted values on due date and price between two methods will increase. It means that more the new orders to be quoted better will be to use simultaneous method of quotation.

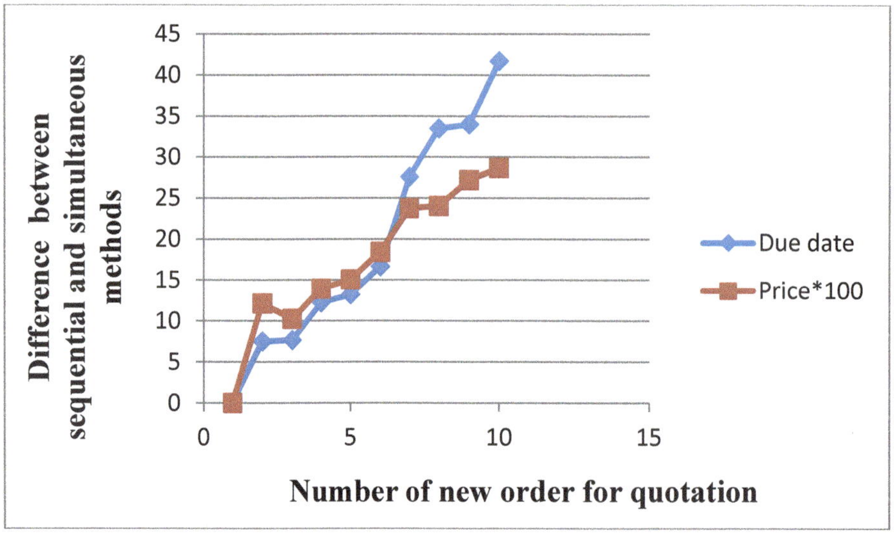

Fig. 1. Relationship between number of new order and difference of quoted values

However, in terms of computation timing, sequential method is better than simultaneous method. This is due to the reason that the number of possible solution/ sequence in simultaneous methods will be more by (n!-1) than by sequential method. For sequential method the major computation timing is needed to carry out step 1 i.e., to manage contingent orders. Otherwise, the possible solution for each new order, at each iteration, will only be 1. However, for simultaneous method it increases exponentially with the increase in the number of new orders. Addition of some constraints for step 2, as in step 1, might helps reduce the no of possible solution considerably which in turn will reduce computation timing for simultaneous method.

5. Conclusions

This paper proposed two steps method to counter the uncertainty caused by contingent orders while preparing quotes for multiple new customers simultaneously. The first step helps manage the effect of contingent orders on quotation being submitted to the new customers. While, second steps help generate quotes for multiple customers simultaneously. Integrated probabilistic and negotiation theories are introduced to offset the effect of contingent orders on quotations. Numerical analysis demonstrated that, except for computational time, proposed method outperform sequential method of quotation on all measure of performances. Therefore, it can be concluded that in contrast to the sequential method, proposed method improves the overall acceptance probability of quotations being submitted to the multiple customers, without diminishing the profit margin of company. This can be attributed to the capability of simultaneous method to reduce the cost and time of production by streamlining the processing of multiple new orders with confirmed and contingent orders. As a future research expansion, the proposed method can be integrated with the simultaneous negotiation mechanism within OA decision framework. Also, as evident from the numerical analysis, even for small to medium size problem the computational time to solve the problem for simultaneous method is high. Therefore, to solve large size industrial problem there is a need to implement intelligent method such as heuristic/ meta-heuristic to reduce computational timing for simultaneous method of quotation.

References

Bac, M. (2001). On creating and claiming value in negotiations. *Group Decision and Negotiation*, *10*(3), 237-251.

Baker, K. R. (1984). Sequencing rules and due-date assignments in a job shop. *Management science*, *30*(9), 1093-1104.

Bertrand, J. W. M. (1983). The effect of workload dependent due-dates on job shop performance. *Management Science*, *29*(7), 799-816.

Cakravastia, A., & Nakamura, N. (2002). Model for negotiating the price and due date for a single order with multiple suppliers in a make-to-order environment. *International journal of production research*, *40*(14), 3425-3440.

Cakravastia, A., & Takahashi, K. (2003). Integrated bidding and manufacturing planning decisions with contingent orders in a make-to-order environment. *Journal of JIMA*, *54*(5), 291-301.

Chaharsooghi, S. K., Honarvar, M., Modarres, M., & Kamalabadi, I. N. (2011). Developing a two stage stochastic programming model of the price and lead-time decision problem in the multi-class make-to-order firm. *Computers & Industrial Engineering*, *61*(4), 1086-1097.

Chand, S., & Chhajed, D. (1992). A single machine model for determination of optimal due dates and sequence. *Operations Research*, *40*(3), 596-602.

Charnsirisakskul, K., Griffin, P. M., & Keskinocak, P. (2006). Pricing and scheduling decisions with leadtime flexibility. *European Journal of Operational Research*, *171*(1), 153-169.

Corti, D., Pozzetti, A., & Zorzini, M. (2006). A capacity-driven approach to establish reliable due dates in a MTO environment. *International Journal of Production Economics*, *104*(2), 536-554.

Easton, F. F., & Moodie, D. R. (1999). Pricing and lead time decisions for make-to-order firms with contingent orders. *European Journal of Operational Research*, *116*(2), 305-318.

ElHafsi, M. (2000). An operational decision model for lead-time and price quotation in congested manufacturing systems. *European Journal of Operational Research, 126*(2), 355-370.

Enns, S. T. (1995). An economic approach to job shop performance analysis. *International journal of production economics, 38*(2-3), 117-131.

Kaminsky, P., & Kaya, O. (2008). Scheduling and due-date quotation in a make-to-order supply chain. *Naval Research Logistics (NRL), 55*(5), 444-458.

Keskinocak, P., Ravi, R., & Tayur, S. (2001). Scheduling and reliable lead-time quotation for orders with availability intervals and lead-time sensitive revenues. *Management Science, 47*(2), 264-279.

Lawrence, S. R. (1995). Estimating flowtimes and setting due-dates in complex production systems. *IIE transactions, 27*(5), 657-668.

Liu, L., Parlar, M., & Zhu, S. X. (2007). Pricing and lead time decisions in decentralized supply chains. *Management Science, 53*(5), 713-725.

Moodie, D. R., & Bobrowski, P. M. (1999). Due date demand management: negotiating the trade-off between price and delivery. *International Journal of Production Research, 37*(5), 997-1021.

Piya, S., Takahashi, K. & Morikawa, K. (2009a), Introducing quotation and negotiation structure in the order acceptance decision model. *International Journal of Operations and Quantitative Management*, 15 (4), 293-318.

Piya, S., Takahashi, K., & Morikawa, K. (2009b), Quotation and capacity planning with contingent orders in make-to-order system. *Proceeding of 20th International Conference on Production Research*, Shanghai, China. (in CD-ROM).

Piya, S., Takahashi, K., & Morikawa, K. (2010). Negotiation with Customer Priority and Dynamic Aspiration Level for Order Acceptance Decision. *Australian Society for Operations Research Bulletin*, 29 (4), 21-47.

Ragatz, G. L., & Mabert, V. A. (1984). A framework for the study of due date management in job shops. *The International Journal of Production Research, 22*(4), 685-695.

Seidmann, A., & Smith, M. L. (1981). Due date assignment for production systems. *Management Science, 27*(5), 571-581.

Shantikumar, J. G., & Sumita, U. (1988). Approximations for the time spent in a dynamic job shop with applications to due-date assignment. *The International Journal of Production Research, 26*(8), 1329-1352.

Slotnick, S. A. (2011). Optimal and heuristic lead-time quotation for an integrated steel mill with a minimum batch size. *European Journal of Operational Research, 210*(3), 527-536.

Sujan, P., Takahashi, K., & Morikawa, K. (2009). Introducing quotation and negotiation structure in the order acceptance decision model. *International Journal of Operations and Quantitative Management*, 15 (4), 293-318.

Van Delft, A., & Nijkamp, P. (1977). *Multi-criteria analysis and regional decision-making* (Vol. 8). Springer.

Van Ooijen, H. P. G., & Bertrand, J. W. M. (2001). Economic due-date setting in job-shops based on routing and workload dependent flow time distribution functions. *International Journal of Production Economics, 74*(1), 261-268.

Watanapa, B., & Techanitisawad, A. (2005a). Simultaneous price and due date settings for multiple customer classes. *European Journal of Operational Research, 166*(2), 351-368.

Watanapa, B., & Techanitisawad, A. (2005b). A genetic algorithm for the multi-class contingent bidding model. *OR Spectrum, 27*(4), 525-549.

Wein, L. M. (1991). Due-date setting and priority sequencing in a multiclass M/G/1 queue. *Management Science, 37*(7), 834-850.

Xianfei, J., Kunpeng, L., & Appa, I.S. (2013). Scheduling and optimal delivery time quotation for the customers with time sensitive demand. *International Journal of Production Economics*, 145(1), 349-358.

Application of multi-stage Monte Carlo method for solving machining optimization problems

Miloš Madić[a*], Marko Kovačević[b] and Miroslav Radovanović[a]

[a]*Faculty of Mechanical Engineering, University of Niš, A. Medvedeva 14, Niš, Serbia*
[b]*Faculty of Electronic Engineering, University of Niš, A. Medvedeva 14, Niš, Serbia*

C H R O N I C L E

Keywords:
Monte Carlo method
Multi-stage
Machining
Optimization
Meta-heuristics

A B S T R A C T

Enhancing the overall machining performance implies optimization of machining processes, i.e. determination of optimal machining parameters combination. Optimization of machining processes is an active field of research where different optimization methods are being used to determine an optimal combination of different machining parameters. In this paper, multi-stage Monte Carlo (MC) method was employed to determine optimal combinations of machining parameters for six machining processes, i.e. drilling, turning, turn-milling, abrasive waterjet machining, electrochemical discharge machining and electrochemical micromachining. Optimization solutions obtained by using multi-stage MC method were compared with the optimization solutions of past researchers obtained by using meta-heuristic optimization methods, e.g. genetic algorithm, simulated annealing algorithm, artificial bee colony algorithm and teaching learning based optimization algorithm. The obtained results prove the applicability and suitability of the multi-stage MC method for solving machining optimization problems with up to four independent variables. Specific features, merits and drawbacks of the MC method were also discussed.

1. Introduction

In recent years, high resource efficiency and optimization of machining processes have been vital for machining companies in order to gain a competitive advantage on the market. Machining processes are highly complex processes affected by a number of input variables, i.e. machining parameters, which in consort have an important role on different process performance characteristics such as quality, cost, productivity, processing time, etc. In real production environment, the selection of machining parameter settings is usually made based on experience and knowledge of a machinist (or a production planner), machining handbooks and manufacturers' recommendations (Rao & Pawar, 2010; Rao & Kalyankar, 2013). Although this is essentially a subjective approach, and chosen values of machining parameters are preferably conservative, or even not near optimal ones, in some cases this approach is sufficient. However, many process parameters and the complex and stochastic nature of the machining processes mean that achieving the optimal performance, even for a highly skilled machinist with a state-of-the-art machine tool, is rarely possible (Rao & Pawar, 2009). The most adverse effect of such approach may lead to product quality deterioration, increase in operation cost and machining time,

* Corresponding author.

decrease in productivity, etc. In order to gain a competitive advantage on the fierce market, get the most from a machine tool and enhance the overall machining efficiency, it is necessary to select machining parameter settings more intelligently. Recent studies indicate that the selection of machining parameter settings is often performed in a two-stage approach consisting of mathematical modeling of a machining performance characteristic and optimization using an optimization method. Mathematical modeling is aimed at establishing mathematical relationships between machining parameters and different performance characteristics. Since many machining optimization problems are complex, highly non-linear and multi-dimensional, meta-heuristic optimization methods have become a preferred trend for solving machining optimization problems.

Meta-heuristic optimization methods conceptually developed to imitate natural phenomena, the biological evolutionary process (genetic algorithm-GA), the physical annealing process (simulated annealing-SA), social behavior of bird flocking (particle swarm optimization-PSO), behavior of real ant colonies (ant colony optimization-ACO), artificial bee colony (artificial bee colony-ABC), have been employed instead of classical optimization methods such as sequential unconstrained minimization technique, feasible direction method, goal programming and non-linear programming. Actually, as noted by Yildiz (2009), the complexity of machining problems creates the requirement for increasingly effective optimization algorithms. Many researchers and authors have found classical optimization methods ineffective for solving machining optimization problems because of the following: a tendency to obtain a local optimal solution (Rao & Pawar, 2010), very complex nature and inability to handle multi-objective problems effectively (Rao & Kalyankar, 2013) and lack of robustness (Rao & Pawar, 2009; Rao & Pawar 2010). As noted by Mukherjee and Ray (2006), researchers and practitioners prefer an alternative cost effective near-optimal (or approximate) solution than the exact optimal one, as it may be extremely difficult to find the exact optimal point in the high dimensional and multimodal search space. Rao et al. (2008) used the PSO algorithm to determine optimal machining parameter settings for an electro-chemical machining (ECM) process and compared its performance with that obtained by other optimization methods. Rao and Pawar (2009) applied the ABC algorithm in order to determine optimal machining parameter values for a wire electric discharge machining (WEDM) process. Rao and Pawar (2010) developed a comparative study of the application of the ABC, SA and PSO algorithm in order to obtain the optimum machining parameter values for various cutting strategies in multi-pass milling process. In the research of Samanta and Chakraborty (2011), the ABC algorithm was applied to search out the optimal machining parameter values combinations of three non-conventional machining processes. The obtained single and multi objective optimization solutions were better compared to optimization solutions obtained by the past researches.

GA was considered by Kilickap et al. (2011) to optimize drilling of AISI 1045 steel. In the study, three machining parameters such as cutting speed, feed rate, and cutting environment were employed. Response surface methodology was used to develop mathematical model and GA was used to determine optimal machining conditions for ensuring the minimum surface roughness. Bhushan et al. (2012) presented the GA approach for optimization of surface roughness in machining of Al alloy SiC particle composite material. A surface roughness prediction model was developed in terms of four machining parameters such as cutting speed, feed rate, depth of cut and tool nose radius. The authors concluded that the GA method outperformed the RSM regarding the best (minimal) surface roughness predicted value. Savas and Ozay (2008) applied GA for optimization of the surface roughness in the process of tangential turn-milling process. The mathematical model consisted of four independent variables, i.e. machining parameters, such as depth of cut, workpiece speed, tool speed and feed rate. Zain et al. (2011) applied GA and the SA algorithm to optimize the machining parameter values of abrasive waterjet (AWJ) machining process. Five machining parameters were considered in the research including traverse speed, waterjet pressure, standoff distance, abrasive grit size and abrasive flow rate. Comparative analysis of optimization solutions indicated that SA provided better solutions than GA. However, recently Yusup et al. (2012) attempted the same optimization problem and observed that the ABC algorithm outperformed both SA and GA. Rao and Kalyankar (2013) applied a newly developed advanced algorithm named 'teaching–learning-based optimization (TLBO) algorithm' for optimization of non-conventional machining processes such as ultrasonic machining (USM), abrasive jet machining (AJM) and WEDM. It was concluded that TLBO has given a considerable improvement in terms of results and convergence over GA, SA, ABC and PSO. Recently, Madić and Radovanović (2014b) proposed the use of pattern search (PS) algorithm, as deterministic direct search method, for solving single- and multi-objective machining optimization problems. While solving several case studies it has been found that the PS algorithm is an efficient optimization method showing good competitive potential against stochastic direct search methods such as meta-heuristic algorithms.

From the literature review, it has been observed that GA, SA, PSO, ABC and ACO have been extensively used for solving machining optimization problems (Yusup et al., 2012). It is observed that meta-heuristic algorithms were identified as effective stochastic optimization methods for solving machining optimization problems. However, the well-known difficulty in determination of optimum algorithm specific controlling parameters such as crossover and mutation rate in the case of GA, initial temperature in the case of SA, inertia weight in the case of PSO, etc., affects the effectiveness of meta-heuristic optimization methods in searching for global optimal solutions (Rao & Kalyankar, 2013). As noted by Bhushan et al. (2012), there is no guideline yet given by the researchers which could be followed in recommending the best combination of parameter values for the best optimal solution. In addition, application of meta-heuristic algorithms implies that the repeatability of optimization results obtained with the same initial condition settings is not guaranteed (Madić et al., 2013). Meta-heuristic optimization methods perform efficient and comprehensive exploration of the optimization search space using the random Monte Carlo (MC) search guided by governing mechanisms which imitate certain strategies taken from nature. GA, SA and recently the ABC algorithm have shown their strength in this respect. Recently, it has been shown that the iterative exhaustive search algorithm has the potential for solving the aforementioned machining optimization problems for a given discrete optimization space (Kovačević et al., 2013). However, to the authors' knowledge there is no application of the MC method for solving machining optimization problems, which is traditionally applied to continuous optimization problems. As noted by Mosegaard and Sambridge (2002), MC methods are powerful optimization tools when searching for globally optimal solutions amongst numerous local optima.

The main motivation behind this paper is to investigate the applicability and efficiency of a conceptually simple MC method for solving machining optimization problems. Although the aforementioned meta-heuristic optimization methods have proved that they can succeed in obtaining good optimization solutions, the average process engineer in real machining environment may not feel familiar with these methods, which require deeper knowledge of artificial intelligence and optimization. Moreover, in some cases, specialized software or programming skills are needed. With an increasing number of meta-heuristic optimization methods and their combinations that have emerged in recent years, it has become very difficult even for researchers to become familiar with all these optimization methods. Because of these reasons, as noted by Besseris (2008), in practice only the simplest of optimization tools are eventually proved workable in the factory. In that sense, this paper is an attempt to investigate the applicability of the MC method for solving machining optimization problems. The proposed optimization procedure was employed to search out the optimal combinations of machining parameters for six machining processes, i.e. drilling, turning, turn-milling, AWJ machining, electrochemical discharge machining (ECDM), and electrochemical micromachining (EMM). The obtained optimization solutions were compared with the optimization solutions obtained by the past researchers using meta-heuristic algorithms.

2. Machining optimization

Optimization of machining processes has been of considerable interest in the literature. The ultimate goal of machining optimization is to select machining parameter values such that the overall machining performance is enhanced. Depending on the criteria, optimization of machining processes should result in an increase in product quality, increase in productivity, decrease in machining cost and time, etc. Mathematically, optimization of machining processes is related to optimization of an objective function, often representing a machining performance, under some machining parameter constraints. The machining optimization problem may be expressed as follows:

$$\begin{aligned} &\text{Minimize}(\text{or maximize})\, f(x) \\ &\text{subject to}: g_i(X) \leq 0, i = 1...m \\ &\qquad X_j^l \leq X_j \leq X_j^u, j = 1...n \end{aligned} \tag{1}$$

where $\boldsymbol{X}$ is the vector of machining parameters, $f(x)$ is the objective function to be minimized (maximized), $g_i(X)$ is the *i-th* functional constraint, and X_j^l and X_j^u are the lower and upper bounds of *j-th* machining parameter X_j.

Generally, objective functions for machining optimization problems can be developed in two ways either analytically or empirically. The development of analytical equations, based on basic laws and principles of a machining process, requires deeper knowledge and understanding of the machining process and is usually quite complex. These equations are more general, but less accurate since they involve some simplifications and approximations in relation to the real machining process and do not take into account any imperfections in the process. Therefore, analytical solutions for describing machining processes are generally not accurate enough for practical usage (Davim, 2001).

Compared with analytical approach, empirical model development does not need deeper knowledge of the phenomena occurring in the machining process, but the development of complex mathematical models and designs implies a more profound knowledge of theory of experimental design and mathematical modeling. The development of empirical models comprises the selection of experimental design, realization of experiment, data collection and development of mathematical equations using different modeling techniques, e.g. regression analysis, artificial neural networks etc. Theory and practice in the field of machining have shown that in most cases the best choice is a mathematical model in the form of polynomials (linear, quasi-linear, square, etc.) (Savas & Ozay, 2008).

3. Monte Carlo Method

The term Monte Carlo method (MC method) was coined in the 1940s by physicists S. Ulam, E. Fermi, J. von Neumann and N. Metropolis under the name "statistical sampling". Since the pioneer studies in the 1940s and 1950s, this method has been applied in almost all areas of science (Yang, 2010). MC methods have been used for a long time, but only in the last few decades, the methods have gained the status of fully rounded numerical methods. As for obtaining a reasonably accurate assessment, the calculation of large number of special cases is needed, as well as appropriate statistical analysis, the efficient application of the MC methods begins with the emergence of high-speed computers able to perform fast computations (Madić & Radovanović, 2014a).

At the heart of any Monte Carlo method there is a random number generator: a procedure that produces an infinite stream: $U_1, U_2, U_3, \ldots \overset{iid}{\sim} Dist$ of random variables that are independent and identically distributed (*iid*) according to some probability distribution *Dist*. When this distribution is the uniform distribution, that is $Dist = U(0,1)$, the generator is said to be uniform random number generator (Kroese et al., 2011). Monte Carlo methods can be divided into two categories: the sampling methods and the optimization methods. Monte Carlo sampling is useful when the space of feasible solutions is to be explored, and measures of resolution and uncertainty of solution are needed. Monte Carlo optimization methods are powerful tools when searching for globally optimal solutions amongst numerous local optima (Mosegaard & Sambridge 2002). The optimization based on Monte Carlo methods can be useful for solving optimization problems with many local optima and complicated constraints, possibly involving a mix of continuous and discrete variables (Kroese et al., 2011). When using the MC method for solving optimization problems there are basically two approaches: single and multi-stage (Dhavlikar et al. 2003; Khayet & Cojocaru 2012). In a multi-stage approach, simulation runs are repeated by modifying the bounds of each independent variable considering the near optimum solution obtained in the previous simulation run, which is illustrated in Fig. 1.

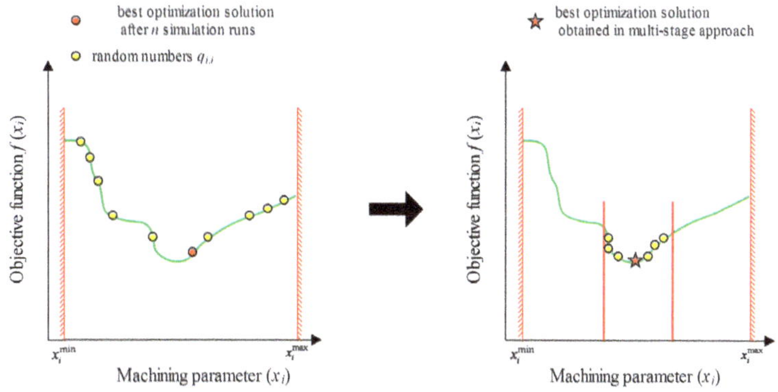

Fig. 1. Illustration of MC optimization procedure in single and multi-stage

The basic idea of the multi-stage MC method for solving optimization problems is given in Fig. 2.

```
Input:
n: Number of independent variables
L[1], L[2], ..., L[n]: lower bounds of independent variables permissible intervals
U[1], U[2], ..., U[n]: upper bounds of independent variables permissible intervals
numOfStages: Number of stages

Output:
Fmax: Maximal mathematical model output value
Qmax: list of independent variables value {Qmax[1], ...., Qmax[n]} that corresponds to maximum mathematical
model output value L[i] ≤ Qmax [i] ≤ U[i], i=1,...,n

Declare:
NmOfSimulations: Number of simulation runs
IntervalSize: Interval size
Fc: Current mathematical objective function value
Qc: List of independent variables values {Qc[1], ...., Qc[n]} that corresponds to current mathematical model output
value L[i] ≤ Qc [i] ≤ U[i], i=1,...,n

for stage = 1 to numOfStages

    /* Single Stage Monte Carlo Method */
    NumOfSimulations = UserInput()
    for i=1 to n do
        R = GenerateRandomNumberFromInterval (0,1)
        Qmax[i] = L[i] + (U[i] + L[i])*R
    end for
    Fmax = ComputeMathematicalModelOutput(Qc)
    for loop = 1 to NumOfSimulations do
        for i = 1 to n do
         R = GenerateRandomNumberFromInterval (0,1)
         Qc[i] = L[i] + (U[i] + L[i])*R
        end for
    Fc = ComputeMathematicalModelOutput(Qc)
    If Fc > Fmax then
       Fmax  = Fc
       Qmax = Qc
    Endif

/* Reduce independent variables permissible intervls */
 for i=1 to n do
   intervalSize = UserInput()
   L[i] = Qmax [i] + intervalSize/2
   U[i] = Qmax [i] + intervalSize/2
 endfor
endfor

Return Fmax, Qmax
```

Fig. 2. Pseudo-code of the multi-stage MC method for solving optimization problems

4. Application examples

The machining optimization application examples considered in this paper are taken from scientific resource bases, such as Springer and Elsevier. Case studies focused on solving machining optimization problems using meta-heuristic algorithms were chosen for comparative analysis of the obtained solutions by the multi-stage MC method. In order to facilitate the validation of obtained optimization solutions, the search for papers was restricted to only those dealing with mathematical models developed using regression analysis. In sum, six optimization case studies were considered - two case studies deal with three, two case studies deal with four and two case studies deal with five independent variables, i.e. machining parameters.

4.1. Application of Monte Carlo method for solving machining optimization problems

In solving machining optimization problems the aim of applying the MC method is to explore the optimization space and identify near optimal values of independent variables (x_i) i.e. machining parameters such that objective function $f(x_i)$ has an extreme value (maximal or minimal) and the constraints for machining parameters $x_i \in \left[x_i^{\min}, x_i^{\max}\right]$ are satisfied. The machining optimization procedure based on the MC method was implemented in the MS Excel software package. The idea behind the MC method was to explore the space of feasible solutions by performing a large number of computational runs using pseudo-random numbers. After the formulation of a machining optimization problem, the first step in optimization procedure involves generation of random numbers $r_{i,j}$ uniformly distributed in the range [0,1] using the function *rand*. Subsequently, in order to satisfy the bounds of machining parameters values, random numbers $r_{i,j}$ were used to generate random numbers $q_{i,j}$ for each machining parameter by using the following equation:

$$q_{i,j} = x_i^{\min} + \left(x_i^{\max} - x_i^{\min}\right) \cdot r_{i,j} \tag{2}$$

In such a way, by using random numbers $q_{i,j}$ and a mathematical model, a number of stochastic computational runs (n) can be performed. When applying the single-stage approach, for each case study different numbers of simulation runs were performed, that is 1000, 2000, 5000 and 10000. Each was repeated 10 times and the best optimization solutions were recorded. In sum, for each case study n = 180000 simulation runs were performed. A simple procedure for computing was implemented in the Excel software package resulting in these computations being performed within few seconds.

In order to improve obtained optimization solutions, the multi-stage approach was implemented. After 180000 simulation runs performed in the first stage the near optimal optimization solution was identified. Subsequently, the bounds of each independent variable were diminished, 10000 simulation runs were repeated and the best optimization solution was identified. Although the multi-stage approach essentially refers to multiple modifications of the bounds of each independent variable, in this paper only one was performed.

4.2. Single objective machining optimization

4.2.1. Case study 1

Sarkar et al. (2006) investigated the influence of applied voltage, electrolyte concentration and inter-electrode gap on material removal rate (MRR), radial overcut (ROC) and thickness of heat affected zone (HAZ) in an ECDM process. Based on the experimental data of Sarkar et al. (2006), Samanta and Chakraborty (2011) developed the following three equations using uncoded values of machining parameters:

$$\begin{aligned} MRR = {} & 4.96423 - 0.20418 \cdot x_1 + 0.09862 \cdot x_2 + 0.00851 \cdot x_3 + 0.00249 \cdot x_1^2 \\ & - 0.00086 \cdot x_2^2 + 0.00039 \cdot x_3^2 - 0.00181 \cdot x_1 \cdot x_2 - 0.00104 \cdot x_1 \cdot x_3 + 0.00125 \cdot x_2 \cdot x_3 \end{aligned} \tag{3a}$$

$$\begin{aligned} ROC = {} & 3.15622 - 0.08019 \cdot x_1 - 0.07678 \cdot x_2 - 0.00356 \cdot x_3 + 0.00069 \cdot x_1^2 \\ & + 0.00048 \cdot x_2^2 + 0.00016 \cdot x_3^2 + 0.00072 \cdot x_1 \cdot x_2 - 0.00026 \cdot x_1 \cdot x_3 + 0.00041 \cdot x_2 \cdot x_3 \end{aligned} \tag{3b}$$

$$\begin{aligned} HAZ = {} & 0.940335 - 0.019541 \cdot x_1 - 0.028638 \cdot x_2 - 0.003122 \cdot x_3 + 0.000147 \cdot x_1^2 \\ & + 0.000242 \cdot x_2^2 + 0.000017 \cdot x_3^2 + 0.000251 \cdot x_1 \cdot x_2 - 0.000017 \cdot x_1 \cdot x_3 + 0.000106 \cdot x_2 \cdot x_3 \end{aligned} \tag{3c}$$

where x_1 = applied voltage, x_2 = electrolyte concentration, x_3 = inter-electrode gap.

In an attempt to obtain optimal machining parameter values for minimization of ROC and HAZ and maximization of MRR, Samanta and Chakraborty (2011) and Rao and Kalyankar (2011) applied the ABC and TLBO algorithm, respectively. The bounds of machining parameters used in optimization procedure as well as optimization solutions are given in Table 1. Optimization solutions obtained using MC methods are also shown in Table 1 as follows.

Table 1

Optimization solutions of case study 4.2.1

Machining parameters and bounds	ABC algorithm			TLBO algorithm			Single stage MC			Multi-stage MC		
	MRR	ROC	HAZ	MRR	ROC	HAZ	MRR	ROC	HAZ	MRR	ROC	HAZ
Applied voltage (V) [50 - 70]	70	50	50	70	50	50	69.973	50.159	50.081	69.999	50.007	50.004
Electrolyte concentration (wt%) [10 - 30]	20	30	24.5	10	30	25	10.299	29.819	24.753	10.002	29.998	24.491
Inter-electrode gap (mm) [20 - 40]	20	20	40	21	20	38	20.598	20.201	38.447	20.017	20.013	39.994
Optimal value	1.3372*	0.05912	0.05409	1.5902	0.0591	0.0541	1.5958	0.0611	0.0542	1.6252	0.05919	0.054092

* Corrected value after Venkata Rao and Kalyankar (2011)

4.2.2. Case study 2

Kilickap et al. (2011) presented a study focused on the influence of machining parameters on the surface roughness obtained in drilling of AISI 1045. The mathematical relationship for correlating the surface roughness and the considered drilling parameters was obtained in the following form:

$$R_a = 4.115 - 0.82767 \cdot x_1 + 8.225 \cdot x_2 + 0.135 \cdot x_3 + 0.0538 \cdot x_1^2 \tag{4}$$

where x_1 = cutting speed, x_2 = feed rate, x_3 = cutting environment. In order to determine optimum drilling parameters for AISI 1045 material at the lowest possible surface roughness value, Kilickap et al. (2011) used GA. The bounds of machining parameters used and comparison of optimization solutions found by GA, PS algorithm (Madić & Radovanović, 2014b) and the MC method are given in Table 2.

Table 2

Optimization solutions of case study 4.2.2

Machining parameters and bounds	GA	PS	Single stage MC	Multi-stage MC
Cutting speed (m/min) [5 - 15]	7.62	7.692	7.813	7.690
Feed rate (mm/rev) [0.1 - 0.3]	0.1	0.1	0.1	0.1
Cutting environment [1 - 3]	1	1	1.016	1
Ra optimal value	1.89	1.8892	1.8958	1.8892

4.2.3. Case study 3

Bhushan et al. (2012) investigated turning of Al Alloy SiC particle composite material using carbide turning inserts. The investigation was focused on the application of GA for finding the optimal machining parameters at which surface roughness has minimal value. Bhushan et al. (2012) developed the following mathematical model relating the machining parameters and surface roughness:

$$R_a = 0.72412 + 0.00324 \cdot x_1 - 0.19694 \cdot x_2 + 4.19915 \cdot x_3 - 0.18753 \cdot x_4 - 0.0000174 \cdot x_1^2 - 3.42419 \cdot x_3^2 + 3.33125 \cdot x_2 \cdot x_3 - 0.56484 \cdot x_3 \cdot x_4 \quad (5)$$

where x_1 = cutting speed, x_2 = feed rate, x_3 = depth of cut and x_4 = tool nose radius. The bounds of machining parameters used in optimization procedure and comparison of optimization solutions found by Bhushan et al. (2012) by using GA to those obtained by using the MC method are given in Table 3. The same optimization problem was solved previously by Madić and Radovanović (2014b) by using the PS algorithm.

Table 3

Optimization solutions of case study 4.2.3

Parameters and objective function	GA	PS	Single stage MC	Multi-stage MC
Cutting speed (m/min) [90 - 210]	207.055	210	206.198	209.867
Feed rate (mm/rev) [0.15 - 0.25]	0.151	0.15	0.152	0.152
Depth of cut (mm) [0.2 - 0.6]	0.201	0.2	0.208	0.2
Tool nose radius (mm) [0.4 - 1.2]	1.199	1.2	1.197	1.2
Surface roughness	1.06509*	1.0498	1.0885	1.0516

*Corrected values

4.2.4. Case study 4

Savas and Ozay (2008) presented the GA approach for optimization of cutting parameters at cylindrical workpieces leading to minimal surface roughness in the tangential turn-milling process. The mathematical model for the prediction of surface roughness considering machining parameters was obtained according to experiment results and is given in the equation below:

$$R_a = \left(0.000008 \cdot x_1^2 - 0.0082 \cdot x_1 + 2.8734\right) \cdot \left(0.00003 \cdot x_2^2 - 0.0135 \cdot x_2 + 1.9924\right) \cdot \left(0.0171 \cdot x_3 + 0.4677\right) \cdot \left(0.2525 \cdot x_4 + 0.4087\right) \cdot 5.3 \quad (6)$$

where x_1 = workpiece speed, x_2 = tool speed, x_3 = feed rate and x_4 = depth of cut.

The optimization solution for turn-milling process obtained by Savas and Ozay (2008) by using GA is given in Table 4. In the same table the bounds of machining parameters used in optimization procedure and optimization solution obtained by the MC method are given. The same optimization problem was solved previously by Madić and Radovanović (2014b) by using the PS algorithm.

Table 4

Optimization solutions of case study 4.2.4

Parameters and objective function	GA	PS	Single stage MC	Multi-stage MC
Workpiece speed (rev/min) [300 - 700]	511.9	512.5	497.716	513.305
Tool speed (rev/min) [150 - 300]	224.9	225	225.164	226.193
Feed rate (mm/min) [3 - 20]	3.2	3	3.055	3
Depth of cut (mm) [0.1 - 1]	0.1	0.1	0.106	0.1
Surface roughness	0.4394*	0.4366	0.4398	0.4366

*Corrected values

4.2.5. Case study 5

Çaydaş and Hasçalik (2008) investigated the relationship between machining parameters and surface roughness in AWJ machining of Al 7075-T6 wrought alloy. Based on experimental data, the following prediction model was developed:

$$R_a = -5.07976 + 0.08169 \cdot V + 0.07912 \cdot P - 0.34221 \cdot h - 0.08661 \cdot d - 0.34866 \cdot m - 0.00031 \cdot V^2 - 0.00012 \cdot P^2 + 0.10575 \cdot h^2 + 0.00041 \cdot d^2 + 0.07590 \cdot m^2 - 0.00008 \cdot V \cdot m - 0.00009 \cdot P \cdot m + 0.03089 \cdot h \cdot m + 0.00513 \cdot d \cdot m \quad (7)$$

where x_1 = traverse speed, x_2 = waterjet pressure, x_3 = standoff distance, x_4 = abrasive grit size and x_5 = abrasive flow rate. Zain et al. (2011) used Eq. (7) as the objective function to determine optimal values of machining parameters so as to minimize surface roughness by using GA and SA. Recently, for solving the same problem, Yusup et al. (2012) and Madić and Radovanović (2014b) attempted the ABC and PS algorithm, respectively. The bounds of machining parameters used in optimization procedure and optimization solutions obtained by past researchers and those obtained by using the MC method are given in Table 5.

Table 5
Comparison of optimization solutions of case study 4.2.5

Optimization approach	Traverse speed (mm/min) [50 - 150]	Waterjet pressure (MPa) [125 - 250]	Standoff distance (mm) [1 - 4]	Abrasive grit size (μm) [60 - 120]	Abrasive flow rate (g/s) [0.5 – 3.5]	Minimal R_a
GA	50.024	125.018	1.636	94.73	0.525	1.554
SA	50.003	125.029	1.486	107.737	0.5	1.533
ABC	50	125	1.550	102.521	0.5	1.522
PS	50	125	1.545	102.494	0.5	1.5223
Single stage MC	50.610	126.443	1.125	103.096	0.506	1.64392
Multi-stage MC	50.030	125.005	1.255	101.778	0.501	1.53351

4.2.6. Case study 6

Munda and Bhattacharyya (2008) presented an approach for investigation of the EMM process using RSM. Five machining parameters, pulse on/off ratio, machining voltage, electrolyte concentration, voltage frequency and tool vibration frequency and two machining performance characteristics such as MRR and ROC were considered in the study. The RSM mathematical models for MRR and ROC prediction on the basis of machining parameters are given in Eqs. (8a) and (8b), respectively:

$$\begin{aligned} MRR = & -1.78917 + 0.111858 \cdot x_1 + 1.36263 \cdot x_2 - 0.0864044 \cdot x_3 + 0.0231122 \cdot x_4 \\ & - 0.00139639 \cdot x_5 - 0.201666 \cdot x_1^2 - 0.0860582 \cdot x_2^2 - 0.000145752 \cdot x_3^2 - 0.000319532 \cdot x_4^2 \\ & + 3.893684 \cdot 10^{-6} \cdot x_5^2 - 0.0704326 \cdot x_1 \cdot x_2 + 0.00838936 \cdot x_1 \cdot x_3 + 0.00275664 \cdot x_1 \cdot x_4 \\ & + 0.00178484 \cdot x_1 \cdot x_5 + 0.00870264 \cdot x_2 \cdot x_3 - 0.00700764 \cdot x_2 \cdot x_4 - 0.00105004 \cdot x_2 \cdot x_5 \\ & + 0.00125437 \cdot x_3 \cdot x_4 + 0.0000247626 \cdot x_3 \cdot x_5 + 0.0000181174 \cdot x_4 \cdot x_5 \end{aligned} \tag{8a}$$

$$\begin{aligned} ROC = & -1.08149 + 1.21039 \cdot x_1 + 0.448639 \cdot x_2 - 0.0821333 \cdot x_3 + 0.0247783 \cdot x_4 \\ & - 0.00258589 \cdot x_5 + 0.0198541 \cdot x_1^2 + 0.0554876 \cdot x_2^2 + 0.00108447 \cdot x_3^2 + 0.000640329 \cdot x_4^2 \\ & + 2.817205 \cdot 10^{-6} \cdot x_5^2 - 0.139966 \cdot x_1 \cdot x_2 - 0.00133867 \cdot x_1 \cdot x_3 - 0.0161759 \cdot x_1 \cdot x_4 \\ & - 5.41335 \cdot 10^{-5} \cdot x_1 \cdot x_5 - 0.00307591 \cdot x_2 \cdot x_3 - 0.0163201 \cdot x_2 \cdot x_4 + 0.000831331 \cdot x_2 \cdot x_5 \\ & + 0.000786541 \cdot x_3 \cdot x_4 + 0.0000725981 \cdot x_3 \cdot x_5 - 6.94181 \cdot 10^{-5} \cdot x_4 \cdot x_5 \end{aligned} \tag{8b}$$

where x_1 = pulse on/off ratio, x_2 = machining voltage, x_3 = electrolyte concentration, x_4 = voltage frequency and x_5 = tool vibration frequency. In order to investigate the applicability and suitability of the ABC algorithm for solving machining optimization problems, Samanta and Chakraborty (2011) used Eqs. (8a) and (8b) as the objective functions to determine optimal values of machining parameters. The bounds of machining parameters used in optimization procedure and optimization results of single objective optimization of the MRR and ROC for the EMM process obtained by the ABC algorithm and MC method are given in Table 6.

As seen from Table 6, optimization of mathematical model for ROC (Eq. 8b) using MC method yielded a negative value as the optimal one. It is recognized that the negative value of ROC is physically meaningless, however, this value is regarded only as the optimal solution of the mathematical model. Although it was observed by Munda and Bhattacharyya (2008) that the mathematical model for ROC prediction is statistically adequate at 97.5% confidence level, it is observed that there is no guarantee that the model will yield meaningful predictions over the entire experimentally investigated range.

Table 6
Optimization solutions of case study 4.2.6

Machining parameters and bounds	ABC algorithm		Single stage MC		Multi-stage MC	
	MRR	ROC	MRR	ROC	MRR	ROC
Pulse on/off ratio [0.5 – 2.5]	0.9168	0.6288	0.912	0.511	0.966	0.501
Machining voltage (V) [2.5 – 4.5]	4.5	3.2639	4.499	2.534	4.498	2.501
Electrolyte concentration (g/l) [10 - 30]	30	18.7930	29.872	20.766	29.996	20.546
Voltage frequency (Hz) [35 - 55]	52.4956	39.4123	54.923	35.663	54.606	35.012
Tool vibration frequency (Hz) [100 - 300]	100	249.4320	108.436	270.838	105.011	270.287
Optimal value	1.47	0	1.4483	-0.2424	1.4584	-0.2649

4.3. Analysis and discussion

As it can be seen from the results in the previous section, the identified near optimal machining conditions are about the same as those selected as optimal by previous researches. Given that the optimization was based on using mathematical models as objective functions, all optimization solutions obtained by the multi-stage MC method can be verified through a trivial calculation using mathematical models given in section 4. Considering stochastic nature of the MC method and meta-heuristic methods, validation and possible improvement of optimization solutions can be performed using exhaustive search algorithm. In our previous work (Kovačević et al. 2013) the possibilities and advantages of applying exhaustive iterative search for solving machining optimization problems and validation of optimization results obtained by other optimization methods were presented.

A deeper analysis of the machining optimization solutions indicates that in the case of mathematical models with up to four independent variables, the optimization solutions obtained by the multi-stage MC method are comparable or even better than solutions obtained by past researchers using meta-heuristic algorithms such as GA, ABC and TLBO. For the case study 5, which deals with five independent variables, it was observed that the optimization solution obtained using the multi-stage MC method is better that the optimization solutions obtained by GA and SA, but poorer than the one obtained by ABC algorithm. A comparative analysis of optimization solutions of case study 4.2.6 in the case of five independent variables reveals that the multi-stage MC method is less effective in high dimensional spaces, and the advantage should be given to the ABC algorithm.

When applying the MC method for solving machining optimization problems it should be noted that a decrease in the variability of obtained solutions was not observed with an increase in the number of simulation runs. Also, it was observed that in some cases e.g. (case study 1) the best optimization solution was obtained after only 5000 simulation runs, whereas in some cases the best optimization solution was obtained after a larger number of simulation runs. In other words, there is no greater certainty that the better optimization solution will be found with an increasing number of simulation runs, which is the case when using meta-heuristics. The impossibility to decrease the variability of obtained solutions and increase certainty of obtaining better optimization solutions with increasing number of simulation runs is a consequence of the fact that the MC method does not use any search guide information and rules, but only random numbers. However, possible improvements of optimization solutions might be expected by a drastic increase in the number of simulation runs.

The optimization solutions presented in this paper indicate that the multi-stage MC method involving few thousand MC computation runs represents a fairly effective and efficient approach for solving multi-dimensional machining optimization problems. Furthermore, the MC based optimization approach is very simple to implement, requires no expert knowledge of optimization theory and artificial intelligence, setting of algorithm parameters and/or defining an initial solution as in the case of using other optimization methods.

The salient advantage of the application of the MC method is that it is possible to obtain a population of optimization solutions. Graphical visualization of the optimization solutions for the case study 4.2.1. (maximization of MRR) obtained using the multi-stage MC method are shown in Figure 3. By plotting these graphs, all acceptable solutions with corresponding values of machining parameters can be easily identified. This is particularly advantageous in machining practice considering different machine/tool constraints and limitations.

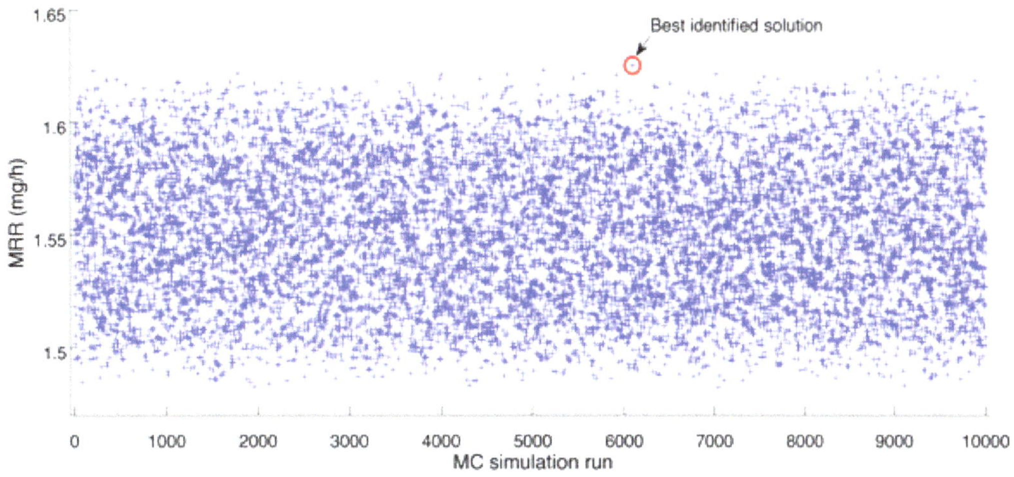

Fig. 3. Optimization solutions obtained using the multi-stage MC method for the case study 4.2.1.

5. Conclusion

This paper presented an investigation of applicability of the multi-stage MC method for solving machining optimization problems. To this aim six machining optimization problems taken from relevant literature were considered and the obtained results were compared with the results obtained by previous researchers using different meta-heuristic optimization methods. The optimization procedure based on the MC method is very simple, both conceptually and in implementation without the need to write a programming code or use specialized software packages, and by the authors' opinion represents an optimization tool being most suited to use by process engineers in real production environment. The salient features of MC include the following: (*i*) solving optimization problems requires no expert knowledge of optimization theory, artificial intelligence, setting of different algorithm parameters and/or defining an initial solution as in the case of using other optimization methods, (*ii*) there is a possibility to find a population of optimization solutions and rank them, (*iii*) the multi-stage approach provides the possibility for an improvement of the optimization solutions. The analysis of the obtained results can be summarized by the following points:

- in the case of mathematical models with up to four independent variables, the optimization solutions obtained by the multi-stage MC method are comparable with the optimization solutions obtained by using other meta-heuristic optimization methods,
- the multi-stage MC method may not be suitable for higher dimensional spaces and the priority for solving machining optimization problems should be given to meta-heuristic optimization methods which have powerful mathematical tools that guide the exploration of optimization space. However, the optimization solutions obtained using the MC method can be used as initial solutions such as in the case of SA,
- the MC method proves its applicability as a very simple optimization method for solving machining optimization problems.

The main scope of our future work will be the application of the multi-stage MC method for solving multi-objective machining optimization problems. Furthermore, attempts will be made to investigate the applicability of a combined optimization approach by integrating the MC method with the brute force algorithm based on an exhaustive iterative search. It is believed that the application of this hybrid approach for solving machining optimization problems in continual search space will outperform the optimization results of other meta-heuristic optimization methods in terms of solution accuracy and computational speed.

Acknowledgements

This work was carried out within the project TR 35034 financially supported by the Ministry of Education and Science of the Republic of Serbia.

References

Besseris, G. J. (2008). Multi-response optimisation using Taguchi method and super ranking concept. *Journal of Manufacturing Technology Management,19*(8), 1015-1029.
Besseris, G. J. (2008). Multi-response optimisation using Taguchi method and super ranking concept. *Journal of Manufacturing Technology Management,19*(8), 1015-1029.
Çaydaş, U., & Hasçalık, A. (2008). A study on surface roughness in abrasive waterjet machining process using artificial neural networks and regression analysis method. *Journal of materials processing technology*, *202*(1), 574-582.

Davim, J. P. (2001). A note on the determination of optimal cutting conditions for surface finish obtained in turning using design of experiments. *Journal of materials processing technology*, *116*(2), 305-308.

Dhavlikar, M. N., Kulkarni, M. S., & Mariappan, V. (2003). Combined Taguchi and dual response method for optimization of a centerless grinding operation.*Journal of Materials Processing Technology*, *132*(1), 90-94.

Khayet, M., & Cojocaru, C. (2012). Artificial neural network modeling and optimization of desalination by air gap membrane distillation. *Separation and Purification Technology*, *86*, 171-182.

Kilickap, E., Huseyinoglu, M., & Yardimeden, A. (2011). Optimization of drilling parameters on surface roughness in drilling of AISI 1045 using response surface methodology and genetic algorithm. *The International Journal of Advanced Manufacturing Technology*, *52*(1-4), 79-88.

Kovačević, M., Madić, M., & Radovanović, M. (2013). Software prototype for validation of machining optimization solutions obtained with meta-heuristic algorithms. *Expert Systems with Applications*, *40*(17), 6985-6996.

Kroese, D. P., Taimre, T., & Botev, Z. I. (2011). *Handbook of Monte Carlo Methods* (Vol. 706). John Wiley & Sons.

Madić, M., Marković, D., & Radovanović, M. (2013). Comparison of meta-heuristic algorithms for solving machining optimization problems. *Facta universitatis-series: Mechanical Engineering*, *11*(1), 29-44.

Madić, M., & Radovanović, M. (2014a). Possibilities of using Monte Carlo method for solving machining optimization problems. *Facta Universitatis, Series: Mechanical Engineering*, *12*(1), 27-36.

Madić M, Radovanović M (2014b) Optimization of machining processes using pattern search algorithm. International Journal of Industrial Engineering Computations, 5(2), 223–234.

Mosegaard, K., & Sambridge, M. (2002). Monte Carlo analysis of inverse problems. *Inverse Problems*, *18*(3), 29-34.

Mukherjee, I., & Ray, P. K. (2006). A review of optimization techniques in metal cutting processes. *Computers & Industrial Engineering*, *50*(1), 15-34.

Munda, J., & Bhattacharyya, B. (2008). Investigation into electrochemical micromachining (EMM) through response surface methodology based approach. *The International Journal of Advanced Manufacturing Technology*,*35*(7-8), 821-832.
Rao, R. V., Pawar, P. J., & Shankar, R. (2008). Multi-objective optimization of electrochemical machining process parameters using a particle swarm optimization algorithm. *Proceedings of the Institution of Mechanical Engineers, Part B: Journal of Engineering Manufacture*, *222*(8), 949-958.
Rao, R. V., & Pawar, P. J. (2009). Modelling and optimization of process parameters of wire electrical discharge machining. *Proceedings of the Institution of Mechanical Engineers, Part B: Journal of Engineering Manufacture*, *223*(11), 1431-1440.
Venkata Rao, R., & Pawar, P. J. (2010). Parameter optimization of a multi-pass milling process using non-traditional optimization algorithms. *Applied soft computing*, *10*(2), 445-456.
Rao, R. V., & Kalyankar, V. D. (2011). Parameters optimization of advanced machining processes using TLBO algorithm. *EPPM, Singapore*, *20*, 21–31
Venkata Rao, R., & Kalyankar, V. D. (2013). Parameter optimization of modern machining processes using teaching–learning-based optimization algorithm. *Engineering Applications of Artificial Intelligence*, *26*(1), 524-531.
Samanta, S., & Chakraborty, S. (2011). Parametric optimization of some non-traditional machining processes using artificial bee colony algorithm. *Engineering Applications of Artificial Intelligence*, *24*(6), 946-957.
Sarkar, B. R., Doloi, B., & Bhattacharyya, B. (2006). Parametric analysis on electrochemical discharge machining of silicon nitride ceramics. *The International Journal of Advanced Manufacturing Technology*, *28*(9-10), 873-881.
Savas, V., & Ozay, C. (2008). The optimization of the surface roughness in the process of tangential turn-milling using genetic algorithm. *The International Journal of Advanced Manufacturing Technology*, *37*(3-4), 335-340.
Yang, X. S. (2010). *Engineering optimization: an introduction with metaheuristic applications*. John Wiley & Sons.
Yıldız, A. R. (2009). A novel particle swarm optimization approach for product design and manufacturing. *The International Journal of Advanced Manufacturing Technology*, *40*(5-6), 617-628.
Yusup, N., Zain, A. M., & Hashim, S. Z. M. (2012). Evolutionary techniques in optimizing machining parameters: Review and recent applications (2007–2011).*Expert Systems with Applications*, *39*(10), 9909-9927.
Yusup, N., Sarkheyli, A., Zain, A. M., Hashim, S. Z. M., & Ithnin, N. (2013). Estimation of optimal machining control parameters using artificial bee colony.*Journal of Intelligent Manufacturing*, 1-10.
Zain, A. M., Haron, H., & Sharif, S. (2011). Genetic algorithm and simulated annealing to estimate optimal process parameters of the abrasive waterjet machining. *Engineering with computers*, *27*(3), 251-259.

9

Process parameter optimization based on principal components analysis during machining of hardened steel

Suryakant B. Chandgude[a*], Padmakar J. Pawar[a] and Mudigonda Sadaiah[b]

[a]*K. K. Wagh Institute of Engineering Education and Research, Nashik-422003, Maharashtra, India.*
[b]*Dr. Babasaheb Ambedkar Technological University, Vidyavihar, Lonere- 402103, Maharashtra, India*

CHRONICLE

Keywords:
End milling
AISI D2
Principal components analysis

ABSTRACT

The optimum selection of process parameters has played an important role for improving the surface finish, minimizing tool wear, increasing material removal rate and reducing machining time of any machining process. In this paper, optimum parameters while machining AISI D2 hardened steel using solid carbide TiAlN coated end mill has been investigated. For optimization of process parameters along with multiple quality characteristics, principal components analysis method has been adopted in this work. The confirmation experiments have revealed that to improve performance of cutting; principal components analysis method would be a useful tool.

1. Introduction

In the world of machining and especially in manufacturing of mould and die components, machining of hardened steel is a vital area for today's scientific research. Mould and die manufacturers prefer this material due to retention of its high strength and good wear resistant property at elevated temperatures. But from mould and die manufacturer's point of view, prime necessity is to reduce machining lead time, improve the quality, and reduce cost of production. For machining of hardened steel, the preferred metal cutting process is end milling. This process is characterised by high metal removal rate, better dimensional accuracy and better surface finish. In practice optimization of every machining process parameter is usually difficult; because it requires simultaneously both machining operation experience and knowledge of mathematical algorithms. The problem of optimization in milling process is complex in nature as multiple objectives and number of constraints has to be considered simultaneously. Moreover, the process planner in practice face problem of optimizing cutting parameters simultaneously for number of mutually conflicting objectives such as machining time, flank wear rate, surface roughness, etc. Considering this fact process parameter optimization in end milling becomes a multi-objective type of problem.

This paper deals with the application of principal components analysis (PCA) method to study the effect of process parameters simultaneously on the performance characteristics during end milling of AISI D2 steel. The results obtained reveals that the surface finish obtained on AISI D2 hardened steel using the proposed approach is close to surface finish in grinding. This leads to elimination of grinding operation after the end milling operation.

2. Literature review

Traditional and non-traditional methods have been applied by various researchers for optimization of process parameters in milling. Traditional methods such as scatter search (Krishna et al., 2006), sequential quadratic programming (Othmani et al., 2011; Shie, 2006) dynamic programming, and method of feasible directions (Othmani et al., 2011) have been employed to the optimization of milling process. However, these methods tend to obtain a local optimum solution and are not suitable for solving complex multimodal problems. Hence, the researchers are now employing non-traditional methods of optimization for solving this class of problems. It is revealed from the literature that researchers had attempted single objective optimization of milling process parameters to achieve desired effects like surface roughness, production rate, machining time, production cost, tool life, and cutting force. Various methods attempted by the researchers for minimization of surface roughness include genetic algorithm (Brezocnik & Kovacic, 2003; Corso et al., 2012), simulated annealing (Corso et al., 2012), and particle swarm optimization (Prakasvidhisarn et al., 2009). The attempts are also made by researchers to maximize the production rate (or minimize machining time) using genetic algorithm (Aggarwal & Xirouchakis, 2012), simulated annealing (Rao & Pawar, 2010), artificial bee colony (Rao & Pawar, 2010), particle swarm optimization (Rao & Pawar, 2010; Gao et al., 2012), harmony search algorithm (Zarei et al., 2009), cuckoo search algorithm (Yildiz, 2012), and teaching learning based optimization algorithm (Pawar & Rao, 2013). Considering cutting force as an objective, researchers employed particle swarm optimization (Farahnakian et al., 2011), for optimization of milling process parameters. Multi-objective optimization for milling process is attempted by researchers using posteriori approaches namely non-dominated sorting genetic algorithm (NSGA) (Wang et al., 2006) and multi-objective particle swarm optimization (MPSO) (Yang et al., 2011).

Although, these methods are successful to determine global optimal process parameter combinations in milling process; their success mainly depends on the accuracy of the mathematical model, and also these methods provide open ended solutions. Hence, in such cases statistical optimization methods may work better. Few researchers tried optimization of surface roughness as a response using grey relational analysis (Yang et al., 2006) and response surface method (Routara et al., 2009). However, it is observed that multi-objective optimization is mostly attempted by earlier researchers using statistical optimization methods. Using response surface method, Premnath et al. (2012) studied the effect of cutting force on the surface roughness simultaneously. Tool life and surface roughness were optimized simultaneously using grey Taguchi method by (Tsao, 2009). Tosun and Pihtili (2010) applied grey relational analysis optimization technique during face milling of 7075 aluminium alloy. Lu et al. (2009) attempted simultaneous optimization of tool life and production rate in end milling operation. Gopalsamy et al., (2009) considered multi-objective optimization aspects of milling process with surface roughness, tool life and machining time as objectives.

AISI D2 is an air hardening, high-carbon, high-chromium tool steel. It has high wear and abrasion resistant properties. It is heat treatable and will offer a hardness in the range 55-62 HRC, and is machinable in the annealed condition. These properties of AISI D2 steel makes it suitable for wide range of applications in the area of tool and die manufacturing. Hence, various researchers have contributed and studied the characterization of AISI D2 tool steel in different environments. The effect of laser parameters such as laser power, speed of workpiece and spot size on tempering depth and hardness value of AISI D2 steel were experimentally investigated by Hindus et al. (2014). Conci et al. (2014) studied the effect of nitrogen potential on the micro-abrasive wear resistance of AISI D2 tool steel. During hard turning of AISI D2 steel using coated carbide insert, the surface roughness analysis was carried out by

Srithar et al. (2014). An experimental study was conducted by Cho et al. (2014) in-order to investigate the surface hardness enhancement of AISI D2 steel by ion nitriding through atomic attrition. Hullapa et al. (2013) explored the use of Magnetic Barkhausen Noise (MBN) to monitor the transformation of austenite to martensite during cooling to sub-zero temperatures. Oliveira et al. (2010) exhibited production and characterization of boride layers on AISI D2 tool steel by thermochemical boriding treatments performed in borax bath. Experimental investigations for machinability during hard turning of AISI D2 cold work tool steel with conventional and wiper ceramic inserts was conducted by Gaitonde et al. (2009). Response surface methodology (RSM) based mathematical models were developed to analyse the effects of depth of cut and machining time on machinability. The quantitative evolution of the residual stress states in surface layers of an AISI D2 steel treated by low energy high current pulsed electron beam was examined by (Zang et al., 2013) using X-ray diffraction technique.

It is thus revealed from the literature that substantial research has been carried out to investigate optimum cutting parameters in end milling. However, very few researchers have attempted to optimize the cutting process parameters using multi-responses simultaneously using principal components analysis. With this understanding, in this study PCA is applied for simultaneous optimization of three important performance measures of the end milling process, which has not been attempted so far by earlier researchers.

3. Principle component analysis

Initially, this technique has been applied to quantify and identify phenomena in social sciences in which it was difficult to directly measure the phenomenal changes. PCA is useful in reduction of data and interpretation of multi-objective sets of data. Currently PCA is finding wide applications in various scientific fields. In this mainly the focus is on correlation analysis of inter-object using linear combinations for each performance measure. To determine optimal combinations during end milling, the algorithm of principal components analysis is given below:

Step 1: Convert the experimental data in to signal-to-noise ratio:

$$\eta_{ij} = -10log\left(\frac{1}{n}\sum_{j=1}^{n} {y_{ij}}^2\right) \tag{1}$$

Step 2: Normalize the signal-to-noise ratio:

$$x_i^*(j) = \frac{x_i^{(o)}(j) - \min x_i^{(o)}(j)}{\max x_i^{(o)}(j) - \min x_i^{(o)}(j)} \tag{2}$$

Step 3: Represent the multi-responses by matrix:

$$X = \begin{bmatrix} x_1(1) & \cdots & x_1(n) \\ \vdots & \ddots & \vdots \\ x_m(1) & \cdots & x_m(n) \end{bmatrix} \tag{3}$$

Step 4: Evaluate the correlation coefficient array:

$$R_{jl} = \frac{COV\left(x_i(j), x_i(l)\right)}{\sigma_{x_i}(j) \times \sigma_{x_i}(l)}, \qquad j = 1,2, \ldots., n, \quad l = 1,2, \ldots \ldots, n \tag{4}$$

where, $COV(x_i(j), x_i(l))$: the covariance of sequences $x_i(j)$ and $x_i(l)$; $\sigma_{x_i}(j)$: the standard deviation of sequence $x_i(j)$; $\sigma_{x_i}(l)$: the standard deviation of sequence $x_i(l)$.

Step 5: Calculate the eigenvalues and eigenvectors:

$$(R - \lambda_k I_m)V_{ik} = 0 \tag{5}$$

Step 6: Obtain the principal components:

$$P_{ik} = \sum_{j=1}^{m} x_i\,(j) \times V_{jk} \tag{6}$$

Step 7: Calculate the Total Principal Component Index:

$$P_i = \sum_{k=1}^{m} P_{ik} \times e(k) \tag{7}$$

where, $e(k) = \frac{eig(k)}{\sum_{k=1}^{m} eig(k)}$ (8)

$eig(k)$ = k^{th} eigenvalue

Step 8: Generate response table and select the optimum levels of cutting parameters:

Control factor	Levels		
	1	2	3
A	$\overline{v_1}$	$\overline{v_2}$	$\overline{v_3}$
B	$\overline{f_{z1}}$	$\overline{f_{z2}}$	$\overline{f_{z3}}$
C	$\overline{a_{e1}}$	$\overline{a_{e2}}$	$\overline{a_{e3}}$
D	$\overline{a_{p1}}$	$\overline{a_{p2}}$	$\overline{a_{p3}}$

$$\overline{v_1} = \frac{(TPCI)_1 + (TPCI)_2 + \cdots + (TPCI)_9}{9} \tag{9}$$

Step 9: Calculate the combined objective function (COF):

$$\begin{matrix} Weighted\ normalized \\ value\ for\ each\ response \end{matrix} = w_i \times \frac{\begin{matrix} Measured\ value\ of\ quality \\ characteristic\ in\ each\ run \end{matrix}}{\begin{matrix} Minimum\ value\ of\ quality \\ characteristic\ in\ the\ data\ set \end{matrix}} \tag{10}$$

w_i = $weightage\ assigned\ to\ the\ quality\ characteristic$

$i = 1,2,\ldots,n$; and n is the number of quality characteristic

$$COF = \sum Weighted\ normalized\ value\ for\ all\ responses\ \ in\ each\ run$$

Step 10: Perform the statistical analysis of variance (ANOVA).

Step 11: Conduct the confirmation run.

4. Experimental Design

In the present study of end milling, the three quality characteristics selected are machining time, flank wear rate, and surface roughness. These three are directly related to quality of the product and the cost. Control factors and levels have been selected from the machining data handbook (Zhuzhou Co. Ltd., 2011) and are shown in Table 1:

Table 1
Control factors and levels

Code	Control Factors	Level 1	Level 2	Level 3
A	v (m min^{-1})	20	70	120
B	f_z (mm tooth^{-1})	0.03	0.04	0.05
C	a_e (mm)	0.3	0.4	0.5
D	a_p (mm)	0.3	0.5	0.7

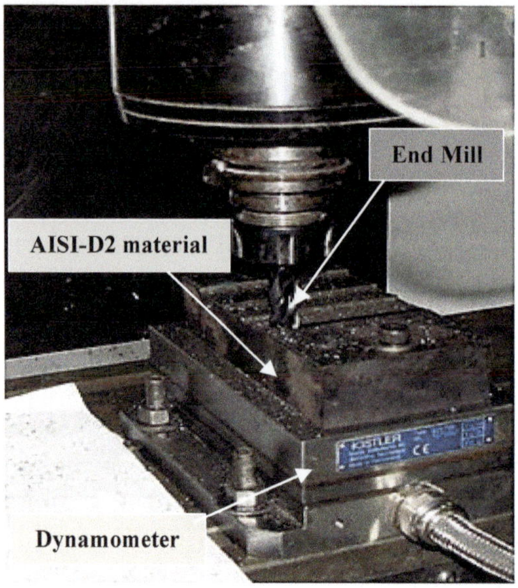

Fig. 1. Experimental set-up: End mill, AISI-D2 hardened steel, and 3-component force dynamometer

The matrix experiments using orthogonal arrays were performed in up-milling mode without any cutting fluid. The factor effects of each level of each control factor can be determined by using the orthogonal arrays. For four factors with three levels L_9 array should be used; however, in our study using PCA for multi-objective type of problem, to get into details in the experiment variation, 27 experiments are performed. To eliminate some of the invisible factors that might contribute to the measured variables, each experiment was performed three times in a random sequence. The workpiece was clamped rigidly on dynamometer which is mounted on table of the machine and a fixed overhang length of 25 mm was selected for the tool. The workpiece was continuously cut for a length of 375 mm, with an approach distance and over travel distance of 5 mm, in each pass in order to minimize tool jerks during the physical contact with the workpiece. The complete experimental setup is shown Fig. 1. The cutting time was recorded using the stop watch and tabulated in Table 2. The surface roughness measurements using R-200 series portable surface analyzer was carried at the end of each experiment. The milled surface being anisotropic in nature, the surface roughness measurement was taken at three fixed points; the first point was in the middle and the other two points on the edge of each milled surface. The mean of three surface roughness readings was recorded in Table 2. To observe and measure the maximum flank wear of the end cutting edges, Nikon microscope with a magnification of 100X was used. The flank wear during each experimental run is shown in Fig. 2.

5. Experimental results and discussion

In this section the detail procedure to find optimal combinations of the process parameters using principal components analysis is discussed.

5.1 Optimal Combination of Process Parameters

The results are shown in Table 2. In the experiments, machining time, flank wear, and surface roughness are considered as quality characteristics. These quality characteristics are continuous and nonnegative, and can be recognized as the smallest-the-better type. The results are substituted into Eq. (1) to obtain the S/N ratios of machining time, flank wear rate and surface roughness. Usually, larger the signal-to-noise ratio, better is the quality characteristic. The calculated S/N ratios are normalized using Eq. (2) and are shown in Table 3.

Table 2

Dataset for measured values of the responses

Expt. No.	Cutting Speed (v) (m min^{-1})	Feed (f_z) (mm tooth^{-1})	Radial Depth of Cut (a_e) (mm)	Axial Depth of Cut (a_p) (mm)	Surface roughness(R_a) (μm)	Machining time (t_m) (s)	Flank Wear (V_B) (mm)
1	20	0.03	0.3	0.3	0.272	388	0.010
2	20	0.04	0.4	0.5	0.302	76	0.014
3	20	0.05	0.5	0.7	0.337	35	0.015
4	20	0.03	0.4	0.7	0.225	372	0.015
5	20	0.04	0.5	0.5	0.200	78	0.009
6	20	0.05	0.3	0.3	0.237	34	0.005
7	20	0.03	0.5	0.5	0.195	370	0.005
8	20	0.04	0.3	0.3	0.252	76	0.015
9	20	0.05	0.4	0.7	0.172	35	0.005
10	70	0.03	0.3	0.3	0.298	359	0.010
11	70	0.04	0.4	0.5	0.145	76	0.005
12	70	0.05	0.5	0.7	0.168	35	0.015
13	70	0.03	0.4	0.7	0.103	368	0.040
14	70	0.04	0.5	0.5	0.128	76	0.005
15	70	0.05	0.3	0.3	0.202	35	0.015
16	70	0.03	0.5	0.5	0.113	369	0.010
17	70	0.04	0.3	0.3	0.130	75	0.010
18	70	0.05	0.4	0.7	0.087	36	0.005
19	120	0.03	0.3	0.3	0.080	363	0.015
20	120	0.04	0.4	0.5	0.160	75	0.005
21	120	0.05	0.5	0.7	0.157	33	0.015
22	120	0.03	0.4	0.7	0.070	366	0.035
23	120	0.04	0.5	0.5	0.083	76	0.015
24	120	0.05	0.3	0.3	0.082	35	0.010
25	120	0.03	0.5	0.5	0.053	365	0.005
26	120	0.04	0.3	0.3	0.043	76	0.010
27	120	0.05	0.4	0.7	0.138	35	0.005

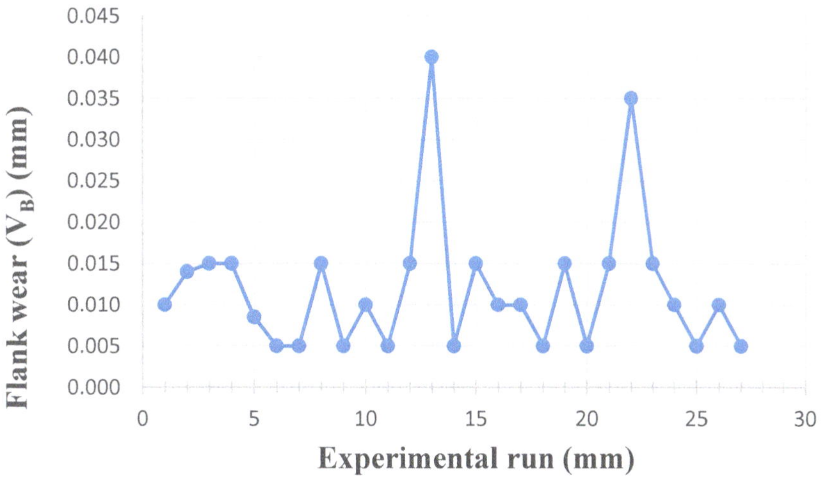

Fig. 2. Flank wear during each experimental run

Correlation coefficients are obtained using Eq. (4). From the correlation coefficient array, the eigenvalues and eigenvectors are calculated using MATLAB. From the eigenvalue properties for symmetric matrices, their eigenvectors are always orthogonal to each other. The corresponding eigenvalues (0.7100, 0.9766, and 1.3134) and the eigenvectors (-0.3604, 0.6922, 0.6253), (0.8910, 0.0570, 0.4505), and (0.2762, 0.7194, -0.6373) of the correlation coefficients are obtained using Eq. (5).

Table 3
Normalized dataset of Signal-to-Noise ratios

Expt. No.	Signal-to-Noise ratio			Normalized Signal-to-Noise ratio		
	R_a	t_m	V_B	R_a	t_m	V_B
1	11.32	-51.777	40.000	0.104	0.000	0.667
2	10.41	-37.616	37.077	0.053	0.657	0.505
3	9.46	-30.881	36.478	0.000	0.970	0.472
4	12.96	-51.411	36.478	0.196	0.017	0.472
5	13.98	-37.786	41.412	0.254	0.650	0.745
6	12.52	-30.630	46.021	0.172	0.982	1.000
7	14.20	-51.352	46.021	0.266	0.020	1.000
8	11.98	-37.559	36.478	0.142	0.660	0.472
9	15.31	-30.881	46.021	0.328	0.970	1.000
10	10.51	-51.102	40.000	0.059	0.031	0.667
11	16.77	-37.559	46.021	0.411	0.660	1.000
12	15.48	-30.881	36.478	0.338	0.970	0.472
13	19.72	-51.305	27.959	0.576	0.022	0.000
14	17.83	-37.559	46.021	0.470	0.660	1.000
15	13.91	-30.756	36.478	0.250	0.976	0.472
16	18.91	-51.329	40.000	0.531	0.021	0.667
17	17.72	-37.501	40.000	0.464	0.663	0.667
18	21.24	-31.005	46.021	0.662	0.964	1.000
19	21.94	-51.198	36.478	0.701	0.027	0.472
20	15.92	-37.501	46.021	0.363	0.663	1.000
21	16.10	-30.238	36.478	0.373	1.000	0.472
22	23.10	-51.258	29.119	0.766	0.024	0.064
23	21.58	-37.616	36.478	0.681	0.657	0.472
24	21.76	-30.881	40.000	0.691	0.970	0.667
25	25.46	-51.234	46.021	0.899	0.025	1.000
26	27.26	-37.559	40.000	1.000	0.660	0.667
27	17.18	-30.881	46.021	0.434	0.970	1.000

Table 4
Principal components and COF for the experimental runs

Expt. No.	Principal Components			TPCI	Weighted Normalized			COF
	PC_1	PC_2	PC_3		R_a	t_m	V_B	
1	0.379	0.393	-0.396	0.045	2.085	3.88	0.660	6.625
2	0.752	0.312	0.166	0.353	2.315	0.76	0.924	3.999
3	0.967	0.268	0.397	0.490	2.584	0.35	0.990	3.924
4	0.236	0.388	-0.234	0.080	1.727	3.72	0.990	6.437
5	0.824	0.599	0.063	0.418	1.535	0.78	0.561	2.871
6	1.243	0.659	0.116	0.561	1.816	0.34	0.330	2.486
7	0.543	0.689	-0.550	0.113	1.497	3.70	0.330	5.522
8	0.701	0.376	0.213	0.382	1.931	0.76	0.990	3.676
9	1.178	0.798	0.151	0.606	1.317	0.35	0.330	1.997
10	0.417	0.354	-0.386	0.045	2.290	3.59	0.660	6.540
11	0.934	0.854	-0.049	0.478	1.113	0.76	0.330	2.198
12	0.845	0.569	0.491	0.601	1.292	0.35	0.990	2.632
13	-0.192	0.515	0.175	0.199	0.793	3.68	2.640	7.108
14	0.913	0.907	-0.032	0.498	0.985	0.76	0.330	2.070
15	0.880	0.491	0.471	0.575	1.548	0.35	0.990	2.883
16	0.240	0.775	-0.263	0.194	0.870	3.69	0.660	5.215
17	0.708	0.752	0.180	0.492	0.998	0.75	0.660	2.408
18	1.054	1.095	0.239	0.712	0.665	0.36	0.330	1.350
19	0.061	0.839	-0.088	0.249	0.614	3.63	0.990	5.234
20	0.953	0.812	-0.060	0.464	1.228	0.75	0.330	2.308
21	0.853	0.602	0.522	0.627	1.202	0.33	0.990	2.517
22	-0.219	0.713	0.188	0.263	0.537	3.66	2.310	6.502
23	0.505	0.857	0.360	0.557	0.640	0.76	0.990	2.390
24	0.839	0.971	0.464	0.719	0.627	0.35	0.660	1.637
25	0.319	1.253	-0.371	0.322	0.409	3.65	0.330	4.384
26	0.513	1.229	0.326	0.665	0.333	0.76	0.660	1.748
27	1.140	0.892	0.180	0.640	1.062	0.35	0.330	1.742

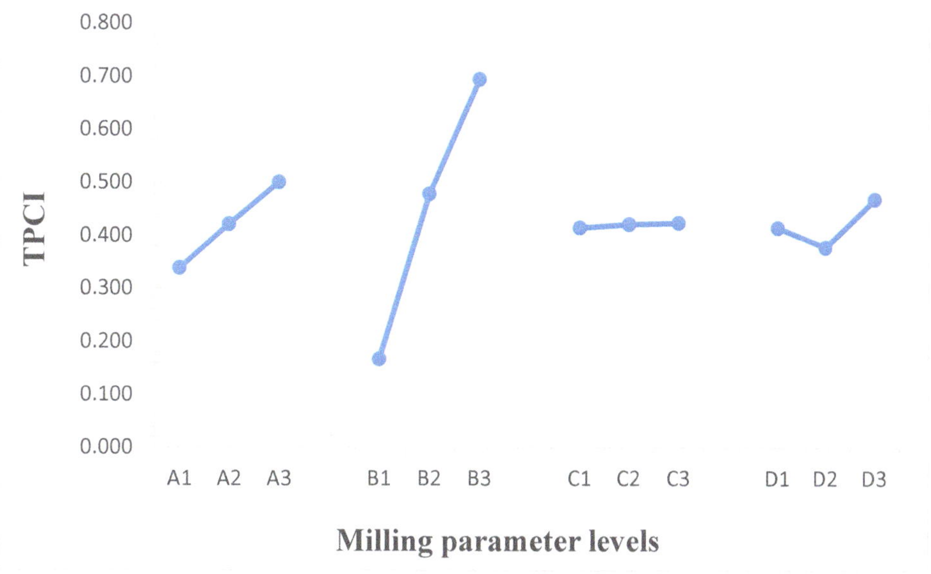

Fig. 3. Response graph of total principal component index (TPCI)

5.2 Obtaining the Principal Components

Principal components analysis is a technique used to transform the correlated variables into linear combinations of uncorrelated variables, which account for most of the variance in the original set of observations. The basic purpose of PCA is to determine the principal components. If the number of linear combinations obtained are *n*, then the number of principal components formed will be less than or equal to *n*. With reference to Table 3 and Eq. (6) the principal component (PC_1) for experiment 1 can be calculated as follows:

$$PC_1 = 0.104 \times (-0.3604) + 0 \times 0.6922 + 0.667 \times 0.6253 = 0.379$$

The present work deals with three responses and therefore, three principal components PC_1, PC_2, and PC_3 are determined. The corresponding values are included in Table 4.

5.3 Calculating Total Principal Component Index (TPCI)

To obtain best optimal combination of factors/levels, TPCI is calculated using Eq. (7) and Eq. (8). Thus, TPCI for experiment number 1 is calculated as:

$$(TPCI)_1 = (PC_1)_1 \times 0.237 + (PC_2)_1 \times 0.326 + (PC_3)_1 \times 0.438 = 0.045$$

All the calculated TPCI values are shown in Table 4 accordingly.

5.4 Generate Response Table for Selection of Optimum Parameters

After calculating the TPCI's for all the experimental run, the next step is to construct the response table. As an illustration in order to calculate the response value for factor A at level 1, using Eq. (9) we get,

$$\overline{v_1} = \frac{(TPCI)_1 + (TPCI)_2 + \cdots + (TPCI)_9}{9}$$

$$\overline{v_1} = \frac{0.045 + 0.353 + 0.490 + 0.080 + 0.418 + 0.561 + 0.113 + 0.382 + 0.606}{9} = 0.3385$$

Using this method of calculation the remaining response values corresponding to the factors and their respective levels are determined. The maximum value of TPCI corresponding to each factor gives the predicted optimum factor level. Fig. 3 shows response graph of TPCI. From response graph best combination consist of the set: A_3 (spindle speed with 120 m min^{-1}), B_3 (feed rate with 0.05 mm $tooth^{-1}$), C_3 (radial depth of cut with 0.5 mm), and D_3 (axial depth of cut with 0.7 mm).

The combined objective function (COF) for all the quality characteristics is calculated using Eq. (10) by assigning the equal weights of 0.33 for each quality characteristic. Table 4 shows the COF values for all the experimental runs.

5.5 Analysis of Variance

To identify significant factors which influence on performance measures, the ANOVA is carried out for COF and the results are given in Table 5. The significant levels (for $\alpha = 0.05$, at 95% confidence level) for each source of variation, associated with the *F*-test are also shown in Table 5. From the principal of *F*-test, if *F* is larger for a specific parameter, the effect on the performance characteristics is greater. In our case from ANOVA table, for feed (factor B), the *F* value is the largest with a total contribution of 80.61 %; which clearly justifies the major effect on the performance measures such as surface roughness, machining time, and flank wear rate. Cutting speed (factor A) was the second significant factor with 5.00 % contribution. The contribution of radial depth of cut (factor C) was 2.50 %, and the contribution of axial depth of cut was found to be 2. 41 %. The contribution due to error was small and clearly signifies that, important factor has not been omitted and high measurement error was not involved.

Table 5

ANOVA result for Combined Objective Function (COF)

Source	Sum of Squares	DOF	Mean Square	F/t	P	Contribution (%)	Remark
Model	79.32	7	11.33	24.81	<0.0001		significant
A-v	4.60	2	2.30	5.04	0.0176	5.00	
B-f_z	74.03	2	37.02	81.04	<0.0001	80.61	
C-a_e	2.30	2	1.15	2.51	0.1075	2.50	
D-a_p	2.22	1	2.22	4.86	0.0400	2.41	
Residual	8.68	19	0.46			9.45	
Cor Total	91.83	26				100	

$R^2 = 0.9014$; $R^2_{Adj} = 0.8650$

5.6 Confirmation Tests

The results of confirmation experiments using the optimal parameters ($A_3B_3C_3D_3$) obtained by PCA with the initial setting level are shown in Table 6. It has been observed that there is a considerable improvement in the results, due to the application of PCA for optimizing all the three responses simultaneously.

Table 6

Results of initial and optimal settings

	Initial settings from machining data handbook	Optimal settings using PCA
Setting level	$v = 110\ m\ min^{-1}$; $f_z = 0.05\ mm\ tooth^{-1}$; $a_e = 0.4\ mm$; $a_p = 0.4\ mm$	$A_3B_3C_3D_3$, i.e. $v = 120\ m\ min^{-1}$; $f_z = 0.05\ mm\ tooth^{-1}$; $a_e = 0.5\ mm$; $a_p = 0.7\ mm$
Surface roughness (R_a) µm	0.635	0.180
Machining time (t_m) s	35	35
Flank wear (V_B) mm	0.033	0.010
COF	7.401	2.391
% improvement in COF = 68		

As shown in Table 6, for the problem under consideration, the machining time is constant for the cutting length of 325 mm. The flank wear is decreased from 0.033 mm to 0.010 mm and the surface roughness is decreased from 0.635 μm to 0.180 μm. Through confirmation tests, it is revealed that the results obtained using the proposed approach shows significant improvement over the existing approach.

6. Conclusions

The application of principal components analysis is an approach for optimization of the cutting process parameters during end milling of AISI D2 hardened steel, using 4-flute flattened solid carbide TiAlN coated end mill with straight shank. Based on the results following conclusions can be drawn:

a) The optimal combination ($A_3B_3C_3D_3$) of the cutting parameters using PCA, as shown in Fig. 2 is the set with; cutting speed (120 m min^{-1}), feed (0.05 mm $tooth^{-1}$), radial depth of cut (0.5 mm), and axial depth of cut (0.7 mm).
b) The corresponding confirmation test shows 70 % improvement in flank wear rate, and 72 % in surface roughness. The overall improvement considering all three objectives is 68 % over existing parameter setting recommended in the machining data handbook.
c) From this study the surface roughness (R_a) value of 0.180 μm obtained during end milling of AISI D2 hardened steel material, based on PCA methodology is acceptable to the mould and die manufacturers. Thus, it is likely that it may be possible to eliminate the grinding operation often carried out after end milling operation.
d) The parameters and their levels considered shows highest effect on flank wear (V_B) with 43.6 % weightage, 32.6 % weightage for machining time (t_m) and 23.7 % weightage for surface roughness (R_a).
e) The results of optimization obtained by the proposed approach provide a ready reference to tool manufacturers as well as to the operators.

It can be concluded that the PCA method is very suitable for solving the flank wear and surface roughness quality problems in milling hardened steel. The cutting forces generated during machining process are important parameters which reflect the machining condition; the cutting forces measured during the experimental run may be helpful in predicting the cutting forces for tool condition monitoring system for future work.

References

Aggarwal, S., & Xirouchakis, P. (2013). Selection of optimal cutting conditions for pocket milling using genetic algorithm. *The International Journal of Advanced Manufacturing Technology*, *66*(9-12), 1943-1958.

Brezocnik, M., & Kovacic, M. (2003). Integrated genetic programming and genetic algorithm approach to predict surface roughness. *Materials and Manufacturing Processes, 18*, 475-491.

Cho, K. T., Song, K., Oh, S. H., Lee, Y. K., & Lee, W. B. (2014). Enhanced surface hardening of AISI D2 steel by atomic attrition during Ion Nitriding. *251*, 115-121.

Conci, M. P., Bozzi, A. C., & Franco Jr, A. R. (2014). Effect of plasma Nitriding potential on tribological bhaviour of AISI D2 cold-worked tool steel. *Wear, 317*, 188-193.

Corso, L. L., Zeilmann, R. P., Nicola, G. L., Missell, F. P., & Gomes, H. M. (2013). Using optimization procedures to minimize machining time while maintaining surface quality. *The International Journal of Advanced Manufacturing Technology*, *65*(9-12), 1659-1667.

Farahnakian, M., Razfar, M. R., Moghri, M., & Asadnia, M. (2011). The selection of milling parameters by the PSO based neural network modeling method. *57*, 49-60.

Gaitonde, V. N., Karnik, S. R., Figueira, L., & Davim, J. P. (2009). Machinability investigations in hard turning of AISI D2 cold work tool steel with conventional and wiper ceramic inserts. *International Journal of Refractory Metals and Hard Materials*, *27*(4), 754-763.

Gao, L., Huang, J., & Li, X. (2012). An effective cellular particle swarm optimization for parameters optimization of a multi-pass milling process. *Applied Soft Computing*, *12*(11), 3490-3499.

Gopalsamy, B. M., Mondal, B., & Ghosh, S. (2009). Optimisation of machining parameters for hard machining: grey relational theory approach and ANOVA. *The International Journal of Advanced Manufacturing Technology*, *45*(11-12), 1068-1086.

Hindus, S. J., Kumar, S. H., Arockiyaraj, S. X., Kannan, M. V., & Kuppan, P. (2014). Experimental investigation on laser assisted surface tempering of AISI D2 tool steel. *Procedia Engineering, 97*, 1489-1495.

Huallpa, E. A., Sanchez, J. C., Padovese, L. R., & Goldenstein, H. (2013). Determining Ms temperature on a AISI D2 cold work tool steel using magnetic barkhausen noise. *Journal of Alloys and Compounds, 577S*, S726-S730.

Krishna, A. G., & Rao, K. M. (2006). Optimization of machining parameters for milling machine operations using a scatter search approach. *International Journal of Advanced Manufacturing Technology, 31*, 219-224.

Lu, H. S., Chang, C. K., Hwang, N. C., & Chung, C. T. (2009). Grey relational analysis coupled with principal component analysis for optimization design of the cutting parameters in high speed end milling. *Journal of Materials Process Technology,* 209, 3808-3817.

Oliveira, C. K., Casteletti, l. C., Neto, L., Totten, G. E., & Heck, S. C. (2010). Production and characterization of boride layers on AISI D2 tool steel. *Vacuum, 84*, 792-796.

Othmani, R., Hbaieb, M., & Bouzid, W. (2011). Cutting parameter optimization in NC milling. *International Journal of Advanced Manufacturing Technology, 54*, 1023-1032.

Pawar, P. J., & Rao, R. V. (2013). Parameter optimization of machining processes using teaching–learning-based optimization algorithm. *The International Journal of Advanced Manufacturing Technology*, *67*(5-8), 995-1006.

Prakasvidhisarn, C., Kunnapapdeelert, S., & Yenradee, P. (2009). Optimal cutting condition determination for desired surface roughness in end milling. *International Journal of Advanced Manufacturing Technology, 41*, 440-451.

Premnath, A., Alwarsmy, T., & Rajmohan, T. (2012). Experimental investigation and optimization of process parameters in milling of hybrid metal matrix composites. *Materials and Manufacturing Processes, 27*, 1035-1044.

Rao, R., & Pawar, P. (2010). Parameter optimization of a multi-pass milling process using non-traditional optimization algorithms. *Applied Soft Computing, 10*, 445-456.

Routara, B. C., Bandyopadhyay, A., & Sahoo, P. (2009). Roughness modeling and optimization in CNC end milling using response surface method: Effect of workpiece material variation. *International Journal of Advanced Manufacturing Technology.*

Shie, J. R. (2006). Optimization of dry machining parameters for high-purity graphite in end milling process by artificial neural networks: A case study. *Materials and Manufacturing Processes, 21*, 838-845.

Srithar, A., Palanikumar, K., & Durgaprasad, B. (2014). Experimental investigation and surface roughness analysis on hard turning of AISI D2 steel using coated carbide inserts. *Procedia Engineering, 97*, 72-77.

Tosun, N., & Pihtili, H. (2010). Grey relational analysis of performance characteristics in MQL millin of 7075 Al Alloy. *International Journal of Advanced Manufacturing Technology,* 46, 509-515.

Tsao, C. C. (2009). Grey-taguchi method to optimize the milling parameters of Alluminium alloy. *International Journal of Advanced Manufacturing Technology,* 45, 41-48.

Wang, Z. G., Wong, Y. S., Rahman, M., & Sun, J. (2006). Multi-objective optimization of high speed milling with parallel genetic simulated annealing. *International Journal of Advanced Manufacturing Technology,* 31, 209-218.

Yang, W., Guo, Y., & Liao, W. (2011). Multi-objective optimization of multi-pass face milling using particle swarm intelligence. *International Journal of Advanced Manufacturing Technology, 56*, 429-443.

Yang, Y. K., Shie, J. R., & Huang, C. H. (2006). Optimzation of dry machinig parameters for high purity graphite in end milling process. *Materials and Manufacturing Processes, 21*, 832-837.

Yildiz, A. R. (2013). Cuckoo search algorithm for the selection of optimal machining parameters in milling operations. *The International Journal of Advanced Manufacturing Technology*, *64*(1-4), 55-61.

Zarei, O., Fesanghary, M., Farshi, B., Saffar, R. J., & Razfar, M. R. (2009). Optimizaton of multi-pass face milling via harmony search algorithm. *Journal of Materials Process Technology, 209*, 2386-2392.

Zhang, K. M., Zou, J. X., Bolle, B., & Grosdidier, T. (2013). Evolution of residual stress states in surface layers of an AISI D2 steel treated by low energy high current pulsed electron beam. *Vacuum*, *87*, 60-68.

Zhuzhou Cemented Carbide Cutting Tools Co., L. (2011). JM series Solid Carbide End Mills. JIS standard, China.

10

Integer batch scheduling problems for a single-machine with simultaneous effects of learning and forgetting to minimize total actual flow time

Rinto Yusriski[*], Sukoyo Sukoyo, T.M.A Ari Samadhi and Abdul Hakim Halim

Department of Industrial Engineering and Management, Institut Teknologi Bandung, Bandung 40132, Indonesia

CHRONICLE

Keywords:
Learning and forgetting effect
Integer batch scheduling
Actual flow time

ABSTRACT

This research discusses an integer batch scheduling problems for a single-machine with position-dependent batch processing time due to the simultaneous effect of learning and forgetting. The decision variables are the number of batches, batch sizes, and the sequence of the resulting batches. The objective is to minimize total actual flow time, defined as total interval time between the arrival times of parts in all respective batches and their common due date. There are two proposed algorithms to solve the problems. The first is developed by using the Integer Composition method, and it produces an optimal solution. Since the problems can be solved by the first algorithm in a worst-case time complexity $O(n2^{n-1})$, this research proposes the second algorithm. It is a heuristic algorithm based on the Lagrange Relaxation method. Numerical experiments show that the heuristic algorithm gives outstanding results.

1. Introduction

The classical theory of scheduling assumes that the processing time of a job is not affected by its position on a schedule (e.g. Morton & Pentico, 1993; Pinedo, 2002; Baker & Trietsch, 2009). However, there are situations where the processing time of a job scheduled at a position can be faster or slower than that at the previous position due to learning and forgetting effects (Wang & Cheng, 2007; Cheng, et al., 2010; Lai & Lee, 2013). Some researchers, i.e. Yang and Chand (2008), and Ji and Cheng (2010) show that one of the crucial factors leading to the learning and forgetting effects is when operators or machines process jobs in batches. The learning effect is caused by the increase of the operator's competence after producing the same parts in a batch repeatedly. Meanwhile, the forgetting effect occurs during a break time between two consecutive batches so that the operator has to learn the operation again when beginning to process the parts in the next batch. Keachie and Fontana (1966) discuss both the effects of learning and forgetting on Economic Order Quantity (EOQ) model. Steedman (1970) proves that the optimal batch sizes in traditional EOQ model have smaller value than that on Keachie and Fontana. Meanwhile, Jaber and Salameh (1995) propose optimal batch sizes based on Economic Production Quantity (EPQ) model by considering learning situation. Other researchers in the same field are Cheng (1991), Cheng (1994), Li and Cheng (1994), Chiu (1997), Chiu et al. (2003), Chiu and Chen (2005),

Chen et al. (2008), Teyarachakul et al. (2008), and Teyarachakul et al. (2011). Although the researchers have found a method to determine the batch sizes, they have not considered on how to schedule the resulting batches on a machine.

Gaweijnowics (1996) and Biskup (1999) are pioneers who discuss job scheduling problems on a machine by considering the learning effect. They assume that job processing time is not a constant value. It may change due to learning and/or deterioration effects. Gaweijnowics (1996) discusses a single-machine scheduling problem to minimize makespan and shows that the optimal schedule should be obtained by scheduling jobs in accordance with the Shortest Processing Time (SPT) rule. In the meantime, Biskup (1999) proves the polynomial solutions for similar problems to that of Gaweijnowics (1996) for two objectives, i.e. minimizing the interval time between completion times of jobs and their common due date, and minimizing the sum of flow times.

Biskup (1999) classifies the learning function into two models, i.e. the position-based learning and the sum-of-processing-times-based learning. The position-based learning assumes that the learning effect increases due to machine-driven and has no or near to zero human interference. The increase of learning effect depends on the position of jobs in a schedule (Cheng et al., 2011). However, the sum-of-processing-times-based learning model assumes that the learning effect increases in line with the number of jobs that the operator has previously completed (Cheng et al., 2009). Kuo and Yang (2006), Anzanello and Foglianto (2010), and Cheng et al. (2013) propose a learning model that combines the sum-of-processing-times-based model and the position-based learning model. Meanwhile, Janiak et al. (2011) propose an operator experience-based learning model and conclude that the learning effect should be restricted by a minimum job processing time called as learning threshold. The readers may get better understanding of learning function models in Janiak et al. (2011), Teyarachakul et al. (2011), and Lai and Lee (2013).

The learning effect is always followed by forgetting effect (Jaber & Bonney, 1996; Jaber, 2011; Nembhard & Uzumeri, 2000). Arzi and Shtub (1997) discuss that any interruption in the course of the learning process generates a forgetting effect. There are various forgetting function models, some of them can be found in Jaber and Boney (1996) and Teyarachakul et al. (2011). Currently, the research on the simultaneous effects of learning and forgetting in the area of scheduling has become an interesting topic for researchers, such as Lai and Lee (2013) and Wu et al. (2014). The researchers deal with single-machine scheduling problems with different objectives. Their research results reveal that in order to minimize both makespan and total completion time, an optimal schedule should be obtained by scheduling jobs in accordance with the SPT rule. Meanwhile, minimizing total weighted completion time can be obtained by arranging jobs in accordance with the Weighted Shortest Processing Time (WSPT) rule. In other case, i.e. to minimize the minimum lateness, maximum tardiness, and total tardiness, the jobs must be sequenced by adopting the Early Due Date (EDD) rule.

This research deals with batch scheduling problems for a single-machine that produces discrete parts where the processing time is affected by learning and forgetting effects simultaneously under a Just-In-Time (JIT) production system. The motivation of research is from a real-life situation, i.e. in the step of inserting components to Printed Circuit Board (PCB). There are several components inserted into PCB by an operator, and then the completed product is placed in a rack. The operator will stop processing and transfer the rack to a temporary warehouse when the number of parts in the rack reaches a certain quantity. After the operator delivers the parts, the operator prepares the process again such as to take the number of new materials from the previous step. The number of parts in one rack can be considered as a batch size. It can also be considered that the interval time when the operator prepares the process as a setup batch. The learning effect occurs because the operator processes the parts repeatedly. Meanwhile, a setup time leads to the forgetting effect.

This research assumes that the arrival time of parts in all batches can be arranged at the time when the machine starts to process, and all finished parts must be delivered exactly at the time coinciding with their common due date. It also assumes that there is a setup time between two consecutive batches. The objective is to minimize the total actual flow time of parts in all batches, as defined by Halim et al. (1994) as the total interval times between the arrival times of parts in all respective batches and their common due date. Halim and Ohta (1994) prove that total actual flow time is effective to minimize total inventory cost and satisfy the due date simultaneously in a Just-In-Time production system. Due to the actual flow time adopt a backward scheduling approach, then the learning effect in this research is shown by the shorter processing time of parts in a batch scheduled at a position than that in another batch scheduled at the next position. On the contrary, the forgetting is shown by longer processing time. The decision variables are the number of batches, batch sizes and the sequence of the resulting batches.

The structure of this paper is as follows. The next section presents a single-machine batch scheduling problems with learning and forgetting effects. The third section shows the problem formulation. The fourth section discusses the solution method and several numerical experiments. Finally, the last is the concluding remarks.

2. Batch Scheduling Problems with Learning and Forgetting Effects

2.1 Batch Scheduling Problems for a Single Machine

Dobson et al. (1987) describe that flow time criteria can be used in batch scheduling problems to minimize setup cost and inventory cost, simultaneously. Flow time in Dobson et al. (1987) is based on the so-called a forward scheduling approach. The researchers assume that all parts have been available since the beginning of the scheduling period (time zero), and all parts in respective batches should be delivered at the completion time of the batches. These assumptions are not always true in many real situations such as in a JIT production system. There are conditions where the completed parts must be delivered exactly at the time coinciding with a due date, and the company is capable of arranging the arrival of parts at the time which the machine starts to process. Halim et al. (1994) propose an objective of actual flow time. The total actual flow time of all parts in a batch is calculated by multiplying the number of parts in a batch with the time interval between the common due date and the arrival time of parts in the batch. For single-machine batch scheduling problems, the constraints that should be considered are as follows: the number of all parts produced equals the demands; the completion time of all batches should not exceed the available time (the interval time from time zero to the due date); the completion of the batch scheduled in the first order (backwardly) must be delivered exactly at the time coinciding with the due date; the batch sizes should be positive value and the number of batches is a positive integer. The decision variables of the research are the number of batches, the number of parts in batches and the sequence of the resulting batches. Halim et al. (1994) solve the problems using the Lagrange Relaxation method. The result shows that the minimum actual flow time is obtained by sequencing the resulting batches in the LPT rule in a backward scheduling approach.

2.2. Processing Time with Learning Effect

A learning effect can be explained as a phenomenon where the processing time of a job at a certain position is shorter than that at the earlier position. It is because the operator's experience increases in line with the number of jobs that the operator has previously completed. Wright (1936) is the pioneer who discusses a learning model the so-called Cumulative Average Power (CAP). The equation of CAP model is as follows.

$$T_{[x]} = T_{[1]}x^{-m} \tag{1}$$

where $m = -\log(\delta)/\log(2)$

It is notated that $T_{[x]}$ is the processing time when producing x-units, $T_{[i]}$ is the initial processing time or the processing time for the unit firstly processed, δ is learning rate and m is learning slope. The δ values are between $0<\delta<1$. However, in manufacturing system, it is between 0.7 and 0.9 (see Jaber & Bonney, 1996).The lower of δ, the greater of the learning effect.

This paper adopts the backward scheduling approach so that the learning effect is shown by a shorter or equal processing time of parts in a batch scheduled at a position than that in another batch scheduled at the next position. The learning function is developed from learning function by Janiak et al. (2011) and CAP model. The batch processing time is calculated on the basis of the maximum values of the learning function and learning threshold as shown in Eq. (2).

$$T_{[i]} = \max\left\{p\left(1+\sum_{k=i}^{N} Q_{[k+1]}\right)^{-m}, v\right\} \tag{2}$$

where $m = -\log\delta / \log 2$, $i=1,2,..,N$

It is assumed that there are N batches, all of which are scheduled on a machine by using the backward scheduling approach. The number of parts in a batch (the batch size) is $Q_{[i]}$ and p stands for the initial processing time, i.e. the processing time of those parts scheduled at the last batch ($Q_{[N]}$). The value of learning rate (δ) determines learning slope (m) according to the formula of Cumulative Average Power (CAP) model. It is also considered that there is a minimum processing time or learning threshold (v) so that the processing time will be constant after having reached a minimum processing time value.

2.3. Processing Time with a Forgetting Effect

A forgetting effect is defined as an increase of job processing time after an interruption during the specific time. It is because an operator must re-learn a process again when he starts to process job after the interruption. Carlson and Rowe (1976) carry out the calculation of processing times with some forgetting effect, well known as a power model as shown in the following equation.

$$\hat{T}_{[x]} = \hat{T}_{[1]} x^{f} \tag{3}$$

It is notated that $\hat{T}_{[x]}$ is a processing time for the x-th unit affected by some forgetting effect, $T_{[i]}$ is the initial processing time or the processing time for the first processed unit, x is the total accumulation of unit that can be produced during interruption time if there is no forgetting effect, and f is a forgetting slope. Jaber and Bonney (1996) assume that an operation process will be interrupted after the operator has produced q number of parts. It is assumed that forgetting parameter is affected by learning slope (m) so that since the operator experience increase in each batch, the forgetting parameter will continuously change too. Jaber and Bonney (1996) put forwards a formula for computing f values as shown in the following equation.

$$f = m(1-m)\log(q)/\log(1+C), \tag{4}$$

where $m = -\log\delta/\log 2$,

$$C = t_B / t_p,$$

$$t_B = T_{[1]}\left[(q+R)^{(1-m)} - u^{(1-m)}\right]/(1-m)$$

$$t_p = t(q) = T_{[1]} q^{(1-m)} / (1-m)$$

The forgetting effect in this research is shown by a longer or equal processing time of parts in a batch scheduled at a position than that in another batch scheduled at the next position in the backward scheduling approach. This paper develops forgetting function for batch processing time based on the Power model in Carlson and Rowe (1976), and assumes that there is a forgetting threshold that restricts the forgetting effect. The formula for calculating batch processing times considers forgetting effect as follows.

$$T_{[i]} = \min\left\{p\left[1+\sum_{k=i}^{N} X_{[k]}/Y_{[i]}\right]^{f_{[i]}}, w\right\} \tag{5}$$

where $X_{[i]} = \lfloor s/T_{[i+1]} \rfloor$, $Y_{[i]} = \lfloor t_B / T_{[i+1]} \rfloor$

Suppose, n units of demand are divided into N batches ($i = 1, .., N$). $T_{[1]}$ is the processing time of the i-th batch. The initial processing time is p, i.e. the processing time of those parts in the batch scheduled in the N-th position using backward scheduling approach, w is the maximum batch processing time or forgetting threshold, f is the forgetting parameter. The forgetting effect occurs due to time break in the form of setup between two consecutive batches. It causes the operator could potentially lose a chance of producing X number of parts. If it is assumed that such interruption takes place during t_B units of time, the operator will undergo a maximum forgetting effect and potentially lose his chance of producing Y number of parts. Thus, the forgetting effect in the time of X should be divided by Y. This paper assumes that an interruption takes place during setup time, so the value of C is equal to X number of parts. It can be formulated as follows.

$$C = X_{[i]} = \lfloor s/T_{[i+1]} \rfloor \tag{6}$$

Applying Eq. (4) and Eq. (6) simultaneously yields.

$$f_{[i]} = \begin{cases} \left\{ m(1-m)\log\left(Q_{[i+1]}\right) / \log\left(1+\lfloor s/T_{[i+1]} \rfloor\right) \right\}, & i = 1,2,...,N-1 \\ 0, \quad i = N \end{cases} \tag{7}$$

2.4. Processing Time with Simultaneous Effect of Learning and Forgetting

Jaber and Bonney (1996), and Arzi and Shtub (1997) show that an interruption of learning process results forgetting effect. It indicates that both learning and forgetting phenomena may happen simultaneously. Based on Eq. (2) and Eq. (5), the model of learning-forgetting functions is shown as follows.

$$T_{[i]} = \min\left\{\max\left\{p\left(\left(1+\sum_{k=i}^{N} Q_{[k+1]}\right)^{-m} + \left(1+\sum_{k=i}^{N} X_{[k]}/Y_{[i]}\right)^{f_{[i]}} - 1\right), v\right\}, w\right\} \tag{8}$$

The trend of batch processing time depends on the height of effect of learning or forgetting. If the learning effect is higher than the forgetting effect, the trend increases. On the contrary, if the forgetting effect is higher than learning effect, the trend decreases.

3. Problem Formulation

There is a single machine that processes n number of parts in N batches, and that needs a setup time between two consecutive batches. It is assumed that a company can arrange the arrivals of the parts as required and deliver the completed parts at the time that is exactly with the common due date. The processing time of parts in a batch depends on its position in the schedule due to the simultaneous effect of learning and forgetting. The objective is to minimize actual flow time. It is then modeled by using the following notations:

Decision variables

- N = number of batches
- $Q_{[i]}$ = the size of batches scheduled at the i-th position, where i=1, ..,N

Parameters

- d = due date
- n = number of demands

s = setup time of batch
p = processing time of N-th batch (initial processing time)
δ = learning rate
m = learning parameter or learning slope
f = forgetting parameter or forgetting slope
v = minimum processing time
w = maximum processing time
t_B = maximum interruption time that results maximum forgetting effect

Dependent variables

$T_{[i]}$ = batch processing time scheduled in the i-th position, i=1, .., N
$B_{[i]}$ = starting time of batch scheduled in the i-th position, i=1, .., N
$X_{[i]}$ = number of parts potentially produced during a setup of batch (s) by a processing time of batch scheduled in the i-th position ($T_{[i]}$), i=1, .., N
$Y_{[i]}$ = number of parts potentially produced during a maximum interruption time (t_B) by a processing time of batch scheduled in the i-th position ($T_{[i]}$), i=1, .., N

The Objective Function

F^a = actual flow time

The mathematical model of the problems can be written as follows.

$$\min\ F^a = \sum_{i=1}^{N}\left(\sum_{j=1}^{i}\left(s+T_{[j]}Q_{[j]}\right)-s\right)Q_{[i]} \tag{9}$$

subject to:

$$\sum_{i=1}^{N} Q_{[i]} = n, \tag{10}$$

$$\sum_{i=1}^{N} T_{[i]}Q_{[i]} + (N-1)s \le d, \tag{11}$$

$$B_{[1]} + T_{[1]}Q_{[1]} = d, \tag{12}$$

$$Q_{[i]} \ge 1 \text{ and integer,} \tag{13}$$

$$1 \le N \le n \text{ and integer,} \tag{14}$$

It is stated in Eq. (9) the objective is to minimize total actual flow time where batch processing time affected by learning and forgetting effects simultaneously. It is explained in Constraint (10) that the produced number of all parts must be equal to demands. It is shown in Constraint (11) that all parts in the batches should be proceeded within the interval time from time zero to the due date. It is stated in Constraint (12) that the completion batch scheduled in the first order must be delivered exactly at the time coinciding the common due date. It is shown in Constraint (13) that batch sizes should be larger than or equal to one and integer. It is explained in Constraint (14) that the number of batches is a positive integer between 1 and the number of demands.

4. The Solution Method

4.1 The Optimal Solution

This research proposes the optimal algorithm based on the Integer Composition method. The solution of the algorithm is searching the optimal solution of total actual flow time for all possible integer batch-size combinations. There are two steps to generate the combinations. The first step is to generate the basic combinations by using the Integer Partition method and then to re-arrange them into a one-to-one correspondence with itself by the Permutation method in the last step. For example, if there is a demand equal to five units, the combinations of batch sizes are showed in Table 1 as follows.

Table 1
The Integer Composition Method Results

Number of batches	Basic Combinations (First Step)	Integer batch-size Combinations (second Step)	Quantity of Combinations	Quantity of Elements
1	[5]	[5]	1	1
2	[4,1], [3,2]	[4,1], [1,4], [3,2], [2,3]	4	8
3	[3,1,1], [2,2,1]	[3,1,1], [1,3,1], [1,1,3], [2,2,1], [2,1,2], [1,2,2]	6	18
4	[2,1,1,1],	[2,1,1,1], [1,2,1,1], [1,1,2,1], [1,1,1,2]	4	16
5	[1,1,1,1,1]	[1,1,1,1,1]	1	5
		Total	16	48

Table 1 shows for 5 unit demands, there are 16 integer batch-size combinations and 48 elements. Shen and Evan (1996) have proved that n units demand will produce 2^{n-1} number of combinations and $(n+1)2^{n-2}$ number of elements. Table 2 shows the examples of the total quantity of batch sizes combinations and total quantity of elements for some number variety of demands.

Table 2
Testing Results for the number variety of demands

Demands (n)	The total quantity of combinations ($f(n)$)	The total quantity of elements
1	1	1
2	2	3
3	4	8
4	8	20
5	16	48
10	512	2,816
15	16,384	131.072
20	524,288	5.505.024
N	2^{n-1}	$(n+1)2^{n-2}$

Table 2 shows that both of the number of combinations and the number of elements increase in line with the increase of the number of demands. Based on Integer Composition method, It is proposed optimal algorithm (P-Algorithm) to solve the problems as follows.

[P-Algorithm]

Step 1: Input parameters n, d, p, s, δ, v, w, and t_B. Continue to Step 2.

Step 2: Generate sets of all integer batch-size combinations by using the Integer Composition Algorithm. Continue to Step 3.

Step 3: For each of the alternative solution batch sizes:

- Step 3.1: Calculate the processing time of batch ($T_{[i]}$) using Eq. (8). Continue to Step 3.2
- Step 3.2: Compute total actual flow time by using Eq. (9) with the constraints which are Eq. (10)-(14). Continue to Step 4.

Step 4: Find the solution that given minimum total actual flow time. **STOP**.

Numerical experiment for a model was conducted by giving the following input data parameters: $n = 5$, $d = 12$, $p = 0.5$, $s = 1$, $\delta = 0.9$, $v = 0.1$, $w = 1.1$, $t_B = 100$. Table 3 presents the solution of the algorithm. Table 3 shows there are 16 number of integer batch-size combinations. The minimum of total actual flow time is 10.35.

Table 3
Testing Results of the P-Algorithm

N	$Q_{[i]}$	$T_{[i]}$	$B_{[i]}$	F^a	Total F^a
1	[5]	[0.5]	[9.5]	[12.5]	12.5
2	[4,1]	[0.45, 0.5]	[0.2, 8.7]	[7.2, 3.3]	10.5
	[1,4]	[0.39, 0.5]	[11.61, 8.6]	[0.39, 13.57]	13.96
	[3,2]	[0.42, 0.5]	[10.73, 8.73]	[3.81, 6.54]	**10.35***
	[2,3]	[0.41, 0.5]	[11.19, 8.69]	[1.62, 9.93]	11.56
3	[3,1,1]	[0.42, 0.45, 0.5]	[10.73, 9.28, 7.78]	[3.81, 2.73, 4.22]	10.75
	[1,3,1]	[0.39, 0.45, 0.5]	[11.61, 9.26, 7.76]	[0.39, 8.23, 4.24]	12.86
	[1,1,3]	[0.39, 0.41, 0.5]	[11.61, 10.2, 7.7]	[0.39, 1.8, 12.89]	15.08
	[2,2,1]	[0.41, 0.45, 0.5]	[11.19, 9.29, 7.79]	[1.62, 5.42, 4.21]	11.26
	[1,1,2]	[0.39, 0.42, 0.5]	[11.61, 9.76, 7.76]	[0.39, 4.48, 8.48]	13.35
	[1,1,2]	[0.41, 0.42, 0.5]	[11.19, 9.77, 7.77]	[1.62, 2.23, 8.47]	12.32
4	[2,1,1,1]	[0.41, 0.42, 0.45, 0.5]	[11.19, 9.77, 8.32, 6.82]	[1.62, 2.23, 3.68, 5.18]	12.72
	[1,2,1,1]	[0.39, 0.42, 0.45, 0.5]	[11.61, 9.76, 8.31, 6.81]	[0.39, 4.48, 3.69, 5.19]	13.75
	[1,1,2,1]	[0.39, 0.41, 0.45, 0.5]	[11.61, 10.2, 8.3, 6.8]	[0.39, 1.8, 7.39, 5.2]	14.78
	[1,1,1,2]	[0.39, 0.41, 0.42, 0.5]	[11.61, 10.2, 8.78, 6.78]	[0.39, 1.8, 3.22, 10.44]	15.85
5	[1,1,1,1,1]	[0.39, 0.41, 0.42, 0.45, 0.5]	[11.01, 10.2, 8.78, 7.33, 5.83]	[0.39, 1.8, 3.22, 4.67, 6.17]	16.25

*: the optimal solution

The optimal number of batches (N) is 2, and batch sizes ($Q_{[i]}$) are 3 and 2. Although the algorithm can generate an optimal solution, it will take a long time to compute and also needs higher physical memory when the results are reported by computer. It is because the problems are solved in polynomial time complexity. The proof which is shown by Lemma and Proposition as follows.

Lemma 1. All feasible integer batch-size combinations in Step 2 can be solved in running time $n2^{n-1}$.

Proof. Suppose there are n unit demands has 2^{n-1} integer batch-size combinations. It is proved by Shen and Evan (1996) that the total number of combinations can be solved in running time $T(n)=n2^{n-1}$. □

Lemma 2. Batch processing time with simultaneous effect of learning and forgetting in Step 3.1 can be solved in running time $T(n)= (n+1)2^{n-2}$.

Proof. Shen and Evan (1996) prove there are $(n+1)2^{n-2}$ elements of integer batch sizes. Batch processing time with simultaneous effects of learning and forgetting can be solved in one-time complexity for each element. Thus, all element batch sizes can be solved in running time $T(n)= (n+1)2^{n-2}$. □

Lemma 3. The quantity of total actual flow time for all combinations in Step 3.2 can be solved in running time $T(n)=2^{n-1}$.

Proof. Since the total number of integer batch-size combinations is equal with 2^{n-1}, the total actual flow time for all combinations can be solved in running time $T(n)=2^{n-1}$. □

Lemma 4. The optimal solution of total actual flow time in Step 4 can be achieved in running time $T(n)=2^{n-1}$.

Proof. Since the quantity of total actual flow time for all combinations is equal with 2^{n-1}, searching the optimal solution of total actual flow time can be solved while running time $T(n)=1$. □

Proposition 1. The running time of batch scheduling problems for a single machine with the simultaneous effects of learning and forgetting to minimize total actual flow time can be calculated using formula $T(n)=(n+1)(2^{n-1})+(n+1)(2^{n-2})+1$ and it can be solved in a worst-case polynomial time complexity $O(n2^{n-1})$.

Proof. Based on Lemma 1, Lemma 2, Lemma 3, and Lemma 4, the running time of the problems can be formulated as follows.

$$T(n)=n\left(2^{n-1}\right)+(n+1)\left(2^{n-2}\right)+\left(2^{n-1}\right)+1$$

$T(n) = O(g(n))$ if and only if there is a positive real number c and a real number n_0 such that:

$$|T(n)|\le c\left|n2^{n-1}\right| \text{ where } n\ge n_0.$$

Thus, the Big-O proved in the following condition as follows.

$$\begin{aligned}\left|n\left(2^{n-1}\right)+n\left(2^{n-2}\right)+2^{n-2}+\left(2^{n-1}\right)+1\right| &\le n\left(2^{n-1}\right)+\left|n2^{n-2}\right|+\left|2^{n-2}\right|+\left|2^{n-1}\right|+|1|\\ &\le n\left(2^{n-1}\right)+n\left(2^{n-1}\right)+n\left(2^{n-1}\right)+n\left(2^{n-1}\right)+n\left(2^{n-1}\right)\le 5\left(n2^{n-1}\right)\end{aligned}$$

If $n_0=1$ then there is exists a positive real number c, that is $c=5$ for all $n\ge n_0$ □

4.2 The Heuristic Solution

Since the problems proved by using Preposition 1, then it can be solved in a worst-case polynomial time complexity $O(n2^{n-1})$, this research develops the heuristic solution based on the Lagrange Relaxation method. The decision variables are the number of batches, batch sizes and the sequence of the resulting batches.

4.2.1. Determining Batch Sizes

The current research assumes that the batch sizes are integer. It is also considered that the processing time of parts in a batch scheduled at a position is different from that at the next position in backward scheduling approach. The learning and forgetting effect are started from the last position of batches to the first. It is proposed a backward computation for determining the batch sizes, that is, the calculation of the batch sizes are started from the last number of index to the first (i=N,..,1).

Halim et al. (1994) propose the formula for calculating the optimal batch sizes for single machine problems by using the Lagrange Relaxation as follows.

$$Q_{[i]}=(n/N)+(1/2)(N+1)(s/t)-(s/t)i,\quad i=1,\dots,N \tag{15}$$

Eq. (15) can be re-written by using the backward computation as shown in the following equation.

$$Q_{[i]}=\left(n-\sum_{k=i+1}^{N}Q_k\right)/i+(1/2)(i+1)(s/t)-(s/t)i,\quad i=N,\dots,1 \tag{16}$$

Substitute t in Eq. (16) by $T_{[i]}$ in Eq. (8) yields:

$$Q_{[i]}=\left(n-\sum_{k=i+1}^{N}Q_k\right)/i+(1/2)(i+1)\left(s/T_{[i]}\right)-\left(s/T_{[i]}\right)i,\quad i=N,\dots,1 \tag{17}$$

where $T_{[i]}=\min\left\{\max\left\{p\left(\left(1+\sum_{k=i}^{N}Q_{[k+1]}\right)^{-m}+\left(1+\sum_{k=i}^{N}X_{[k]}/Y_{[i]}\right)^{f_{[i]}}-1\right),v\right\},w\right\}$

Applying Constraint (13) and Eq. (17) simultaneously yields:

$$Q_{[i]}=\max\left\{\left\lceil\left(n-\sum_{k=i+1}^{N}Q_k\right)/i+(1/2)(i+1)\left(s/T_{[i]}\right)-\left(s/T_{[i]}\right)i\right\rceil,\ 1\right\},\quad i=N,\dots,1 \tag{18}$$

4.2.2 Determining the Maximum Number of Batches

The optimal number of batches cannot be determined precisely because batch processing time depend on its position on a schedule. The maximum number of batches can be calculated by assuming that all parts in all batches are processed on a machine by the minimum value of batch processing time. According to Constraints (11) and (14), the maximum number of batches (Nmax) can be calculated as follows.

$$N\max = \min\left\{\left\lfloor \left(d - nT_{\min}\right)/s \right\rfloor + 1, n\right\} \tag{19}$$

T minimum can be relaxed using the formula:

$$T\min \approx \max\left\{p\left(1+n\right)^{-m}, v\right\} \tag{20}$$

Eq. (20) shows that minimum batch processing time is found when all parts are proceeded together and no forgetting effect. Substitute Tmin in Eq. (19) by Eq. (20) yields:

$$N\max = \min\left\{\left(\left\lfloor \left(d - n\max\left\{p\left(1+n\right)^{-m}, v\right\}\right)/s \right\rfloor + 1\right), n\right\} \tag{21}$$

Based on Eq. (14) and Eq. (21), it also can be observed that the optimal value of total actual flow time is within the closed interval [1, Nmax]. Lemma says as follows.

Lemma 5. If the objective function of total actual flow time is convex in a closed interval [1, Nmax], then the optimal value of total actual flow time is within the interval.

Proof. The optimal number of batches should be convex in a closed interval [1, Nmax] if and only if Hessian Matrix is a definite positive. f: $R^n \rightarrow R$ convex function $\leftrightarrow$ $H_f(x)$ is a definite positive.

Based on Eq. (8), $T_{[i]}$ is the function of Q, s, t_B, f, and δ. It is defined dependent variable $T_{[i]}$ as a parameter, i.e $T_{[i]}$ as t. Eq. (10) can be re-written as follows.

$$f\left(x\right) = F^a = s\sum_{i=1}^{N} iQ_{[i]} + (1/2)t\sum_{i=1}^{N} Q_{[i]}^2 + (1/2)t\left(\sum_{i=1}^{N} Q_{[i]}\right)^2 - s\sum_{i=1}^{N} Q_{[i]}$$

Gradient: $\partial F / \partial Q = si + tQ_{[i]} + t\left(\sum_{i=1}^{N} Q_{[i]}\right) - s$

$$\partial F / \partial N = sNQ_{[N]} + (1/2)tQ_{[N]}^2 - sQ_{[N]}$$

Hessian Matrix: $H_f\left(x\right) = \nabla^2 f\left(x\right) = \begin{pmatrix} t & 0 \\ 0 & sQ_{[N]} \end{pmatrix}$

Using Silvester Criterion yields: $\det\begin{vmatrix} t & 0 \\ 0 & sQ_{[N]} \end{vmatrix} \rightarrow tsQ_{[N]} > 0$

Hessian Matrix is a definite positive. Therefore, the objective function is convex. The optimal number of batches is within the interval of [1, Nmax]. □

4.2.3 Determining Sequence of the Resulting Batches

Proposition 2. Minimizing total actual flow time in batch scheduling problems on single-machine with processing time affected by simultaneous effects of learning and forgetting can be obtained by sequencing the resulting batches by using the LPT rule in the backward scheduling approach.

Proof. There are two feasible schedules for N batches. The first schedule places batch of (i)-th position at the (i)-th position and batch of (i+1)-th position to the (i+1)-th position. The last schedule is only different from the first schedule in batch of (i)-th position at the (i+1)-th position and batch of (i+1)-th position at the (i)-th position. Both of them are arranged by using backward scheduling approach. If F^{a1} and F^{a2} are the value of total actual flow time for the first schedule and the second respectively, the following is obtained

$$F^{a1} - F^{a2} = \left(T_{[i]}Q_{[i]}Q_{[i+1]} + sQ_{[i+1]}\right) - \left(T_{[i+1]}Q_{[i+1]}Q_{[i]} + sQ_{[i]}\right)$$

The value of total actual flow time at the first schedule will be less than or equal to that value at the second schedule if only if

$$\left(T_{[i]}Q_{[i]} + s\right)Q_{[i+1]} \leq \left(T_{[i+1]}Q_{[i+1]} + s\right)Q_{[i]} \text{ or } T_{[i]} + s/Q_{[i]} \leq T_{[i+1]} + s/Q_{[i+1]}$$

The value of total actual flow time at the left side will be less than the right side if the batch size at the left side is greater than the right side. Therefore, minimizing total actual flow time can be obtained by arranging the batches in order to the LPT rule in the backward scheduling approach.

$$\left(T_{[1]}Q_{[1]} + s\right)/Q_{[1]} \leq \left(T_{[2]}Q_{[2]} + s\right)/Q_{[2]} \leq \ldots \leq \left(T_{[N]}Q_{[N]} + s\right)/Q_{[N]} . \square$$

Based on the Lagrange Relaxation method, it is proposed heuristic algorithm (H-Algorithm) to solve the problems as follows.

[H-Algorithm]

Step 1: Input parameters $n, d, p, s, \delta, v, w,$ and t_B. Continue to Step 2.

Step 2: Find Nmax by Eq. (21). Determine j as an index for the number of batches (j= 1, …, Nmax). Continue to Step 3.

Step 3: Begin with j = 1. Determine $Q_{[1]} = n$, $T_{[1]} = p$. Continue to Step 4.

Step 4: Calculate the value of $M = np$. If $M \leq d$, then continue to Step 5, otherwise **STOP** because the demand cannot be scheduled.

Step 5: Calculate total actual flow time for j =1 by Eq. (9). Determine the value of $F_{[1]}$ as F^*. Continue to Step 6.

Step 6: Determine the index of the number of batches j=j+1. If $j < N_{max}$, continue to Step 7, otherwise determine F^* as the best solution and then **STOP**.

Step 7: Determine N=j. Set $Q_{[N+1]}$=0, $T_{[N+1]}$=0. Continue to Step 8.

Step 8: Calculate the processing time of batch ($T_{[i]}$) using Eq. (8) and the batch sizes ($Q_{[i]}$) using Eq. (18). Continue to Step 9.

Step 9: Based on Proposition 2, sequence the resulting batches in a non-increasing size of $Q_{[i]}$ in backward scheduling approach. Continue to Step 10.

Step 10: Calculate $M = \sum_{i=1}^{N} T_{[i]}Q_{[i]} + (N-1)s$. If $M \leq d$, continue to Step 11, otherwise determine F^* as the best solution and then **STOP**.

Step 11: Compute total actual flow time for j by Eq. (9). Continue to Step 12.

Step 12: Compare the value of total actual flow time $F_{[j]}$ and F^*. If $F_{[j]} < F^*$, determine $F_{[j]}$ as F^* and then return to Step 6, otherwise determine F^* as the best solution and then **STOP**.

Numerical experiment test was conducted by giving the following input data parameters: n = 5, d = 12, p = 0.5, s = 1, δ = 0.9, v = 0.1, w = 1.1, t_B = 100. The result of processing by the proposed algorithm yields Nmax = 5. Table 4 presents the solution of the algorithm. Table 4 shows that the solution searching method begins with the number of batches (N) that equals 1. The result of total actual flow time is 12.5. The search of the solution is increased for N=2. The result of the total actual flow time is 10.35. The total actual flow time for N=2 is compared with N=1, and it is found that the total actual flow time for N=2 is smaller than N=1.

Table 4
Testing Results of the H-Algorithm

N	i	$Q_{[i]}$	$T_{[i]}$	$B_{[i]}$	Total F^a
1	1	5	0.5	9.5	12.5
2	1	3	0.42	10.73	**10.35***
	2	2	0.5	8.73	
3	1	3	0.42	10.73	10.75
	2	1	0.45	9.28	
	3	1	0.5	7.78	

*: the best solution

Thus, the total actual flow time for N=2 is temporarily best solution. The search of the solution continues for N=3. The result of the total actual flow time is 10.75. The total actual flow time for N=3 is compared with N=2, and it is found that the total actual flow time for N=2 is smaller than N=3. Thus, the algorithm is stopped by determining that the total actual flow time is 10.35 as the best solution obtained at N=2. For this case, the best solution of the H-Algorithm is equal to the optimal solution of the P-Algorithm.

4.3 Comparison test

The tests are continued by testing the optimality of both proposed algorithms. They are carried out by comparing the solution of the P-Algorithm and the H-Algorithm. The P-Algorithm and the H-Algorithm are coded in C++, and the results are reported by using computer processor Intel Core i3 with 6 GB of Random Access Memory (2 GB usable in 64-bit of platform) and 1 GB of Graphical Processor Unit. The comparison tests have been done on more than 100 problems. The results of comparison test show that the H-algorithm works adequately for any numbers of demands. However, it does not guarantee to produce an optimal solution. Meanwhile, the P-Algorithm guarantee to produce the optimal solution but it is only adequate for demands less than 23 units (small size problems). If there are more than 23 unit demands, the solution can not be found because of out-of-memory (o.o.m). In order to make well understanding of the behavior of comparison tests, Table 5 shows the examples of comparison tests for 10 problems where input parameters are generated randomly.

Table 5
The result of comparison test between the H-Algorithm and the P-Algorithm

Problems		H-Algorithm				P-Algorithm			
No	Demand	Total Actual Flow Time minimum	Time to compute (minute)	Memory usage	Optimal Sequencing Rule	Total Actual Flow Time minimum	Time to compute (minute)	Memory usage	Optimal Sequencing Rule
1	4	10.956	< 0.1	1 KB	LPT	10.956	< 0.1	4 KB	LPT
2	6	26.219	< 0.1	2 KB	LPT	26.219	< 0.1	13 KB	LPT
3	8	53.939	< 0.1	2 KB	LPT	53.939	< 0.1	56 KB	LPT
4	10	70.179	< 0.1	3 KB	LPT	70.179	< 0.1	244 KB	LPT
5	13	74.992	< 0.1	8 KB	LPT	74.970	< 0.1	2.2 MB	LPT
6	16	167.152	< 0.1	2 KB	LPT	167.152	0.52	20 MB	LPT
7	18	206.009	< 0.1	5 KB	LPT	206.009	0.9	85 MB	LPT
8	20	191.752	< 0.1	9 KB	LPT	191.752	0.42	370 MB	LPT
9	22	218.225	< 0.1	6 KB	LPT	218.225	1.79	1.57 GB	LPT
10	24	320.485	< 0.1	2 KB	LPT	N.a.N	N.A	o.o.m	LPT
Average variance solution of total actual flow time						0.2%			

Table 5 shows that for some cases, the total actual flow time of the H-Algorithm has different results from the optimal solutions produced by the P-Algorithm. The average difference is about 0.2%. It proves that the H-Algorithm does not guarantee to produce an optimal solution. However, the H-Algorithm is very useful to solve the problems. In addition, the optimal solution that is generated by the P-Algorithm and the H-Algorithm becomes the proof of the Proposition 2 that the minimum of total actual flow time can be obtained by sequencing the resulting batches using the LPT rule in the backward scheduling approach. It also shows that the H-Algorithm takes time to compute, and uses memory less than the P-Algorithm.

5. Concluding Remarks

The current research deals with single-machine batch scheduling problems with the simultaneous effect of learning and forgetting to minimize total actual flow time. The problems are proved by using the Integer Composition method can be solved in a worst-case polynomial time complexity $O(n2^{n-1})$. This paper proposes two algorithms to solve these problems. The first algorithm produces the optimal solution that is developed by using the Integer Composition method. The second algorithm is the heuristic algorithm that is developed based on the Lagrange Relaxation method. Although the heuristic algorithm does not guarantee to produce an optimal solution, numerical experiments on randomly generated examples show that the heuristic algorithm gives outstanding results. The model introduced in this paper can be extended to discuss batch scheduling problems with learning and forgetting effects, as well as deterioration simultaneously. In a real system, it is also known that the changes in processing times occur not only on the operators but also on the machines.

Acknowledgments

The authors wish to thank the editors and the referees for their constructive comments and suggestions.

References

Anzanello, M.J. & Foglianto, F. S. (2010). Scheduling learning dependent jobs in customized assembly lines. *International Journal of Production Research*, 48, 6683-6699.

Arzi, Y. & Shtub, A. (1997). Learning and forgetting in mental and mechanical tasks: A comparative study. *IIE Transactions*, 29 759-768.

Baker, K.R. & Trietsch, D. (2009). *Principles of sequencing and scheduling*. John and Wiley & Sons.

Biskup, D. (1999). Single machine scheduling with learning considerations. *European Journal of Operational Research*, 115, 173-178.

Carlson, J. G. & Rowe, A. J. (1976). How much does forgetting cost?, *Industrial Engineering*, 8 40-47.

Chen, C.K., Lo, C.C. & Liao, Y.X. (2008). Optimal lot size with learning consideration on an imperfect production system with allowable shortages. *International Journal of Production Economics*, 113, 459-469.

Cheng, T.C.E, Lee, W.C. & Wu, C. C. (2010). Scheduling problems with deteriorating jobs and learning effects including proportional setup times. *Computers & Industrial Engineering*, 58, 326–331.

Cheng, T.C.E. (1991). An EOQ model with learning effect on set-ups. *Production and Inventory Management*, 32, 83-84.

Cheng, T.C.E. (1994). An economic manufacturing quantity model with learning effect. *International Journal of Production Economics*, 33, 257-264.

Cheng, T.C.E., Kuo, W.H. & Yang, D.L. (2013). Scheduling with a position-weighted learning effect based on sum-of-logarithm-processing-times and job position. *Information Sciences*, 221, 490–500.

Cheng, T.C.E., Lai, P.J., Wu, C.C. & Lee, W.C. (2009). Single-machine scheduling with sum-of-logarithm-processing-times-based learning considerations. *Information Sciences*, 179, 3127–3135.

Cheng, T.C.E., Wu, W.H., Cheng, S.R. & Wu, C.C. (2011). Two-agent scheduling with position-based deteriorating jobs and learning effects. *Applied Mathematics and Computation*, 217, 8804–8824.

Chiu, H.N. & Chen, H.M. (2005). An optimal algorithm for solving the dynamic lot-sizing model with learning and forgetting in set-ups and production. *International Journal of Production Economics*, 95, 179-193.

Chiu, H.N. (1997). Discrete-time varying demand lot-size models with learning and forgetting effect. *Production Planning Control*, 8, 484-493.

Chiu, H.N., Chen, H.M. & Weng, L.C. (2003). Deterministic-time varying demand lot-size models with learning and forgetting in set-ups and production. *Production Operation Management*, 12, 120-127.

Dobson, G., Karmakar, U. S. & Rummel, J. (1987). Batching to minimize flow times on one machine. *Management Science*, 33, 784-799.

Gaweijnowics, S. (1996). A note on scheduling on a single processor with speed dependent on a number of executed jobs. *Information Processing Letters*, 57, 297-300.

Halim, A.H. & Ohta, H. (1994). Batch Scheduling Problems to minimize inventory cost in the shop with both receiving and delivery Just-In-Time. *International Journal of Production Economics*, 33, 185-194.

Halim, A.H., Miyazaki, S. & Ohta, H. (1994). Batch scheduling problems to minimize actual flow times of component through the shop under JIT environment. *European Journal of Operational Research*, 72, 529-544.

Jaber, M.Y. (2011). *Learning Curves: Theory, model, and applications*, Taylor & Francis Group, US.

Jaber, M.Y. & Bonney, M. (1996). Production break and the learning curve: The forgetting phenomenon. *Applied Mathematical Modelling*, 20, 162-169.

Jaber, M.Y. & Salamah, M.K. (1995). Optimal lot sizing under learning considerations: Shortages allowed and back ordered. *Applied Mathematical Modeling*, 19, 307-310.

Janiak, A., Lichtensteind, M. & Rusoń, A. (2011). Scheduling jobs with a linear model of simultaneous ageing and learning effects. *Decision Making in Manufacturing and Services*, 5, 37–48.

Ji. M. & Cheng, T.C.E. (2010). Batch scheduling of simple linear deteriorating jobs on a single machine to minimize makespan. *European Journal of Operational Research*, 202, 90-98.

Keachie, E.C. & Fontana, R.J. (1966). Effect of learning on optimal lot size. *Management Science*, 13, B102-B108.

Kuo, W.H. & Yang, D.L. (2006), Minimizing the total completion time in a single-machine scheduling problem with a time-dependent learning effect. *European Journal of Operational Research*, 174, 1184-1190.

Lai, P.J. & Lee, W.C. (2013). Single machine scheduling with learning and forgetting effect. *Applied Mathematical Modelling*, 37, 4509-4516.

Li, C.L. & Cheng, T.C.E. (1994). An economic production quantity model with learning and forgetting considerations. *Production and Operations Management*, 3, 118-132.

Morton, T.E. & Pentico, D.W. (1993). *Heuristic scheduling systems*. John Wiley & Sons, Inc.

Nembhard, D.A. & Uzumeri, M.V. (2000). Experiential learning and forgetting for manual and cognitive tasks. *International Journal of Industrial Ergonomics*, 25, 315–326.

Pinedo, M. (2002). *Scheduling: Theory, algorithms, and systems*. Prentice Hall, New Jersey.

Shen, H. & Evans. D. J. (1996). A new method for generating integer compositions in parallel. *Parallel Algorithms Applications*, 9, 101-109.

Steedman, I. (1970). Some improvement curve theory. *International Journal Production Research*, 8, 189-206.

Teyarachakul, S., Chand, S. & Ward, J. (2008). Batch sizing under learning and forgetting: Steady state characteristics for the constant demand case. *Operations Research Letters*, 36, 589-593.

Teyarachakul, S., Chand, S. & Ward, J. (2011). Effect of learning and forgetting on batch sizes. *Production and Operations Management*, 20, 116-128.

Wang, X. & Cheng, T.C.E., (2007). Single machine scheduling with deteriorating jobs and learning effects to minimize the makespan. *European Journal of Operational Research*, 178, 57 -70.

Wright, T.P. (1936). Factors affecting the cost of airplanes. *Journal of Aeronautical Sciences*, 3, 122-128.

Wu, C.H., Lai, P. J. & Lee, W. C. (2014). A note on a single machine scheduling with sum of processing time based on learning and forgetting effects. *Applied Mathematical Modelling*, 39, 415-424.

Yang, W.H. & Chand, S. (2008). Learning and forgetting effect on a group scheduling problem. *European Journal of Operational Research*, 187, 1033-1044.

11

Application of imperialist competitive optimization algorithm in power industry

Mohammad Shahrazad and Amir Hessam Alikhanzadeh*

Department of Electrical Engineering, Brunel University, London, UK

CHRONICLE

Keywords:
Imperialist Competitive Algorithm
FACTS Devices
Optimization
Evolutionary Algorithms
Total Transfer Capability

ABSTRACT

In future electricity industry transferring high quality of power is essential. In this case, using Flexible AC Transmission System (FACTS) devices is inevitable. FACTS devices are used for controlling the voltage, stability, power flow and security of transmission lines. Therefore, finding the optimal locations for these devices in power networks is necessary. There are several varieties of FACTS devices with different characteristics, deployed for different purposes. Imperialist Competitive (IC) algorithm is a recently developed optimization technique, applied in power systems. IC algorithm is a new heuristic approach for global optimization searches based on the concept of imperialistic competition. In this paper, an IEEE 4-bus system is deployed as a case study in order to demonstrate the results of this novel approach using MATLAB.

1. Nomenclature

θ	Approaching angle of a colony towards an empire,
γ	Gamma (open parameter),
ζ	Zeta (constant co-efficient),
β	Direction co-efficient,
B	Total Line Charging Susceptance (p. u.),
d	Distance,
Bus_I	Bus number (positive integer),
Bus_Type	Bus Type (1 = PQ, 2 = PV, 3 = ref, 4 = isolated),
F_Bus	"from" Bus Number,
P_D	Real Power Demand (MW),
P_G	Real power output (MW),
Q_D	Reactive Power Demand (MVar),
Q_G	Reactive power output (MVar),
V_A	Voltage Angle (Degrees),
V_M	Voltage Magnitude (p. u.),
R	Resistance (p. u.),

* Corresponding author.

T_Bus	"to" Bus Number,
Tap	Transformer off nominal turns ratio,
TTC	Total Transfer Capability,
X	Reactance (p. u.).

2. Introduction

In both present and future power systems, transmission network plays significant role in transporting electricity within large distances over national and international borders (Krishans et al., 2007). In recent years by integrating Renewable Energy Sources (RES) in power systems a range of issues such as efficiency, controllability and security have become significantly important. According to the electricity demand growing, these parameters are developing and every year it can be observed that new technologies and methods are introduced in order to minimize the losses and increase the quality of power networks. Principally, power system can transfer the maximum electricity if network can be controlled accurately. In this case, by deploying FACTS devices the transfer capability in power systems can be enhanced.

Rapid developments in power electronics have made different types of FACTS for different purposes. These devices are flexible and assist system operator to improve the power transfer and stability (Belsnes et al., 2011). Considering the existence of high competition in the electricity market, using FACTS devices is essential to achieve the maximum power transfer. In this case, both the *amount* and *type* of FACTS devices should be computed to reach the maximum efficiency within a stable network. There are several advantages using FACTS devices in power networks such as (Saravanan et al., 2007):

- Power transfer control,
- Stability control,
- Voltage collapse prevention,
- Power quality improvement,
- Power factor improvement,
- Voltage profile improvement,
- Dynamic stability improvement,
- Load-ability of the transmission lines enhancement, and
- Decreasing in the reactive power losses as a result of increasing the active power transfer.

Allocating and estimating the types and sizes of FACTS devices are the most important area in the transmission network. Indeed, with deploying optimization techniques, maximum efficiency from power networks can be obtained without installing new generators (Saravanan et al., 2007). Based on the FACTS devices' prices and their financial aspects, it is essential to allocate these devices in the optimal locations. For this purpose, initially a mathematical model of the FACTS devices should be considered and then, by using an optimization technique they can be allocated in optimal places. Allocating the FACTS devices is a combinatorial optimization in which both the location and size of these devices should be determined to obtain maximum savings in the power systems investments (Chansareewittaya & Jirapong, 2010). The aim of this study is to find the best and optimum solution for the proposed problems. In this case a range of evolutionary algorithms have been deployed, which help to achieve the solution. More details are provided in Section 3. Classic mathematic methods have got a range of disadvantages (Hong-Bo & Shuxia, 2008), since some of these methods consider local optimum as a final answer of the problem.

A novel optimization method called Imperialist Competitive (IC) method is introduced in Section 4. This optimization algorithm is based on the human evolution. The key feature of this evolutionary algorithm comparing to other techniques such as Genetic Algorithm (GA) and Particle Swarm

Optimization (PSO) is that it finds the optimal placing for the FACTS devices alongside the type and amount for each of them (Shahrazad & Zobaa, 2013).

Section 5 presents different objective functions with various numbers of constraints that have been taken into consideration in order to illustrate the application of IC algorithm in power industry. It is obvious that by using different objective functions different solutions can be obtained. This is because of the behavior of each objective function. Introducing a range of constrains such as: voltage profile and FACTS compensator sizes is compulsory in the optimization techniques. IC algorithm has essentially proved that it is faster than the other optimization techniques such as GA and PSO. This algorithm can approach the solution with less number of iterations. In this study the IEEE 4-bus system (Deep & Bansal, 2009) is analyzed and results are presented in Section 6. Finally, Section 7 presents the concluding remarks and future steps.

3. Study Background

The aim of deploying power electronic devices in power industry is to enhance the capability of the system in terms of transferring high quality of power to the demand side. FACTS are based on the power electronics converters. With this regard, they are able to create rapid adjustment to control the electrical components. These devices are able to join and connect to the power systems in parallel, series, or a combination of them. Installing extra transmission lines is not always an appropriate solution to increase the power transfer capability. By applying FACTS devices the power transfer and thermal limits of the transmission lines can be controlled in order to assure they are working within their limits.

One of the major parameters, which has special role in operation of both present and future electricity highway systems (EHS), is the power transfer capability (Chansareewittaya & Jirapong, 2010). Power transfer capability can be defined as an ability of the power network in terms of transferring the power in a specific condition without disturbing the constraints and security of the power systems. This condition depends on the physical constraints and financial issues of the transmission systems. Regarding physical constraints, major issues are related to the security of the power systems. Power transfer constraints can be categorized as a thermal limit of the transmission lines, voltage constraints at bus bars and stability of the power systems (Avramenko, 2005). In this case, power transfer capability of the system has a direct relationship with the economic aspect of the network.

3.1 Available Transfer Capability

Available Transfer Capability (ATC) is one of the major functions of the power transfer capability. Principally, it can be defined as the available power transferred in the transmission lines for achieving more profit. This parameter can be obtained by deducting the Transmission Reliability Margin (TRM), Capacity Benefit Margin (CBM), from Total Transfer Capability (TTC) (Rajabi-Ghahnavieh et al., 2009). ATC can be identified as follows:

$$ATC = TTC - TRM - Existing\ Transmission\ Commitments\ (including\ CBM) \quad (1)$$

Transfer capacity and transfer capability have different expressions in the power systems. Transfer capacity represents the loading of the transmission network components such as thermal limit, which is constant and depends on the physical characteristics of the electrical components. However, transfer capability covers mostly the whole system. For maintaining security of supply, this parameter should be calculated and it is essential to assure that this value does not exceed a certain amount (Takasaki, 2006). The power transfer capability between two areas, connected by transmission lines, is greater than the power transfer ability between those areas. This is because of the existence of both physical and electrical limitations of the components in the power network. In spite of transfer capacity, transfer capability is an oriented vector. This means the amount of power transfer capability from area A to B is not necessarily equal to the amount of power transfer capability from B to A. FACTS devices can enhance the power transfer capability. In this case, it is vital to examine the impact of FACTS devices in the market during the total transfer capability assessment.

To obtain the total transfer capability all initial conditions of the power system such as estimating the load, distributing the load between the generators, schedule of transfer and operating point of the system should be identified. Total transfer capability is determined according to thermal, voltage, and stability constraints. These constraints can be changed due to different situation such as breakdown of components.

3.2 FACTS

FACTS devices are not able to prevent the fault in the system; however, they are able to reduce the inconvenient condition, which occurs due to any fault in the system. For example, cutting off one load from system causes line voltage raise. In this case, FACTS compensators eliminate the extra voltage from the lines and make them stable. Future smart grid operation needs high quality of power. Using FACTS devices decreases the losses and creates high quality of power to the customers. In general FACTS can be divided into four categories:

- Parallel devices
- Series devices
- Series-Series devices
- Series-Parallel devices

List of FACTS devices introduced in this paper are as follows (Tabatabaeiet al., 2011):

3.2.1 SVC

Static VAR Compensator (SVC) is an electrical component deployed in high voltage transmission networks. SVC is the parallel compensator, which can be used as a capacitor and variable inductor according in power networks. In fact, SVC is the components that can control the voltage. Therefore, SVC is a kind of generator with zero active power, which compensates voltage at the connected bus-bar.

3.2.2 TCSC

Thyristor-Controlled Series Compensator (TCSC) is a series compensator component, which is used as a capacitor and inductor element in the transmission lines. This device increases the dynamic stability of the power transmission lines and improves the load sharing between the transmission networks.

Table 1 illustrates the characteristics of various types of FACTS devices in order to provide a general overview of their applications in the power systems.

Table 1.
FACTS Devices Specifications

Type	Power Flow Control	Voltage Control	Dynamic Stability	Transient Stability
SVC	✓	✓✓✓	✓✓	✓
STATCOM	✓	✓	✓✓	✓✓
TCSC	✓✓	✓✓✓	✓✓	✓✓✓
UPFC	✓✓✓	✓✓✓	✓✓	✓✓

3.3 Evolutionary Algorithms

Optimization method generally is divided into two categories; local optimization and global optimization techniques (Minasian & Bird, 2013). In general, metaheuristic optimization techniques tend to obtain the global solution; however, they are not always successful or efficient according to the size of the problem. In this case, they may achieve a local optimal solution. These algorithms include GA, PSO, Simulated Annealing, and etc. These algorithms have widely applied in the power systems in order to determine the optimal number and location of FACTS devices (Cai et al., 2004; Majumdar et al., 2006). As presented in (Gerbex et al., 2001) once the number of FACTS devices reaches to the optimal value, installing more FACTS cannot improve the load-ability of the power system.

According to (Panda & Padhy, 2008) both PSO and GA are inspired by nature, and they can be deployed in optimizing the parameters of a FACTS-based controller. However, PSO has better evolutionary characteristics as it can achieve to the final parameter values in fewer generation comparing to the GA. Considering the computational time, GA can have better performance and achieve to the optimal solution with less iterations.

In fact, many optimization algorithms are based on the nature process. However, considering the social and historical evolution of human have not been considered yet. In fact, this process can be defined as an algorithm that can be faster than other optimization algorithms like GA. More details will be provided in the Section 3. This type of algorithm has considered some part of social-politic of human and by converting this process as a mathematical equation different problems can be solved. This algorithm is also deployed for designing controllers for industrial systems, designing intelligent systems, and finding solution for management problems. Table 2 presents the advantages of IC algorithm over the other evolutionary algorithms such as GA and PSO.

Table 2
FACTS Devices Specifications

Method	Criteria					
	Speed of Convergence	Global Search Ability	Number of Iterations	Solution Reliability	Cost Function Value	Accurancy of Optimal Value
GA	✓✓	✓✓	✓✓	✓✓✓	✓✓✓	✓✓✓
PSO	✓✓✓	✓✓	✓✓✓	✓✓	✓✓	✓✓
IC	✓✓✓	✓✓	✓✓	✓✓	✓✓	✓✓✓

4. Methodology

This algorithm starts with initial selections, which are called *countries*. They can be the answers for the proposed problem. These countries are as the same as chromosomes in the genetic algorithm and particles in the particle swarm optimization. All these countries are divided into two main groups, *imperialist* and *colony* (Atashpaz-Gargari & Lucas, 2007). After several iterations the algorithm will converge and obtain optimum solution for the problem. This algorithm is based on human evolution history. In this study, the IC algorithm is implemented on a case study with six different analysis and all related results are demonstrated in Section 6.

4.1 Imperialist Competitive Algorithm Concept

This algorithm is based on assimilation, imperialistic competition and revolution (Shahrazad & Zobaa, 2013). IC starts with initial random answer, which in this case called country. Some of these countries are selected in term of less cost function and other main positive parameters. This group is called *imperialist* and the rest of other countries called *colonies*. Imperialists will grab the colonies according to their power. After initialization, imperialist competition will be in process. In this case, if one imperialist could not success in competition or increasing its own power, it will be eliminated automatically from the competition. Therefore, each empire will not be collapse unless it grabs more colonies. In this process some of the empires will grow and become stronger and some of them collapse and will be eliminated from the competition. This process results in developing empires. Consequently, only one empire will remain and all other colonies will join the last empire. In this case, a kind of convergence can be achieved.

4.2 Imperialist Competitive Algorithm Implementation

The countries in the IC algorithm can be defined as follow (Rajabioun et al., 2008):

$$Country = [p_1, p_2, p_3 \dots p_N] \tag{2}$$

Considering the social-political aspects, the countries can be defined with several features such as culture, language, structure of economy, and etc. Fig. 1 illustrates this performance clearly.

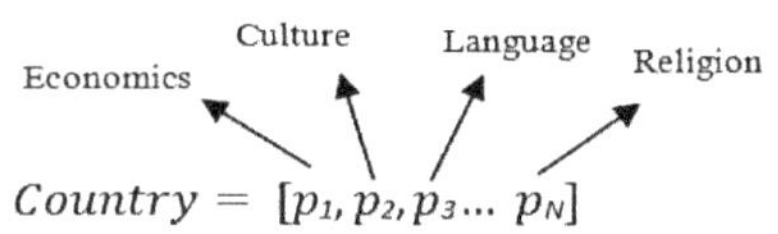

Fig. 1. Characteristics of the Function Variables

During optimization problem, the IC algorithm is looking for the best country with the best features. Realistically, the algorithm needs to find the country with less cost function. Cost of these countries is defined as below:

$$Cost = f\,(country) = f\,(p_1, p_2, p_3 \dots p_N) \tag{3}$$

Accordingly, this algorithm looks for the best country and then by making assimilation and imperialistic competition different colonies will be grabbed. Initially, $N_{country}$ countries are created and N_{imp} countries from the best countries which have less cost functions are selected as imperialists. The rest, N_{col}, are the colonies distributed between the empires. Having all the empires, the cost function can be defined as follow:

$$C_n = c_n - max\,\{c_i\} \tag{4}$$

where, c_n is the cost function of nth imperialist, $max\,\{c_i\}$ is the maximum cost between the imperialists, and C_n is the normalized cost for the imperialist. Therefore, the empires with more cost are weak empires and will have less normalized costs. According to normalized cost, normalized power of each empire can be defined as follow:

$$p_n = \left| \frac{Cn}{\sum_{i=1}^{Nimp} Ci} \right| \tag{5}$$

On the other hand, power of one empire depends on the colonies in its authority. Hence, the initial number of colonies of an empire will be:

$$N.C._n = round\,\{p_n, N_{col}\} \tag{6}$$

where $N.C._n$ is the initial colonies of one empire, N_{col} is the total colonies that exist, and $round$ is the function which gives an integer value in case of having a decimal value. With considering the initial position of the all empires, the IC starts. At this stage, the empires try to gather all their colonies in term of different states. Fig. 2 indicates a schematic movement of a colony towards an empire.

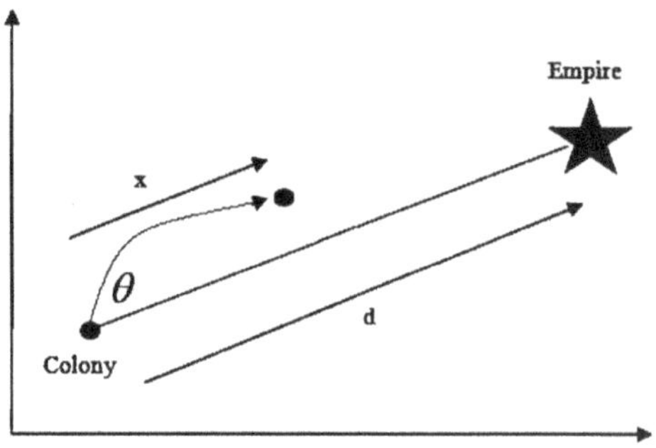

Fig. 2. Movement Process

According to Fig. 2, it can be seen that the colony starts to move towards the empire by x and reaches the new position. The total distance between empire and colony is showed by d. The value X is also a random value that follows the below relation:

$$X \sim U\ (0, \beta \times d) \tag{7}$$

where β is a parameter greater than 1 and close to 2 according to different analysis and simulations. In this case, the β value can be assumed to be equal to 2 in order to achieve better results. When $\beta > 1$, it can be seen that the colony moves towards the empire in different sides (Rajabioun et al., 2008). By investigating historical data it can be observed that during the assimilation between the imperialists all the expected policies were not according to their favorite plans and sometimes there was a deviation (Rajabioun et al., 2008). This deviation can be define as θ. With this regard, the colonies movement towards the empires might be done with deviation angle as illustrated in Fig. 3.

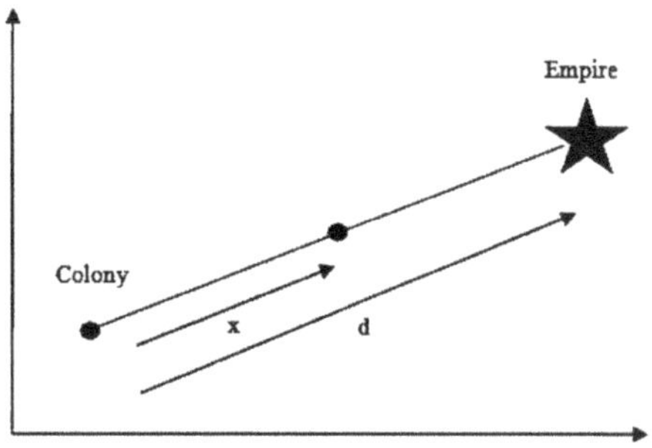

Fig. 3. Colony Movement towards an Empire with Angle

θ is also a random value and follows the below relation:

$$\theta \sim U\ (-\gamma, \gamma) \tag{8}$$

γ is the open parameter. By selecting high value of γ the colony searches more places around the empire and by choosing small value it will make short distance colony to reach the empire. $\pi/4$ has been selected as the optimum value (Shahrazad & Zobaa, 2013). It is important to note that in some cases during the colonies movements towards empires, the colonies may obtain better places than empires. This means that they reach points where the cost function is lower than the empires'. In this condition, colony and empire swap their places and the algorithm continues with new empire. This time the new empire starts to do the assimilation and grab the colonies.

The total power of imperialist is equal to the power of empire plus some percentage of its colonies power. Hence, the total power of the imperialist is defined as follow:

$$T.C._n = Cost\ (imperialist_n) + \zeta\ mean\ \{Cost\ (colonies\ of\ empire_n)\} \quad (9)$$

where, $T.C._n$ is the total cost of nth empire and ζ is a value between 0 to 1. Assuming zero value for ζ causes the total cost of imperialist becomes equal to the cost of empire itself. However, in optimal cases it can be assumed that $\zeta = 0.05$ to achieve reasonable results (Rajabioun et al., 2008).
If any empire could not increase its power and loses the competition, it will be eliminated during the imperialist competition. This elimination will happen gradually.

Hence, the algorithm assumes one empire is about to collapse. In this case, the algorithm selects couple of weakest colonies and makes competition between the empires. The competition between the empires to grab the colony considering the total cost of empires can be modeled. First, the algorithm calculates the total normalized cost:

$$N.T.C._n = max\ \{T.C._i\} - T.C._n \quad (10)$$

where $N.T.C._n$ is the total normalized cost of nth empire. Accordingly, $N.T.C._n$ is equal to the total cost of one empire and $N.T.C._n$ is equal to the total power of that empire. Therefore, an empire with a less cost function is the strongest one. Considering the total normalized cost, the probability of each empire in terms of grabbing the colony can be defined as:

$$P_{pn} = \left|\frac{N.T.C.n}{\sum_{i=1}^{Nimp} N.T.C.i}\right| \quad (11)$$

The algorithm needs a mechanism such as roulette wheel in the GA in order to allocate the colony according to the empire power. In this study a new mechanism to implement such an algorithm is introduced. One of the advantages of IC algorithm over the GA is that the IC algorithm has got less cost function because it eliminates complicated computation required in the roulette wheel in the GA. In fact, it deletes the *cumulative distribution function* (CDF) and only needs *probability density function* (PDF). Based on the probability of grabbing each empire, the algorithm distributes the related colony. *P* is a vector that can be defined as (Rajabioun et al., 2008):

$$P = [Pp_1, Pp_2, Pp_3, \dots, Pp_{N(imp)}] \quad (12)$$

This vector includes all the probability of the possession values of the empires. Vector *R* can be created to the same size of vector *P*. The arrays of this type of vector have a random value between 0 and 1.

$$r_1, r_2, r_3, \dots, r_{N(imp)} \sim U(0,1), R = [r_1, r_2, r_3, \dots, r_{N(imp)}] \quad (13)$$

Hence, *D* vector can be created as follow:

$$D = P - R = [D_1, D_2, D_3, \dots, D_{N(imp)}] = [Pp_1 - r_1, Pp_2 - r_2, Pp_3 - r_3, \dots, Pp_{N(imp)} - r_{N(imp)}] \quad (14)$$

The colony is transferred to the empire, which has higher value in the *D* vector. This part of algorithm is terminated with grabbing the colony by one of the empire. As it was mentioned earlier, during the imperialist competition, weak empires collapse gradually and their colonies are shared between other strong empires. In this case, only one empire will remain and all the colonies will be under its authority and the algorithm will be converged. Fig. 4 presents a general workflow of the IC algorithm.

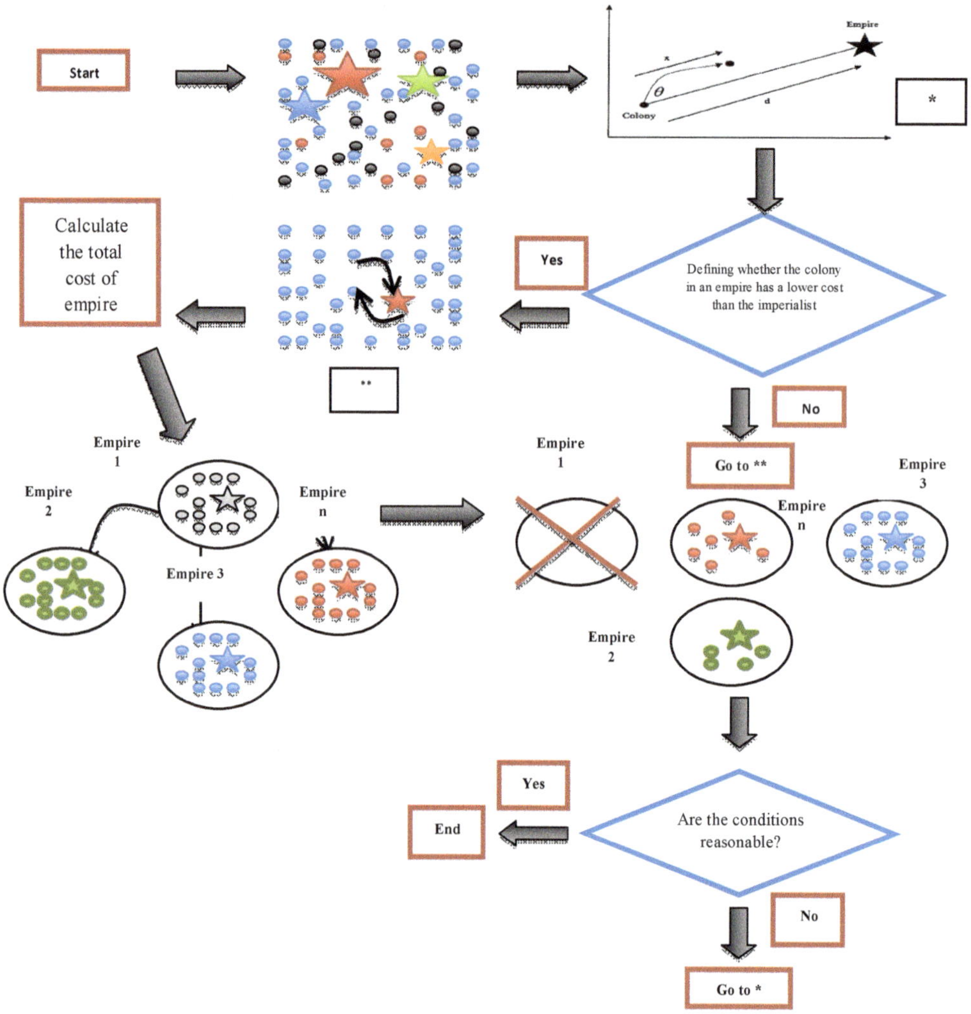

Fig. 4. Schematic Diagram of the IC algorithm

5. Case Study Related Objective Functions Optimizations

The IEEE 4-bus system has been considered as a case study. Fig. 5 illustrates a schematic of this network.

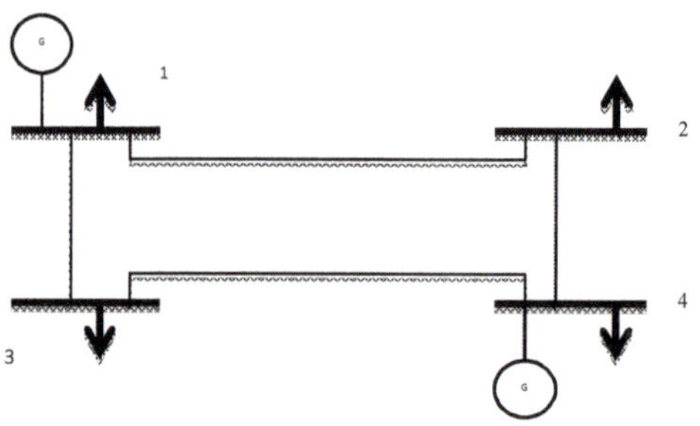

Fig. 5. Single Diagram of the IEEE 4-Bus System

In this case, three objective functions have been involved. As mentioned previously, the size and cost of implementing FACTS devices play significant roles in deployment of these devices in power industry. The objective functions can be defined as follow:

- Obj_1: Minimizing voltage deviation
- Obj_2: Minimizing the FACTS size
- Obj_3: Minimizing the implementation cost

These three objective functions can define as:

$$Obj_1 = V_d = \sum_{k \in \Omega}(V_k - V_{(ref)k})^2 \quad (15)$$

where, J_1 is the voltage deviation and V_k is the voltage magnitude.

$$Obj_2 = \sum_{j=1}^{N_{units}} \eta_j \quad (16)$$

where, J_2 is the FACTS size, η_j is the size in $MVar$ of FACTS and N_{units} is the number of FACTS that should be allocated.

$$Obj_3 = IC = C \times S \quad (17)$$

where, IC is the minimum installation cost in $US\$$ and C is the installation cost function in $US\$/kVar$. According to the data applied from Siemens AG (Hong-Bo & Shuxia, 2008) the installation costs of TCSC and SVC are:

$$C_{TCSC} = 0.0015S^2 - 0.71305S + 153.75 \quad US\$/kVar \quad (18)$$
$$C_{SVC} = 0.0003S^2 - 0.3051S + 127.38 \quad US\$/kVar \quad (19)$$

where, S is the operating range of the FACTS devices, which can be defined as:

$$S = |Q_{post}| - |Q_{initial}| \quad (20)$$

In this case Q_{post} is the reactive power in the transmission line after installing FACTS devices and $Q_{initial}$ is the reactive power in transmission line before installing FACTS devices both in $MVAr$. Furthermore, other constraints should be defined as:

$$|V_i|_{min} \leq |V_i| \leq |V_i|_{max} \quad (21)$$
$$S_{ij} \leq S_{ijmax}, \delta_{ij} \leq \delta_{ijmax} \quad (22)$$

where, $|V_i|_{min}$, $|V_i|_{max}$ are minimum and maximum voltage magnitudes at bus i, S_{ij} is the apparent power of the line i to j, S_{ijmax} is the thermal limit of that line, and δ_{ij} is the difference of phase angels between buses i and j. In this study, a range of assumptions have been made as follows:

- The thermal rating has been considered to be equal to 250 MVA.
- Minimum and maximum reactive power outputs of the generator have been assumed to be equal to -999 and 999 $MVar$ respectively.
- Minimum and maximum voltage magnitudes have been assumed to be equal to 0.95 and 1.05 pu, respectively.
- The S_{base} has considered being equal to 100 MVA.

The total power loss equation is given as:

$$P_l = \sum [V_i^2 + V_j^2 - 2V_iV_j \cos(\delta_i - \delta_j)] Y_{ij} \cos\varphi_{ij} \quad (23)$$

where δ_i is the angle of the voltage at bus i, Y_{ij} and φ_{ij} are the magnitude and angle of the admittance of the transmission line between i to j, respectively. For computing the TTC the real power generations at bus i, P_{Gi}, and both real and reactive load powers at bus i, P_{Di} and Q_{Di}, which are in function of λ, should be calculated as follow:

$$P_{Gi} = P_{Gi} + (1 + \lambda k_{Gi}) \quad (24)$$
$$P_{Di} = P_{Di} + (1 + \lambda k_{Di}) \quad (25)$$
$$Q_{Di} = Q_{Di} + (1 + \lambda k_{Qi}) \quad (26)$$

where, k_{Gi} and k_{Di} are the constants used to show the rate of changes in the load with various λs. The total transfer capability can be defined as (Shahrazad et al., 2013):

$$TTC = \sum_{i=1} P_{Di}(\lambda_{max}) - \sum_{i=1} P^0{}_{Di} \quad (27)$$

In this case, $P_{Di}(\lambda_{max})$ is the sum of the load in sink area where: $\lambda = \lambda_{max}$. On the other hand $P^0{}_{Di}$ is the sum of load where: $\lambda = 0$. In fact, λ is the parameter for increasing load or generation. When $\lambda = 0$, there is no power transfer between the areas and when $\lambda = \lambda_{max}$ there is maximum power transfer.

6. Experimental Results

In this section the installation of FACTS devices in terms of single multiple, cost of FACTS devices, load flow, convergence characteristics and P-V curves are investigated. In this case, the number of initial counties is 4 and maximum number of iterations is considered to be 300. Six analyses are simulated and presented in this paper. Table 3 presents the specifications of the considered case study.

Table 3
Case Study Resistance, Reactance and Susceptance

F Bus	T Bus	R (pu)	X (pu)	1/2B (pu)	Tap
1	1	0.01008	0.0504	0.05125	0
1	3	0.00744	0.0372	0.03875	0
2	4	0.00744	0.0372	0.03875	0
3	4	0.01272	0.0636	0.06375	0

Table 4 indicates a summary of the load flow. In this case, the program analyzes the IEEE 4-bus system using quadratic Newton-Raphson load flow algorithm. Fig. 6 presents the P-V curve before and after installing any FACTS devices. The P-V curve indicates the loading of the power system. To obtain the P-V curve in the power system, load is increased slightly and in each step voltage magnitude is analyzed.

Table 4
Load Flow Summary

F_Bus	T_Bus	P_{ij} (pu)	Q_{ij} (pu)	S_{ij} (pu)	Loss P (pu)	Loss Q (pu)
	-	1.368	0.835	0.160	-	-
1	2	0.386	0.222	0.446	0.0022	-0.089
	3	0.981	0.612	1.156	0.0103	-0.235
	-	-1.7	-1.053	1.99	-	-
2	1	-0.384	-0.3123	0.4955	0.0022	-0.0893
	4	-1.31	-0.741	1.50	0.0171	0.00806
	-	-2	-1.239	2.352	-	-
3	1	-0.9708	-0.6356	1.1604	0.01031	-0.02356
	4	-1.029	-0.603	1.193	0.0183	-0.03441
	-	2.38	1.318	2.7208	-	-
4	2	1.332	0.749	1.528	0.01715	0.00806
	3	1.047	0.569	1.192	0.0183	-0.0344

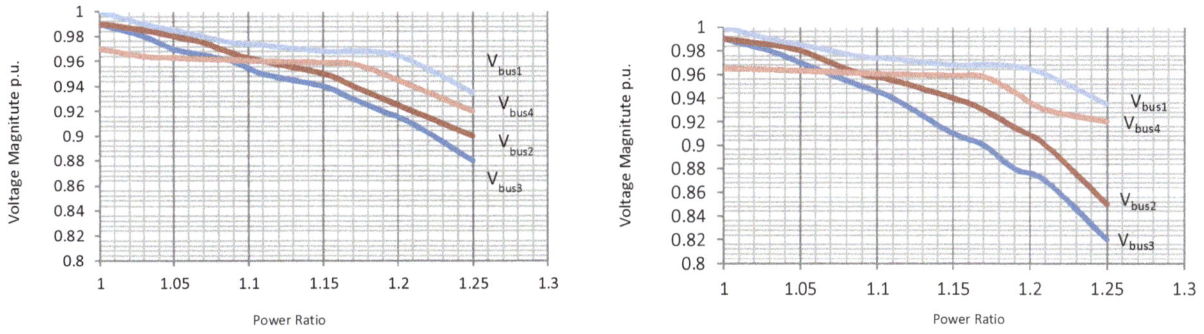

(a) (b)

Fig. 6. P-V Curves Before (a) and After (b) Installing FACTS Devices

According to the behavior of the slope, the voltage collapse point can be obtained; therefore, voltage stability can be analyzed. Fig. 7 illustrates the cost function graph for the SVC and TCSC devices. In this case TCSC has more load-ability characteristics and SVC has less installation cost.

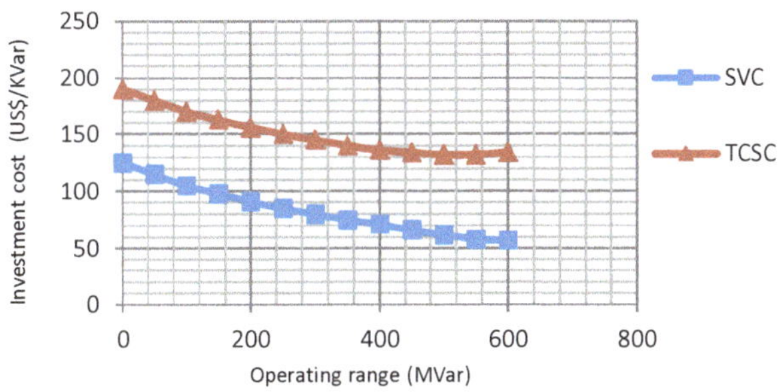

Fig. 7. Investment Cost Curves

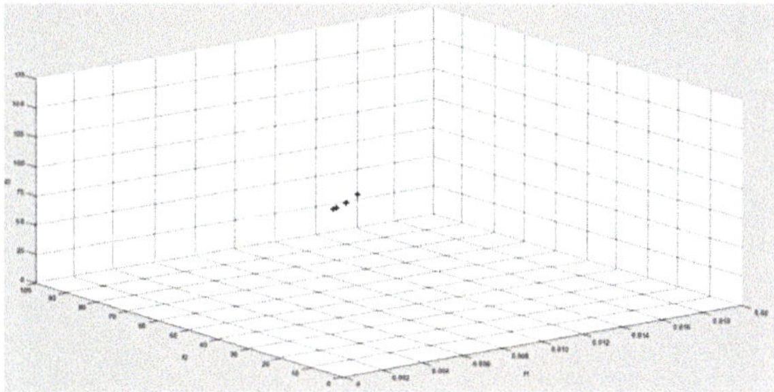

Fig. 8. Solutions' Positions after 4 Iterations

Voltage deviation is one of the major issues, which all industries must also consider. Voltage level can be divided from milliseconds to few seconds. Therefore, installing the FACTS devices in different places can solve the voltage sag problem and supply high quality power in the network. Fig. 8 presents the location of four countries considered in the IC algorithm after four iterations. This paper considers TCSC as a series component in transmission lines and SVC as a parallel component at bus-bars. In this regard, six analyses are performed as follows:

- Allocating one TCSC as a single element,
- Allocating couple of TCSCs,
- Allocating one SVC as a single element,
- Allocating couple of SVCs,
- Allocating a TCSC and SVC at the same time,
- Allocating couple of TCSCs and SVCs at the same time.

It can be identified that by allocating FACTS devices optimally, the losses can be reduced and the system load-ability can be enhanced. Furthermore, it can be observed that by increasing the number of FACTS devices in the transmission system, the installation cost will increase significantly.

6.1 Allocating one TCSC as a Single Element

In this analysis only one TCSC has been taken into consideration. Results are presented in Table 5. It can be identified that the amount of losses has been reduced; furthermore, both active and reactive powers flowing through the allocated place were increased. Generally imperialist competitive has a good convergence characteristic. This is because of the fact that the algorithm searches the answer by initializing the problem in different groups of area. The speed of convergence is indicated in Fig. 9. It can be understood that the algorithm can be converged after 5 iterations and the optimal place for allocating the FACTS device will be between busses 1-3.

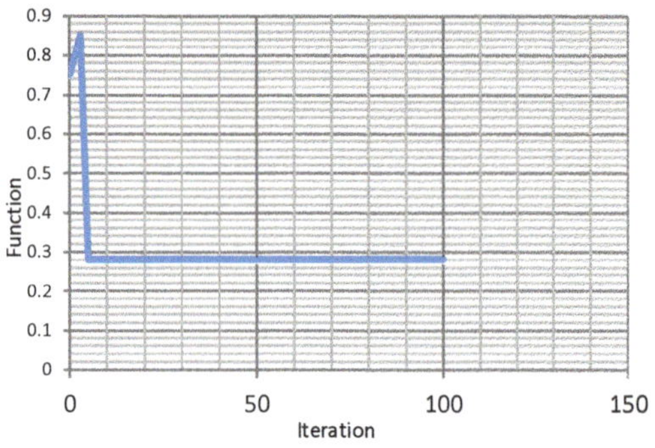

Fig. 9. Convergence Speed with One TCSC

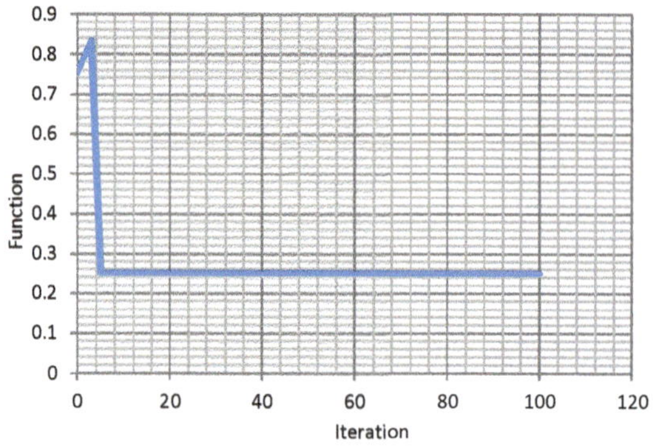

Fig. 10. Convergence Speed with Couple of TCSCs

Table 5
Case Study with One TCSC

Case	Type of FACTS	FACTS Location		P_{ijb} (pu)	Q_{ijb} (pu)	P_{ija} (pu)	Q_{ija} (pu)	X_{TCSC} (pu)	Ic (×10^6US$)
		F_Bus	T_Bus						
Single Type	**TCSC**	1	3	0.98	0.612	1.0072	0.603	-0.64	0.0029

6.2 Allocating Couple of TCSCs

In this analysis couple of TCSCs have been considered. According to Table 6, it can be identified that the amount of losses were decreased significantly however the cost function increased comparing to installing one TCSC. In this case, the system load-ability increased. Furthermore, both active and reactive powers flowing through the allocated place were increased. In total, two places between busses 1-3 and 3-4 were found, respectively. It is important to note that the installation cost decreases by increasing load-ability of the power network. Speed of convergence with installing two FACTS devices is indicated in Fig. 10. In this case, after 5 iterations the algorithm has been converged.

Table 6
Case Study with Couple of TCSCs

Case	Type of FACTS	FACTS Location		P_{ijb} (pu)	Q_{ijb} (pu)	P_{ija} (pu)	Q_{ija} (pu)	X_{TCSC} (pu)	Ic (×10^9US$)
		F_Bus	T_Bus						
Single Type	**TCSC**	1	3	0.98	0.612	1.0085	0.59	-0.66	0.0194
		3	4	-1.029	-0.603	-1.036	-0.583	-0.32	

6.3 Allocating one SVC as a Single Element

This sub-section demonstrates the analysis with installing one parallel SVC in the power network. Subsequently, deploying this FACTS device it can be observed that the installation cost decreased; however, the installation cost by SVC is more than TCSC. In this analysis the power losses also decreased. Both active and reactive powers flowing through the transmission lines are also increased. Furthermore, the load-ability of the system with installing the SVC is enhanced. The optimal place for allocating the FACTS device is bus 3 (Table 7). After 9 iterations the algorithm can be converged (Fig. 11).

Table 7
Case Study with One SVC

Case	Type of FACTS	FACTS Location		P_{ijb} (pu)	Q_{ijb} (pu)	P_{ija} (pu)	Q_{ija} (pu)	X_{TCSC} (pu)	Ic (×10^6^US$)
		F_Bus	T_Bus						
Single Type	SVC	3		-2.00	-1.239	-2.052	-1.201	-0.48	0.0033

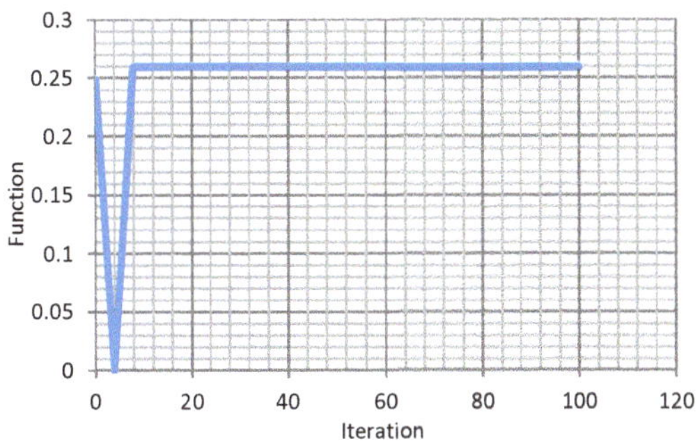

Fig. 11. Convergence Speed with One SVC

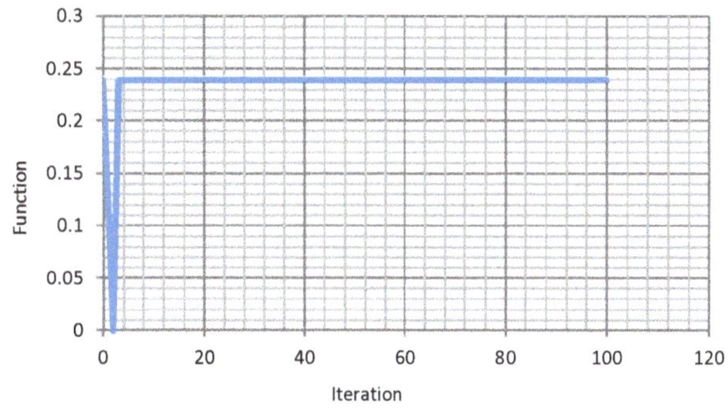

Fig. 12: Convergence Speed with Couple of SVCs

6.4 Allocating Couple of SVCs

In this analysis couple of SVCs in the power network have been considered. It can be identified that because of installing more FACTS devices in the transmission network, the installation cost is increased. System load-ability is enhanced and losses are decreased (Table 8). Total installation cost is less than allocating TCSCs in the transmission lines with same number of FACTS devices. The optimal places for allocating FACTS devices are at busses 1 and 3. The IC algorithm is converged after 5 iterations (Fig. 12).

Table 8
Case Study with Couple of SVCs

Case	Type of FACTS	FACTS Location		P_{ijb} (pu)	Q_{ijb} (pu)	P_{ija} (pu)	Q_{ija} (pu)	X_{TCSC} (pu)	Ic (×10^9^US$)
		F_Bus	T_Bus						
Single Type	SVC	1		1.368	0.835	1.40	0.78	-0.62	0.018
		3		-2.00	-1.239	-2.031	-1.14	-0.502	

6.5 Allocating a TCSC and SVC at the Same Time

This sub-section demonstrates an analysis where different types of FACTS devices are allocated in the power system. With this regard a SVC as a parallel component and a TCSC as a series component have been installed at a bus-bar and on transmission line, respectively. It can be realized that total installation cost increased sharply and total losses are decreased while the system load-ability has increased significantly. Correspondingly, both active and reactive powers flowing through the transmission lines are increased (Table 9). In this case, the optimal place for allocating one SVC is bus 1, and the best place for allocating one TCSC is between busses 1 and 3. After 13 iterations the IC algorithm can approach the solution (Fig. 13). Hence, by allocating the SVC and TCSC the TTC can increase and this results in reducing the need for constructing new transmission lines.

Table 9
Case Study with a TCSC and SVC

Case	Type of FACTS	FACTS Location		P_{ijb} (pu)	Q_{ijb} (pu)	P_{ija} (pu)	Q_{ija} (pu)	X_{TCSC} (pu)	Ic (×10^6^US$)
		F_Bus	T_Bus						
Multi Type	TCSC	1	3	0.98	0.612	0.9908	-0.410	-0.50	8.92
	SVC	1		1.36	0.835	1.405	-0.249	-0.382	

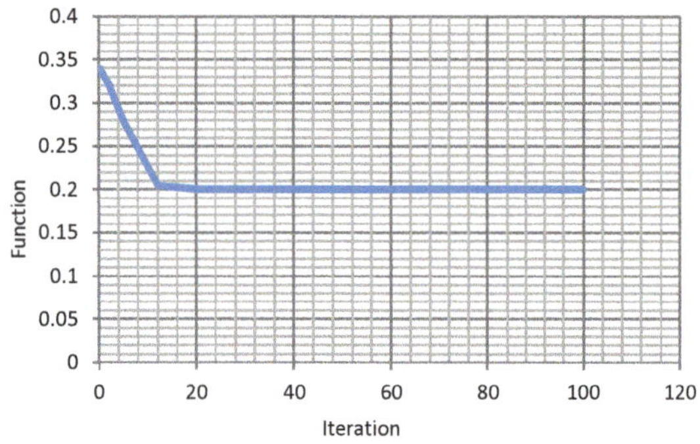

Fig. 13. Convergence Speed with a SVC and TCSC

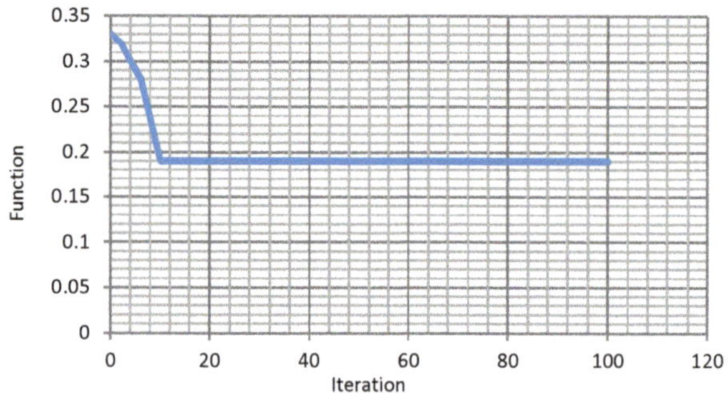

Fig. 14: Convergence Speed with couple of SVC and TCSC Combination

6.6 Allocating Couple of TCSCs and SVCs at the Same Time

In this sub-section the algorithm performs optimal analysis for allocating a different types of FACTS devices. In this case, one TCSC and two SVCs have been taken into consideration. As illustrated in Table 10 the installation cost sharply increased. Moreover, the total losses are decreased; on the other hand, the system load-ability increased. Furthermore, by deploying a combination of FACTS devices more optimal results can be achieved over using single type of them. With this respect, the IC algorithm finds 3 optimal places. The parallel FACTS devices can be allocated at busses 1 and 3, and TCSC will be placed between busses 1 and 3. The algorithm approached the optimal solution after 10 iterations (Fig. 14).

Table 10
Case Study with Couple of TCSCs and SVCs

Case	Type of FACTS	FACTS Location		P_{ijb} (pu)	Q_{ijb} (pu)	P_{ija} (pu)	Q_{ija} (pu)	X_{TCSC} (pu)	Ic (×10^{9}US$)
		F_Bus	T_Bus						
Multi Type	TCSC	1	3	0.98	0.612	0.99	0.325	-0.57	5.92
	SVC	1		1.36	0.835	1.4	-0.54	-0.42	
		3		-2.0	-1.239	-2.04	-1.173	-0.63	

7. Conclusion

FACTS devices are the most common components used to increase the stability and load-ability of the power networks. Evolutionary algorithms are deployed to allocate FACTS devices in optimum place lead to decrease the losses and improve the voltage profile. Imperialist competitive algorithm is used in this paper to find optimum places of the compensators. Different scenarios are discussed in this paper to demonstrate different approaches. According to the results, it can be identified that using more number of FACTS devices in transmission lines increases the installation cost sharply. Hence, considering the financial terms of FACTS devices are desirable. Imperialist competitive is the latest optimization method recently used in power industry, which can improve both accuracy and speed of the convergence.

Acknowledgement

The authors would like to thank the anonymous referees for constructive comments on earlier version of this paper.

References

Atashpaz-Gargari, E., & Lucas, C. (2007, September). Imperialist competitive algorithm: an algorithm for optimization inspired by imperialistic competition. *IEEE Congress o Evolutionary Computation*, 2007. CEC 2007. (pp. 4661-4667).

Avramenko, V. N. (2005, June). Power system stability assessment for current states of the system. In Power Tech, 2005 IEEE Russia (pp. 1-6). IEEE.

Belsnes, M. M., Warland, G., & Wolfgang, O. (2011, June). Simulating equilibrium prices in oligopoly power markets. *IEEE Trondheim in PowerTech,* 2011 (pp. 1-8). IEEE.

Cai, L. J., Erlich, I., & Stamtsis, G. (2004, October). Optimal choice and allocation of FACTS devices in deregulated electricity market using genetic algorithms. *In Power Systems Conference and Exposition*, 2004. IEEE PES(pp. 201-207). IEEE.

Chansareewittaya, S., & Jirapong, P. (2010, November). Power transfer capability enhancement with multitype FACTS controllers using particle swarm optimization. In TENCON 2010-2010 IEEE Region 10 Conference (pp. 42-47). IEEE.

Deep, K., & Bansal, J. C. (2009, December). Optimization of directional overcurrent relay times using Laplace Crossover Particle Swarm Optimization (LXPSO). *In Nature & Biologically Inspired Computing, 2009. NaBIC 2009. World Congress on* (pp. 288-293). IEEE.

Gerbex, S., Cherkaoui, R., & Germond, A. J. (2001). Optimal location of multi-type FACTS devices in a power system by means of genetic algorithms. *Power Systems, IEEE Transactions on*, 16(3), 537-544.

Hong-Bo, S., & Shuxia, L. (2008, October). The Comparison Between Genetic Simulated Annealing Algorithm and Ant Colony Optimization Algorithm for ASP. In Wireless Communications, Networking and Mobile Computing, WiCOM'08. *4th International Conference on* (pp. 1-6). IEEE.

Krishans, Z., Kutjuns, A., & Kalnins, M. (2007, July). Method of transmission power networks reliability estimation. *In Power Tech, 2007 IEEE Lausanne* (pp. 1996-1999). IEEE.

Majumdar, S., Chakraborty, A. K., & Chattopadhyay, P. K. (2009, December). Active power loss minimization with FACTS devices using SA/PSO techniques. *International Conference on Power Systems, 2009*. ICPS'09. (pp. 1-5). IEEE.

Minasian, A. A., & Bird, T. S. (2013). Particle Swarm Optimization of Microstrip Antennas for Wireless Communication Systems. *IEEE Transactions on Antennas and Propagation*, 61(12), 6214-6217.

Panda, S., & Padhy, N. P. (2008). Comparison of particle swarm optimization and genetic algorithm for FACTS-based controller design. *Applied Soft Computing*, 8(4), 1418-1427.

Rajabi-Ghahnavieh, A., Fotuhi-Firuzabad, M., Shahidehpour, M., & Feuillet, R. (2009). Optimal allocation of available transfer capability in operating horizon. *IEEE Transactions on Power Systems*, 24(2), 967-975.

Rajabioun, R., Hashemzadeh, F., Atashpaz-Gargari, E., Mesgari, B., & Salmasi, F. R. (2008, July). Identification of a MIMO evaporator and its decentralized PID controller tuning using colonial competitive algorithm. In be presented in IFAC World Congress.

Saravanan, M., Slochanal, S., Venkatesh, P., & Abraham, J. (2007). Application of particle swarm optimization technique for optimal location of FACTS devices considering cost of installation and system loadability. *Electric Power Systems Research*, 77(3), 276-283.

Shahrazad, M., & Zobaa, A. F. (2013). Optimal allocation of FACTS devices in distribution networks using Imperialist Competitive Algorithm. *International Review of Electrical Engineering*, 8(5).

Tabatabaei, N. M., Aghajani, G., Boushehri, N. S., & Shoarinejad, S. (2011). Optimal location of FACTS devices using adaptive particle swarm optimization mixed with simulated annealing. *International Journal on Technical and Physical Problems of Engineering (IJTPE)*, (7), 60-70.

Takasaki, M. (2006, May). Power transfer capability enhancement with UPFC under circumstances of uncertain power flow pattern. *In Transmission and Distribution Conference and Exhibition, 2005/2006 IEEE PES (pp. 659-665). IEEE.*

12

An ordered heuristic for the allocation of resources in unrelated parallel-machines

André Serra e Santos[a*], Ana Maria Madureira[b] and Maria Leonilde R. Varela[a]

[a]*Department of Production and Systems - School of Engineering - University of Minho (EEUM/UM), Portugal*
[b]*GECAD - Knowledge Engineering and Decision Support Research Center - School of Engineering – Polytechnic of Porto (ISEP/IPP), Portugal*

CHRONICLE

Keywords:
Scheduling
Makespan
Unrelated Parallel Machines
MCT
MOMCT

ABSTRACT

Global competition pressures have forced manufactures to adapt their productive capabilities. In order to satisfy the ever-changing market demands many organizations adopted flexible resources capable of executing several products with different performance criteria. The unrelated parallel-machines makespan minimization problem (Rm||Cmax) is known to be NP-hard or too complex to be solved exactly. In the heuristics used for this problem, the MCT (Minimum Completion Time), which is the base for several others, allocates tasks in a random like order to the minimum completion time machine. This paper proposes an ordered approach to the MCT heuristic. MOMCT (Modified Ordered Minimum Completion Time) will order tasks in accordance to the MS index, which represents the mean difference of the completion time on each machine and the one on the minimum completion time machine. The computational study demonstrates the improved performance of MOMCT over the MCT heuristic.

1. Introduction

Operations scheduling is a decision process that defines strategies from orders release down to detailed programming (Baker & Trietsch, 2009; Blazewicz et al., 2001; Pinedo, 2012). Therefore, for determining how an order will be executed it is necessary to decide when an order should be processed and which resource should be used for each task. Moreover, the establishment of a production sequence indicating the tasks' execution order has also to be considered in order to optimize one or more performance measures. Consequently, scheduling tries to answer two main questions (Baker & Trietsch, 2009; Pinedo 2012; Xhafa, & Abraham, 2008):

(1). How should tasks be distributed between the available resources in order to optimize the performance measure(s)?

(2). In which sequence should the tasks be processed to optimize the performance measure(s)?

According to these two main questions, Operations Scheduling can be divided into two main phases, the allocation phase that answers the first question and the sequencing phase that answers the second one. There is a strong relation between both phases and it is often hard to clearly distinguish between them.

* Corresponding author.

In problems where the resources are in parallel, the allocation phase is usually addressed before determining the sequence in which the tasks will be processed. In problems with more complex implementations, it is harder to separate the allocation phase from the sequencing one, since there is a correlation between them.

In complex production implementations, the problem constraints and the number of available tasks can make scheduling operations extremely hard through exact methods (Baker & Trietsch, 2009; Blazewicz et al., 2001; Pinedo, 2012). Therefore, many different kinds of contributions exist for dealing with these more complex scenarios, namely web-based systems and platforms, which would integrate a variety of scheduling approaches and methods, mainly based on heuristic, for solving different types of scheduling problems. Some examples are described by Varela et al. (2003, 2008, 2012, 2014), which include several distinct kinds of exact and heuristic scheduling methods. This paper proposes an ordered approach to the *MCT* heuristic for unrelated parallel-machines *makespan* minimization problem (*Rm*||*Cmax*) (Pinedo, 2012; Varela et al., 2008), based on the performance limitations of the random like order in which *MCT* allocates tasks. The proposed *MOMCT* is a constructive heuristic, which would construct a schedule from an empty solution by adding parts to the solution (Briceño, 2008).

In this paper, resources are identified as machines, which execute jobs or tasks, and where the elementary part of the job or task only requires a single resource is referred as an operation. The scheduling problem will be represented by the Graham's classification as described by Pinedo (2012) and Varela et al. (2008). This classification is represented by three fields, **α|β|γ** with the first field representing the machine environment, the second field the process characteristics and constraints and the third field the problem optimization criterion. This paper is divided into five sections. Section 2 defines the *makespan* minimization problem in parallel-machines (*Pm*||*Cmax*), in parallel-machines with different speeds (*Qm*||*Cmax*) and in unrelated parallel-machines (*Rm*||*Cmax*). Section 3 presents the literature review and four heuristics for the *Rm*||*Cmax* problem. In section 4, the proposed *Modified Ordered Minimum Completion Time* heuristic is described. The computational study and a discussion of the achieved results is presented in section 5. Lastly, the paper presents some conclusions.

2. Problem Definition

The *makespan* minimization problem in parallel-machines (*Pm*||*Cmax*) is known to be NP-hard. It can describe a machine environment where there are *m* similar machines in parallel. It is equivalent to the PARTITION problem (Pinedo 2012). *Pm*||*Cmax* is solved by the priority rule *LTP* (Longest *Processing Time).* The problem (*Pm*||*Cmax*) is formulated as seen bellow. In this problem the optimization criterion is to minimize the *makespan* (*Cmax*), which is subjected to three constraints, where *n* is the number of tasks and *m* the number of machines. The first (1) forces each task to be allocated to only one machine; the second (2) defines the *makespan* as the completion time of the last task to leave the system. The third (3) forces the decision variables to be non-negative. *xij* are the decision variables that define which tasks (j) are allocated to each machine (i), p_{ij} is the processing time of task *j* in machine *i.*

min C_{max}

subject to

$$\sum_{i=1}^{m} x_{ij} = 1, \qquad j = 1, \cdots, n \tag{1}$$

$$\sum_{j=1}^{n} p_j x_{ij} \leq C_{\max}, \qquad i = 1, \cdots, m \tag{2}$$

$$x_{ij} \geq 0 \tag{3}$$

The *makespan* minimization problem in parallel-machines with preemptions (*Pm|prmp|Cmax*) can be solved exactly since it is possible to divide the execution of tasks among the various machines. *Pm|prmp|Cmax can* also be solved by the priority rule *LRPT* (Longest *Remaining Processing Time).* The problem (*Pm|prmp|Cmax*) is formulated as seen bellow. In the problem the optimization criterion is to minimize the *makespan* (*Cmax*), which is subjected to four constraints. Eq. (4) forces each task to be executed, Eq. (5) makes sure that each task execution is less than or equal to the *makespan.* Eq. (6) ensures that the processing time in each machine is less than or equal to the *makespan.* Eq. (7) forces the decision variables to be non-negative.

min C_{max}

subject to

$$\sum_{i=1}^{m} x_{ij} = p_j, \qquad j = 1, \cdots, n \tag{4}$$

$$\sum_{i=1}^{m} x_{ij} \le C_{\max}, \qquad j = 1, \cdots, n \tag{5}$$

$$\sum_{j=1}^{n} x_{ij} \le C_{\max}, \qquad i = 1, \cdots, m \tag{6}$$

$$x_{ij} \ge 0 \tag{7}$$

The *makespan* minimization problem in unrelated parallel- machines (*Rm||Cmax*) is a particular case of *makespan* minimization problem in parallel-machines (*Pm||Cmax*) (Pinedo, 2012). In this problem, the machine has different characteristics, which means, each machine can have a different processing time for each task. Unrelated parallel-machines can describe problems in workshops where there are different machines that can be used to execute the same tasks, which is usual when considering an upgrade to the installed capacity, or grid-computing problems where computers with different configurations are shared between various users. The problem (*Rm||Cmax*) formulation can be seen bellow.

min C_{max}

subject to

$$\sum_{i=1}^{m} x_{ij} = 1, \qquad j = 1, \cdots, n \tag{8}$$

$$\sum_{i=1}^{n} p_{ij} x_{ij} \le C_{\max}, \qquad i = 1, \cdots, m \tag{9}$$

$$x_{ij} \ge 0 \tag{10}$$

In this problem the optimization criterion is to minimize the *makespan* (*Cmax*), which is subjected to three constraints, where *n* is the number of tasks and *m* the number of machines. Eq. (8) forces each task to be allocated to only one machine, Eq. (9) defines the *makespan* as the completion time of the last task to leave the system. Eq. (10) forces the decision variables to be non-negative. Finally, x_{ij} are the decision variables that define which tasks (j) are allocated to each machine (i), p_{ij} is the processing time of task *j* in machine *i*.

3. Literature Review

Due to the complexity of the $R_m||C_{max}$ problem, many heuristics have been developed. Some of the most used algorithms, presented in (Ibbara et al. 1997), such as the, *MCT Min-Min* and *Min-Max* heuristics are efficient methods to approach the problem. Braun et al. (2001) presented and compared *OLB* (*Opportunistic Load Balance*), *MET* (*Minimum Completion Time*), *MCT*, *Min-Min*, *Min-Max*, *Duplex*, *GA* (*Genetic Algorithms*), *SA* (*Simulated Annealing*), *GSA* (*Genetic Simulated Annealing*), *Tabu Search* and *A**. In the computational study, *GA* had the best performance and *Min-Min* the best results between the constructive heuristics. *OLB* and *MET* constructive heuristics had the worse results. Other approaches (Briceño et al., 2012; Chaturvedi & Sahu 2011; Pfund et al., 2004) to the problem are the Suffrage heuristics that allocates the tasks that would suffer more when not allocated to their favorite machine first, *KPB* (*K-Percent Best*) and *SWA* (*Switching Algorithm*) that combine the *MET* and *MCT* heuristics, *WQ* (*Work Queue*) that allocates task to the machine with the minimum work load. *DFPLTF* (*Dynamic Fastest Processor to Largest Task First*) and *RR* can be used to approach the dynamic $R_m||C_{max}$ problem. Sivasankaran et al. (2010) proposed a local search heuristic. The proposed heuristic consists of two phases: in the first tasks are allocated to the machines and in the second they are shifted from one machine to another. Sugavanam et al., (2007) compared *SA* to *GRASP* (*Greedy Randomized Adaptive Search Procedure*). The computational study found that the *SA* performed better than the *GRASP*. Some research is focused on mixed approaches, which in the first phase use integer programing and in the second phase use heuristics methods. Hariri et al. (1991) proposed the *DA* (*Descent Algorithm*) and Glass et al. (1994) proposed an altered *GA*, which applies the *DA* to each population solution, the *GDA* (*Genetic Descent Algorithm*) is then compared with *DA*, *SA*, *TS* and *GA*. In the computational study the *GDA* outperformed the *GA*, with results similar to those of *SA* and *TS*. The *LP/Roundup* heuristic is proposed by Lin et al. (2011), the heuristic solves the LP and then allocates the task to the machines where they have the highest fraction. In the computational study, the *LP/Roundup* was compared with *MCT*, *LP/MCT*, *LP/M* and *GA*. *LP/Roundup* performed better than the other constructive heuristics, *GA* achieved the best results but at the expense of more computational time.

Lenstra et al. (1990) proposed an approximation algorithm based on LP-relaxation and Martello et al. (1997) proposed lower bounds based on Lagrangian relaxations to obtain new exact and approximation algorithms. Ebenlendr et al. (2014) and Verschae et al. (2014) proposed a special case of the $R_m||C_{max}$ problem. In this case each task needs to be allocated to at most two machines and with the similar execution times (*Restrictive Assignment*), with a two phase *LP/Rounding* approach. Vakhania et al. (2014) proposed another special case of the $R_m||C_{max}$ problem, where each task can have two possible execution (p and q) times. The authors presented a polynomial-time algorithm for $q=2p$ and then modified to deliver approximated solutions for the $p\neq 2p$ problem.

In this section, three heuristics for $R_m||C_{max}$ will be presented, along with an illustrative example with three machines (m=3) and four tasks (n=4), seen in Table 1.

Table 1
Illustrative Example

	Machine 1	Machine 2	Machine 3
Task 1	100	120	260
Task 2	100	170	270
Task 3	200	300	400
Task 4	160	80	300

3.1. Minimum Execution Time

The *MET* (*Minimum Execution Time*) heuristic uses the execution time to allocate tasks. This does not take into consideration tasks already allocated, what can result in an inefficient solution when several tasks are allocated to one machine (Braun et al., 2001; Briceño et al., 2012; Ibbara et al., 1997). The MET algorithm is:

1. Unmapped task are placed into a list in a random sequence;
2. The first task is allocated to the minimum execution time machine;
3. Unmapped task are placed into a list in a random sequence;
4. Steps II, III and IV are repeated until all tasks have been mapped.

In the illustrative example, *MET* found a solution (Fig.1) with *makespan* of 400 *t.u*, with the allocation of tasks 1, 2 and 3 into machine 1; the remaining task (task 4) was allocated into machine 2. The solution is far from the optimal solution that would have a *makespan* of 260 *t.u.*

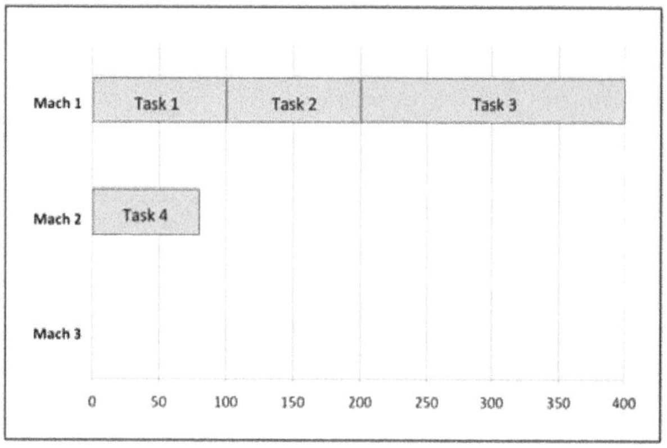

Fig. 1. MET Solution

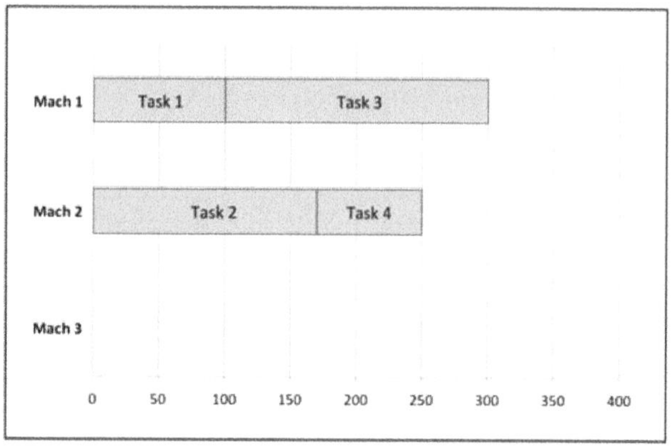

Fig. 2. MCT Solution

3.2.Minimum Completion Time

The *MCT* (*Minimum Completion Time*) heuristic uses the completion time to allocate tasks; this takes into consideration tasks already allocated, in opposition to *MET* heuristics (Braun et al., 2001; Briceño et al., 2012; Ibbara et al., 1997). The *MCT* algorithm is:

1. Unmapped tasks are placed into a list in a random sequence;
2. The first is allocated to the minimum completion time machine;
3. The task selected in step 2 is removed from the tasks list;
4. Ready times are updated with assigned task execution time;
5. Steps II, III and IV are repeated until all tasks have been mapped.

In the illustrative example, *MCT* found a solution (Fig. 2) with *makespan* of 300 *t.u*, with the allocation of tasks 1 and 3 to machine 1 and tasks 2 and 4 to machine 2.

3.2.Min-Min

The *Min-Min* heuristic follows a two-phased approach. First, the machines with the minimum completion time for each task are determined, then, between the task/machine pairs determined in the first step the one with lowest completion time is allocated. This approach makes this heuristic indifferent to the sequence in which the tasks are allocated (Braun et al., 2001; Briceño et al., 2012; Chaturvedi & Sahu 2011; Gupta & Singh, 2012) The *Min-Min* algorithm is:

1. Unmapped tasks are placed into a list in a random sequence;
2. For each listed task, the machine that provides the minimum completion time is identified;
3. For each task/machine pair found in step 2, the one with the minimum completion time is identified;
4. The task determined in step 3 is removed from the task list and allocated to the selected machine;

5. Machine ready times are updated with the assigned task execution time;
6. Steps 2, 3, 4 and 5 are repeated until all tasks have been mapped

In the illustrative example depicted in Fig. 3, *Min-Min* obtained a solution with *makespan* of 380 *t.u.*, with the allocation of tasks 1 and 2 into machine 1 and tasks 3 and 4 allocated into machine 2. The obtained solution is far from the optimal solution that would have a *makespan* of 260 *t.u.* There are several other heuristics that use a similar two-phase approach, such as *Max-Max* or *Max-Min* heuristics.

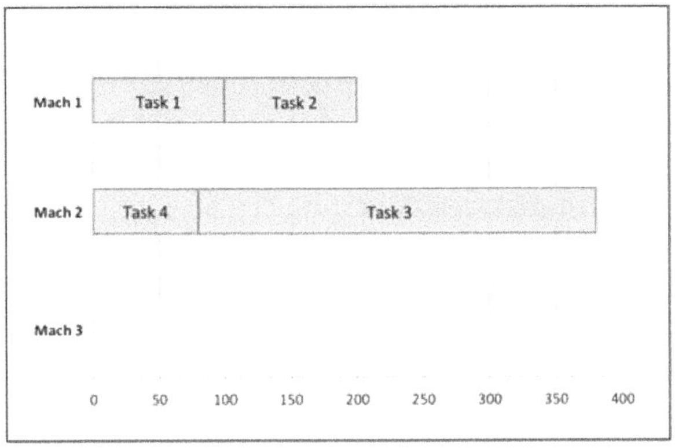

Fig. 3. Min-Min Solution

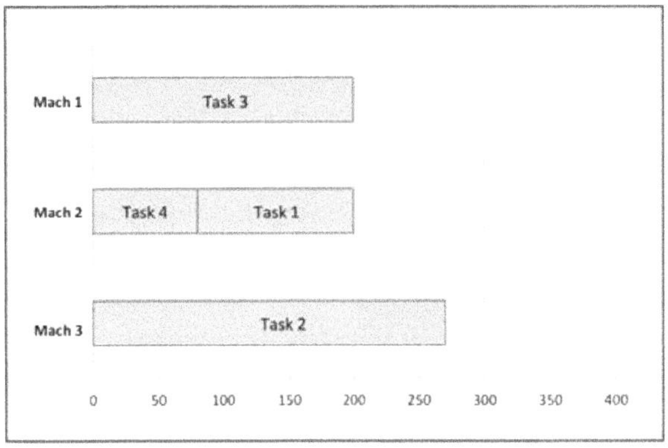

Fig. 4. Suffrage Solution

3.3.Suffrage

Suffrage finds the minimum and second minimum completion time for each task. When there are several tasks with the minimum completion time in the same machine, the one that would suffer more by being allocated to the machine with the second best completion time is allocated to the minimal completion time machine (Briceño et al., 2012; Gupta & Singh, 2012; Paranhos & Brasileiro 2003). *Suffrage* allocates first the tasks that would have more impact in the *makespan*, when they are not allocated to their favorite machine. The *Suffrage* algorithm is:

1. Unmapped tasks are placed into a list in a random sequence;
2. While there are still unmapped tasks;
   i. For each machine find the tasks that have the minimum completion time on that machine;
      a. If there is only one task competing for that machine, allocate it to that machine and remove it from the tasks list;
      b. If there is more than one competing task, allocate the one with the highest Suffrage value and remove it from the task list;
   ii. Machine ready times are updated with the assigned task execution time

The suffrage value can be calculated as seen in Eq. (11), where *2ndMCTM* represents the second smallest completion time and *1stMCTM* the smallest completion time machine.

$$S = 2^{nd} MCTM - 1^{st} MCTM \qquad (11)$$

In the illustrative example seen in Fig. 4, *Suffrage* obtained a solution with *makespan* of 270 *t.u*, in three iterations. Task 3 is allocated to machine 1, task 1 and 4 to machine 2 and task 2 to machine 3.

4. Modified Ordered Minimum Completion Time

The *MCT* heuristic, performance depends on the random like order in which tasks are allocated. In this paper, an ordered approach to the *MCT* heuristic is proposed. Santos et al. (2014) proposed an ordered approach to *MCT*. The *Ordered Minimum Completion Time* (*OMCT*) will determine a sequence of allocation before using the *MCT* algorithm to map tasks. The sequence of allocation will be determined by the standard deviation of the completion times of all tasks and the Suffrage value. The weighted sum of the standard deviation of the completion time on each machine and the Suffrage index will allocate first tasks with the biggest dispersion of their completion time while also taking into consideration the fact that each task is more likely to be allocated to its preferred machines. In the computational study, the *OMCT* heuristic performed better than *MCT*.

In this paper, a *Modified Ordered Minimum Completion Time* (*MOMCT*) is proposed. In this ordered approach to *MCT*, tasks are sequenced in non-increasing order of the MS (Mean Suffrage) value. MS is the difference between the mean of the completion time of a variable number of machines (2° minimum completion time machine, 3° minimum completion time machine, ..., m° minimum completion time machine) and the minimum completion time machine. This does not assume that the tasks will be allocated to their preferred machine. The number of machines considered for the calculation of MS value should be determined, experimentally. The number of machines use in MS should take in consideration how likely are tasks to be allocated into their favorite machines. Several factors can contribute to this, the number of machines (m), the number of tasks (n), the dispersion execution time of tasks between machines (p_j) and the dispersion of execution time of machines between tasks (p_i). The *MOMCT* algorithm is described as follows:

1. Calculate the MS value for each task;
2. Tasks are placed into a list in a decreasing order of their SM value;
3. The first task is allocated to the minimum completion time machine;
4. The task selected in step 3 is removed from the tasks list;
5. Machine ready times are updated with the assigned task execution time;
6. Steps 3, 4 and 5 are repeated until all tasks have been mapped.

The MS can be calculated as seen in Eq. (12), where a is determined experimentally, between the values of [$2; m$]. $i^{nd}MCTM$ represent the completion time on the i^{nd} minimum completion time machine and $1^{st}MCTM$ the completion time on the minimum completion time machine.

$$MS = \frac{\sum_{i=2}^{\alpha} iMCTM}{\alpha - 1} - 1^{st}\,MCTM \qquad \alpha \in [2;m] \tag{12}$$

In the illustrative example of the *MOMCT* heuristic (Fig. 8), with a=3, task 3 is allocated to machine 1, task 2 and task 4 are allocated to machine 2 and task 1 is allocated to machine 3. *MOMCT* heuristic obtained the optimal solution with a *makespan* of 260 *t.u.*

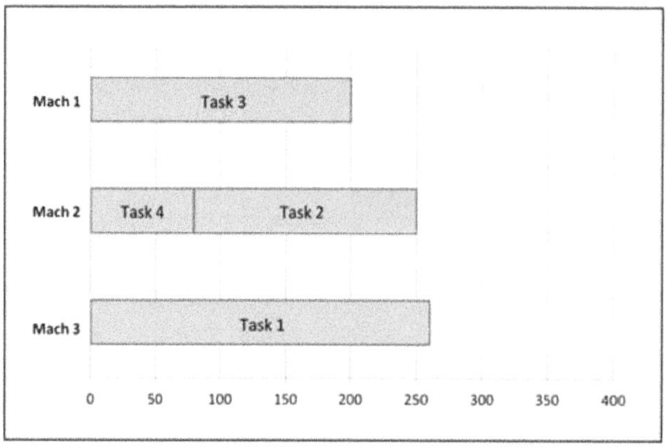

Fig. 5. MOMCT Solution

5. Computational Study

The MCT and MOMCT heuristics were implemented in C in Microsoft Visual Studio 2012 to be compared and evaluated. The computational tests were performed on a MacBook Air with an Intel® Core™ 2 @ 1.60 GHz processors, 4 GB of RAM memory and with Windows 7 64-bit as the operative system with all the available updates. Since the academic benchmark problems as it has been used by multiple authors and in diverse areas, it was considered an effective evaluation method. Both heuristics were tested on benchmark instances. *MOMCT* and *MCT* heuristics performance was evaluated with three groups of 20 instances with variable problem size (No. of Machines X No. of Tasks: 2X5,2X6, 2X7, 2X8, 2X9, 3X5, 3X6, 3X7, 3X8, 3X9, 4X5, 4X6, 4X7, 4X8, 4X9, 5X5, 5X6, 5X7, 5X8 and 5X9) retrieved from Sivasankaran et al. (2010).

5.1.MCT and MOMCT

Both heuristics were compared on their deviation from the optimal *makespan*, since this allows to overview the performance of each heuristic across all problems. Since there are 60 problems, the deviation from the optimal *makespan* can be assumed to follow a normal distribution, according to Central Limit Theorem (*CLT*).

In the computational study *MOMCT* heuristic performed better than the *MCT* heuristic. The bar chart from Fig. 6 depicts the deviation of heuristic obtained solutions from the optimal solution. *MOMCT* found the optimal solution in 30 (50%) instances while *MCT* only found the optimal solution in 21 (35%) instances.

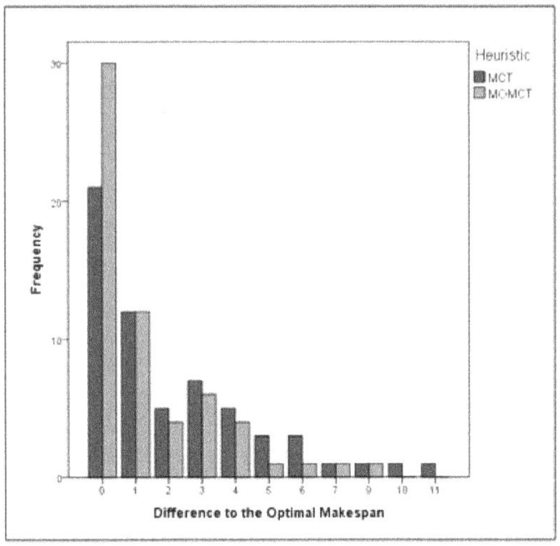

Fig. 6. Bar chart of the solutions of *MCT* and *MOMCT*

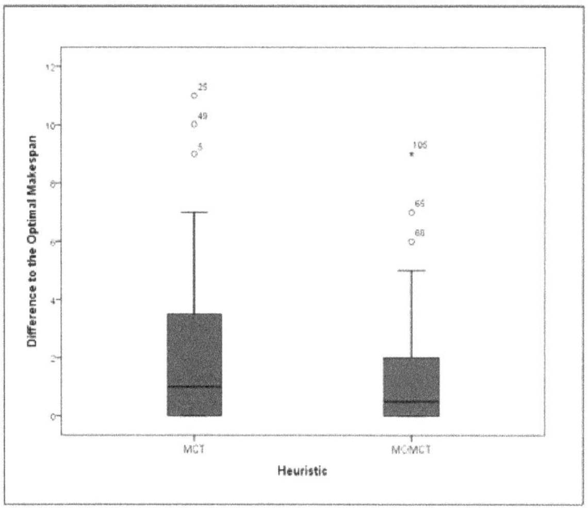

Fig. 7. Boxplot of the solutions of *MCT* and *MOMCT*

The boxplot from Fig. 7 permits to conclude about the advantage of *MOMCT* heuristic, in terms of minimization of *makespan* in the resolution of the analyzed instances. This boxplot allows the analysis of location, dispersion, median and asymmetry of data of the *MOMCT* and *MCT* obtained solutions when compared to the optimal solution of the problems.

To evaluate the significance of the obtained results the Student's t-test for independent samples has been used, considering the hypotheses:

$H_0: \mu_{MCT} - \mu_{OMCT} \geq 0$

$H_1: \mu_{MCT} - \mu_{OMCT} < 0$

Table 2 shows the result of the Student's t-test. It is not possible to assume equal variances (p-value of 0.030) and the student's t-test has a p-value of 0.022 (0.044/2). From the Student's t-test, it is possible to conclude that the MOMCT performs better in problems with more machines, with 95% of confidence level ($p=0.005<\alpha$).

Table 2
Student's T Test Results for *MCT* and *MOMCT*

	Levene's Test for Equality of Variances		t-test for Equality of Means				
	F	Sig.	t	df	Sig. (2-tailed)	Mean Difference	Std. Error Difference
Equal variances assumed	4.828	0.030	2.035	118	0.044	0.867	0.426
Equal variances not assumed			2.035	108.810	**0.044**	0.867	0.426

5.2. Number of Machines

The *MOMCT* heuristics performance appeared to depend on the number of machines. In the computational tests *MOMCT* seemed to perform better in problems with more machines. This is expected since the *MOMCT* heuristic was implemented to sweep all possible *a* values and due to the fact that *MOMCT* does not assume that the tasks will be allocated to their preferred machine taking more machines in consideration when ordering the tasks. The performance of *MOMCT* with 2/3 and 4/5 machines, where compared on their deviation from the makespan of the optimal solution.

In the computational study *MOMCT* heuristic performed better with more machines (4 and 5). The bar chart from Fig. 8 depicts the deviation of the heuristic obtained solutions from the optimal solution, with 2/3 and 4/5 machines. With 4/5 machines *MOMCT* found the optimal solution in 17 (56.7%) instances while with 2/3 machines it only found the optimal solution in 13 (43.3%) instances.

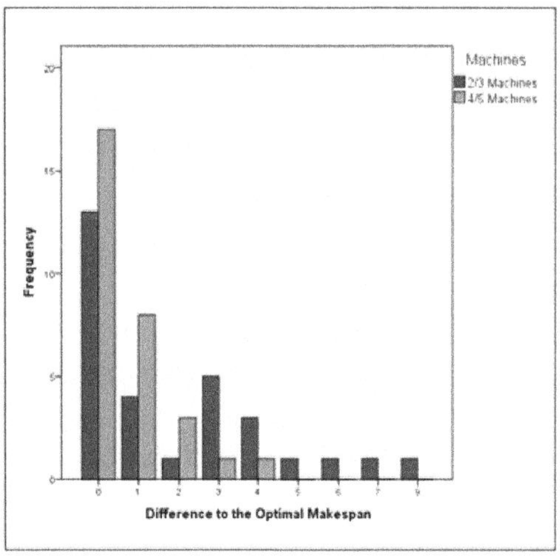

Fig. 8. Bar chart of the solutions for 2/3 and 4/5 machines

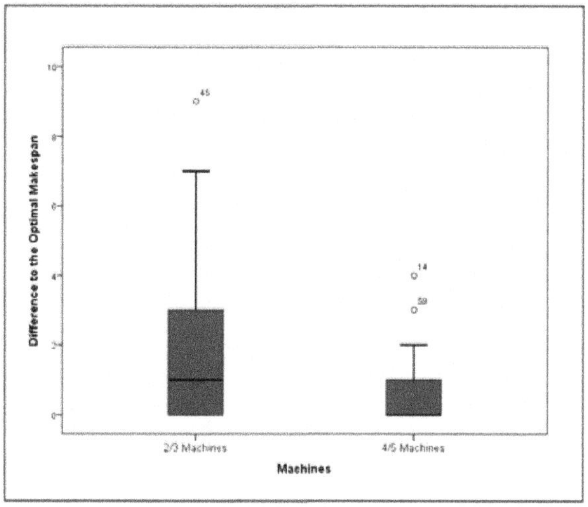

Fig. 9. Boxplot of the solutions for 2/3 and 4/5 machines

The boxplot from Fig. 9 permits to conclude that the *MOMCT* heuristic performs better with 4/5 machines than with 2/3 machines, in terms of minimization of makespan. The boxplot allows the analysis of location, dispersion, median and asymmetry of *MOMCT* with a variable number of machines.

To evaluate the significance of the obtained results the Student's t-test for independent samples has been used, considering the hypotheses:

H_0: $\mu_{2/3} - \mu_{4/5} \geq 0$

H_1: $\mu_{2/3} - \mu_{4/5} < 0$

Table 3 shows the result of the Student's t-test. It is not possible to assume equal variances (p-value of 0.000) and the Student's t-test has a *p-value* of 0.005 (0.010/2). The null hypothesis H_0, that considers the non-existence of difference between the performances on 4/5 and 2/3 machines was rejected ($p=0.005<\alpha$). This permits to conclude, with 95% of confidence that the *MOMCT* heuristic performs better in problems with more machines.

Table 3
Student's T Test Results for 2/3 and 4/5 machines

	Levene's Test for Equality of Variances		t-test for Equality of Means				
	F	Sig.	t	df	Sig. (2-tailed)	Mean Difference	Std. Error Difference
Equal variances assumed	19.975	0.000	-2.696	58	0.009	-1.300	0.482
Equal variances not assumed			-2.696	38.910	**0.01**	-1.300	0.482

6. Conclusions

In this paper, the *makespan* minimization problem in unrelated parallel-machines was analyzed. Some of the heuristics used to solve the problem were described and compared and a new heuristic was proposed. *MCT* performance depends on a random like order in which tasks are allocated. The possibility of ordering tasks before applying the *MCT* was studied in the proposed heuristic.

The *MOMCT* heuristic orders tasks in accordance to the difference between the mean of the completion time of a variable number of machines and the completion time on the minimum completion time machine. This does not assume that the tasks would be allocated to their preferred machine as several machines can be used to determine in which order tasks would be allocated to machines. Experimental analysis was performed in order to validate the performance of the proposed *MOMCT* heuristic. From the obtained results, it is possible to conclude that *MOMCT* found better solutions than *MCT* with a deviation of 3.00%. *MOMCT* found the optimal solution in 30 (50%) instances while *MCT* only found the optimal solution in 21 (35%) instances. From the obtained results it was possible to conclude with statistical evidence, through the Student's t-test, that the *MOMCT* was more effective than the *MCT* heuristic, with a 95% confidence level. The proposed heuristic performance seems to be dependent on the number of machines in the problem. In the computational study *MOMCT* found the optimal solution in 17 (56.7%) for 4/5 machines and 13 (43.3%) for 2/3 machines. From the Student's t-test it is possible to conclude that the *MOMCT* performs better in problems with more machines, with 95% of confidence level ($p=0.005<\alpha$). This is expected since the *MOMCT* does not assume that the tasks will be allocated to their preferred machine and will sweep all possible *a* values to determine the allocation sequence.

Future work should focus on more computational tests that will compare the performance of the proposed heuristic with other heuristics described in the literature. Moreover, additional emphasize will be given to compare the performance of the proposed heuristic when dealing with problems of different sizes and the formulation of how the performance of *MOMCT* depends on the number of machines used to order tasks in problems with different characteristics will also be further explored.

Acknowledgements

This work is supported by FEDER Funds through the "Programa Operacional Factores de Competitividade - COMPETE" program and by National Funds through FCT "Fundação para a Ciência e a Tecnologia" under the project: FCOMP-01-0124-FEDER-PEst-OE/EEI/UI0760/2011 and PEst-OE/EEI/UI0760/2014.

References

Baker, K. R., & Trietsch, D. (2009). *Principles of Sequencing and Scheduling* (First Edition). Wiley.

Blazewicz, A.J., Ecker, K. H., Pesh, E., Schmidt, G., & Weglarz, J. (2001). *Scheduling Computer and Manufacturing Process* (Second Edition). Springer.

Braun, T. D., Siegel, H. J., Beck, N., Bölöni, L. L., Maheswaran, M., Reuther, A. I., ... & Freund, R. F. (2001). A comparison of eleven static heuristics for mapping a class of independent tasks onto heterogeneous distributed computing systems. *Journal of Parallel and Distributed computing, 61*(6), 810-837.

Briceño, L. D., Siegel, H. J., Maciejewski, A. A., & Oltikar, M. (2012). Characterization of the iterative application of makespan heuristics on non-makespan machines in a heterogeneous parallel and distributed environment. *The Journal of Supercomputing, 62*(1), 461-485.

Ebenlendr, T., Krčál, M., & Sgall, J. (2014). Graph balancing: A special case of scheduling unrelated parallel machines. *Algorithmica, 68*(1), 62-80.

Chaturvedi, A. K., & Sahu, R. (2011). New heuristic for scheduling of independent tasks in computational grid. *International Journal of Grid and Distributed Computing*, 4(3), 25-36.

Fujimoto, N., & Hagihara, K. (2004). A Comparison among grid scheduling algorithms for independent coarse-grained tasks. *SAINT Workshops*, 674-680.

Glass, C. A., Potts, C. N., & Shade, P. (1994). Unrelated parallel machine scheduling using local search. *mathematical. Computer Modeling*, 20(2), 41-52.

Gupta, K., & Singh, M. (2012). Heuristic based task scheduling in grid. *International Journal of Engineering and Technology*, 4(4), 254-260.

Hariri, A., & Potts, C. (1991). Heuristics for scheduling unrelated parallel-machines. *Computers and Operations Research*, 18(3), 323–331.

Ibbara, O., & Kim, C. (1977). Heuristic algorithms for scheduling independent tasks of no identical processors. *Journal of Association for Computing Machinery*, 24(2), 280-289.

Lenstra, J. K., Shmoys, D. B., & Tardos, E. (1990). Approximation algorithms for scheduling unrelated parallel machines. *Mathematical Programming*, 46(1-3), 256-271.

Lin, Y. K., Pfund, M. E., & Flower, J. W. (2011). Heuristics for minimizing regular performance measures in unrelated parallel machine scheduling problems. *Computer & Operations Research*, 38(6), 901-916.

Martello, S., Soumis, F., & Toth, P. (1997). Exact and approximation algorithms for makespan minimization on unrelated parallel machines. *Discrete Applied Mathematics*, 75(2), 169-188.

Paranhos, D., Cirne, W., & Brasileiro, F. V. (2003). Trading cycles for information: Using replication to schedule bag-of-tasks application on computational grids. *International Conference on Parallel and Distributed Computing*, 2790, 169-180.

Pfund, M., Flower, J. W., Jatinder, & Gupta, J. N. D. (2004). A survey of algorithms for single and multi-objective unrelated parallel-machine deterministic scheduling problems. *Journal of the Chinese Institute of Industrial Engineers*, 21(3), 230-241.

Pinedo, M. L. (2012). *Scheduling Theory, Algorithms, and Systems* (Fourth Edition). Springer.

Serra e Santos, & A., Madureira, A. M. (2014). Ordered minimum completion time heuristic for unrelated parallel-machines. Proceedings of the *9th Iberian Conference on Information Systems and Technologies* (CISTI*2014*), 695-700.

Sivasankaran, P., Sornakumar, T., & Panneerselvam, R. (2010). Efficient heuristic to minimize makespan in single machine scheduling problem with unrelated parallel machines. *Intelligent Information Management*, 2, 188-198.

Sugavanam, P., Siegel, H. J., Maciejewski, A. A., Oltikar, M., Mehta, A., Pichel, R., ... & Pippin, A. (2007). Robust static allocation of resources for independent tasks under makespan and dollar cost constraints. *Journal of Parallel and Distributed Computing*, *67*(4), 400-416.

Vakhania, N., Hernandez, J. A., & Werner, F. (2014). Scheduling unrelated machines with two types of jobs. *International Journal of Production Research*, 52(13), 3793-3801.

Varela, L. R., & Ribeiro, R. A., (2003). Evaluation of simulated annealing to solve fuzzy optimization problems, *Journal of Intelligent & Fuzzy Systems*, 14(2), 59-71.

Varela, M.L.R., Aparício, J.N., & Silva, S.C. (2003), A web-based application for manufacturing scheduling. Proceedings of the *International Conference on Intelligent Systems and Control* (*IASTED*), 400- 405.

Varela, M. L. R., & Carmo-Silva, S. (2008). An ontology for a model of manufacturing scheduling problems to be solved on the web. *Innovation in Manufacturing Networks, 8th IFIP International Conference on Information Technology for Balanced Automation Systems*. Azevedo, A. (Ed.); Springer, 197-204.

Varela, M. L. R., Barbosa, R., & Putnik, Goran (2012). Experimental platform for collaborative inter and intra cellular fuzzy scheduling in an ubiquitous manufacturing system, Proceedings of the *First International Conference on Virtual and Network Organizations Emergent Technologies and Tools* (*ViNOrg'11*), 227-236.

Varela, M. L. R., Putnik G. D., & Cruz-Cunha M. M., (2012) Web-based technologies integration for distributed manufacturing scheduling in a virtual enterprise. *International Journal of Web Portals*, 4(2), 19-39.

Varela, M. L. R., Putnik, G. D, & Ribeiro, R. A. (2012). A web-based platform for collaborative manufacturing scheduling in a virtual enterprise. *Information and Communication Technologies for the Advanced Enterprise an international journal*, 2, 87-108.

Varela, M. L. R., & Ribeiro, R. A. (2014), Distributed manufacturing scheduling based on a dynamic multi-criteria decision model. *Studies in Fuzziness and Soft Computing*, 317, 81-93.

Verschae, J., & Wiese, A. (2014). On the configuration-LP for scheduling on unrelated machines. *Journal of Scheduling*, 17, 371-383.

Xhafa, F., & Abraham, A. (2008). *Metaheuristics for scheduling in industrial and manufacturing applications series: Studies in computational intelligence,* 128, Springer.

Design and simulation of CONWIP in the complex flexible job shop of a Make-To-Order manufacturing firm

Giovanni Romagnoli*

Department of Industrial Engineering, University of Parma, Italy

CHRONICLE

Keywords:
CONWIP
Simulation
Shop floor
Complex manufacturing system
Flexible job shop
Make-to-order

ABSTRACT

This paper presents a methodology for the design and integration of CONWIP in a make-to-order firm. The approach proposed was applied directly to the flexible job shop of a real manufacturing firm in order to assess the validity of the methodology. After the description of the whole plant layout, attention was focused on a section of the shop floor (21 workstations). The CONWIP system deals with multiple-product families and is characterized by path-type cards and a pull-from-the-bottleneck scheme. The cards release strategy and a customized dispatching rule were created to meet the firm's specific needs. After the simulation model of the present state was built and validated, the future state to be implemented was created and simulated (i.e. the CONWIP system). The comparison between the two systems achieved excellent results, and showed that CONWIP is a very interesting tool for planning and controlling a complex flexible job shop.

1. Introduction

The market of products having high added value has undergone a profound change in the past decades. Consumers require a variety of products, product quality, quick and reliable delivery times, and short lead times. On the other hand, managers and shareholders ask for minimal work-in-process (WIP) and maximum utilization of resources. For these reasons, modern manufacturing systems are becoming more and more familiar with concepts like flexibility, quality and adaptability to consumers' demands, particularly when operating in a make-to-order (MTO) environment (Sultana & Ahmed, 2014).

Unfortunately, the above mentioned goals (i.e. variety, quality, short lead times, minimal WIP and maximum utilization of resources) appear to be conflicting: one of the main objectives of modern production planning and control systems is to find a compromise solution amongst them (Lengyel et al., 2003). The first systems that tried to attain such a solution are the well-known master production schedule (MPS), material requirement planning (MRP) and manufacturing resource planning (MRPII) (Gibson et al., 1995). They have been (and they still are) excellent long term planning, scheduling and control tools (Vollman et al., 1997), but they have also revealed a series of problems which are often associated with push systems (Lambrecht & Decaluwe, 1988; Ramsay et al., 1990; Adetunji &

* Corresponding author.

Yadavalli, 2012). Pull systems, on the other hand, instead of scheduling the start of jobs, authorize them. The best known pull system is kanban (the Japanese word for card), described and analysed in depth by Hall (1981), Ohno (1988) and Monden (1998). MRP is generally considered to be applicable to many more manufacturing firms than kanban is. However, kanban seems to produce superior results when it can be applied (Spearman et al., 1990). Hodgson and Wang (1991) also assert that push and pull control strategies have different advantages and disadvantages, and that combinations of push and pull strategies may be easier to implement and may achieve better results than either pure push or pure pull. After this study, several researchers tried to combine these two types of control systems (Flapper et al., 1991; Larsen & Alting, 1993; Villa & Watanabe, 1993, to name a few). Spearman et al. (1990) proposed a hybrid production planning and control system called CONWIP (CONstant Work In Process or, as some researchers argue, CONtrolled Work In Process), with the objective of sharing the positive aspects of pull systems with the wide applicability of push systems. Although this system is relatively recent, Framinan et al. (2003) report that the acronym CONWIP was firstly coined in 1988 and its basic mechanics dates as back as 1963.

Since 1990, CONWIP has been the object of research from many points of view, especially by comparing it with scheduling and control systems such as MRP, JIT and synchro-MRP (for a literature review and description of synchro-MRP see Bertolini et al., 2013). According to Benton and Shin (1998), the studies in literature that simultaneously address push/pull systems (MRP/JIT) can be classified into comparison studies and integration studies, and their approach may be either conceptual (broader and more general) or analytical (narrower and more specific); in analytical comparison and integration studies, methods like simulation and mathematical analysis are the most frequently used. Amongst these studies, Geraghty and Heavey (2004), Yang et al. (2007), Koulouriotis et al. (2010) and Chong et al. (2013) have focused on advantages (or disadvantages) of a CONWIP system and on his potential improvements. Sharma and Agrawal (2009) used AHP algorithm to compare kanban, CONWIP and a hybrid system proposed by Bonvik et al. (1997). Other studies of CONWIP simulate a whole supply chain, such as the works by Rubiano Ovalle and Crespo Marquez (2003), Özbayrak et al. (2006), Pettersen & Segestedt (2009), or even extend the use of CONWIP to Project Management (Anavi-Isakow & Golany, 2003). Some papers also evaluate performances of CONWIP by studying card setting and control (Framinan et al., 2006; Renna, 2010; Braglia et al., 2011) or simulate a simple production system (Huang et al., 1998; Duri et al., 2000). However, though most of the papers claim that CONWIP is superior to both MRP and JIT (Roderick et al., 1994; Huang et al., 1998; Pettersen & Segerstedt, 2009), all the previously mentioned studies suffer from one specific limitation: they fail to address the integration of CONWIP into the shop floor of a real complex firm. Preliminary work on this matter was undertaken by Li (2010), although his main focus is the coordination of layout change and quality improvement, rather than general guidelines for the integration of CONWIP into MTO systems. In order to attempt to fill this gap, this paper presents a simulative study for the design and integration of CONWIP into the flexible job shop of an MTO firm, operating with a general job shop configuration. The study is developed in cooperation with a well-known firm, world leader in the manufacturing and assembling of oil hydraulic accessories.

According to Stevenson et al. (2005), different shop floor configurations are available for MTO industries, namely (i) pure flow shop, (ii) general flow shop, (iii) general job shop and (iv) pure job shop. Key differences among these configurations are the direction of material flow and the degree of customization (for further details on flow routeing matrices see Enns, 1995). Besides, Mahdavi et al. (2010) suggest that a flexible job shop is characterized by the presence of a set of workstations in which an operation may be performed by any machine/assembly station of the work centre.

The remainder of the paper is organized as follows: in Section 2 we will present and briefly describe the proposed approach. This approach was directly implemented in an industrial case of relevance which sets a good example and assesses the validity of our approach. The case study is presented in Section 3: after a description of the firm, subsections 3.1 to 3.11 scrupulously follow and apply the guidelines of the proposed approach. Finally, Section 4 describes and analyses the results of the simulation and Section 5 reports conclusions and possible future developments.

2. The proposed approach

The proposed approach for the design and integration of CONWIP into the flexible job shop of a real manufacturing system is similar to the one proposed in Bertolini and Romagnoli (2013), and structured according to Table 1.

Table 1
Steps of the proposed approach

Step	Guideline
1	Classify workstations and machines
2	Group jobs in families
3	Connect every family with an average value of Cycle Time (CT) and Sample Standard Deviation (SSD)
4	Choose characteristics of the CONWIP system
5	Define the cards release strategy
6	Define a dispatching rule
7	Build the simulation model of the current state
8	Validate the simulation model
9	Build the simulation model of the future state
10	Simulate the future state
11	Analyze results
12	Possibly implement the simulated future state

The first step of the approach consists in the identification and classification of workstations and machines, and in most cases can be directly achieved from the shop floor. Afterwards, the investigation moves from workstations and machines to jobs, and it is often necessary to group jobs into families. This last statement is particularly appropriate for job shops, and one of the instructions on how to group jobs is to attribute a single routing to each family, when possible. Every job family must then be connected with an average value of CT and SSD, prior to the selection of the general characteristics of the CONWIP system to be implemented (steps 3 and 4). Strict attention must be paid to the cards release strategy (step 5), i.e. a general characteristic of the CONWIP system that much too often defines the behaviour of the future state. Together with the cards release strategy, it is also important to define a dispatching rule, so as to sort jobs as they arrive at machines. Afterwards, the simulation model must be built and validated, i.e. the as is state of the system must be reproduced by a simulation model that is an accurate representation of the real system. Once the simulation model of the as is state has been built and validated, the future state model can be built and its critical parameters can be determined. Finally, the future state is simulated (step 11) and, after an analysis of the results it has produced, possibly implemented (step 12).

3. Industry application

The firm we collaborated with is a joint-stock company that designs and produces hydraulic pumps and motors. Attention was focused on the assembly phase of oil hydraulic pumps. Main objectives of this company are the high quality of end products, high flexibility in product design and manufacturing, and a strong commitment to short delivery times. In order to meet these goals, rigorous testing is carried out on every product and several customizations are available to customers. The company produces through a mix of *job production* and *batch production* methods, and its manufacturing philosophy follows an assemble-to-order production approach for 1/3 of the products, whilst the remaining 2/3 follow a MTO approach (and, rarely, an engineer-to-order approach). Despite its complexity, as often happens in real enterprises, the main production approach of the firm is MTO.

Given the wide variety of manufacturing methods and philosophies, as well as the need for high production flexibility, the company's organization follows a flexible general work shop layout, where the following characteristics can be found:

- each job passes through a different number of workstations;
- on some WSs, operations may be performed by any machine/assembly station of the work centre;
- the routing may be different from job to job;
- CTs on each machine are often different from job to job;
- the workflow is not unidirectional, though a dominant flow direction exists.

In the present configuration, production planning and control is completely based on a push system (MRP) with EDD dispatching rule and expediting. With a classic MRP system, the firm exercises direct control on throughput and indirect control on WIP; for this reason, in order to ensure the maximum production rate, every station is provided with a certain *load* (i.e. a certain amount) of parts to be processed, so as to avoid starvation, even if this condition highly increases the WIP. As a matter of fact, the current situation relies on an operator who daily verifies all the orders that must be dealt with and short lists the priority ones.

An analysis of the flow times of any product code shows how, with an average flow time of 10 working days, effective flow times range from 1 to 100 days. A first attempt at controlling flow times was made in early 2007 by adopting assembly kits (set of all parts needed to assemble a given product). This solution became necessary so as to allow a fixed position assembly line to deal with a wide range of different products.

The plant layout of the company is divided into different areas: (i) **machining and quality control area**, (ii) **acceptance and storage area**, (iii) **assembling and testing area**, (iv) **finished product storage area.** Our attention focused on the assembling and testing area (iii), core business of the company.

3.1 Classify workstations and machines

The workstations (WSs) of the assembling and testing area of the plant are reported and classified in Fig. 1. Note that the organization of Fig. 1 is mainly illustrative, because product routings are often not linear (i.e. not from top to bottom) and they seldom call at every zone of the assembling and testing area. For these reasons, as well as for the sake of comprehensibility, materials and information flow are not reported in Fig. 1.

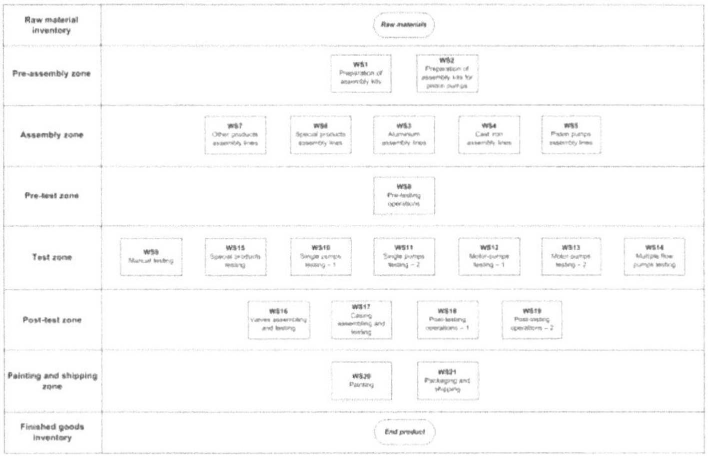

Fig. 1. workstations (WSs) of the assembling and testing area and their classification

The attention of the study was directed to the assembling and testing area because this is the core business of the firm itself. As Fig. 1 shows, this area can be divided into six zones between raw material inventory and finished goods inventory: (i) pre-assembling, (ii) assembling, (iii) pre-testing, (iv) testing, (v) post-testing, (vi) painting and shipping.

The **pre-assembly zone** (2 WSs) is the junction point between the inbound stock point and the assembly zone: here the list of products to be manufactured is supplied by the MRP, and the workers, i.e. 4 people per shift on two 8-hour shifts a day, fill the assembly kits according to the Bill Of Materials.

The **assembly zone** (5 WSs) of the plant works on a single 8-hour shift and roughly assembles 2,600 products per day. The assembly operations vary according to the product variant; they can be carried out on a fixed-position assembly layout or an unpaced multi-model assembly line (for a classification of assembly lines see Scholl, 1999). This zone is directly fed by the pre-assembly zone and the assembly lines are divided as follows:

- aluminium – 4 mixed-model unpaced assembly lines. Each line counts 3 assembly stations (i.e. 3 workers) that perform all of the assembly operations;
- cast iron – 2 mixed-model unpaced assembly lines. Each line counts 3 assembly stations;
- piston pumps – 3 fixed-position multi-model assembly lines. Each line counts 2 workers that pre-assemble products (if necessary) and carry out the entire assembly process;
- special products – 6 fixed-position multi-model assembly lines. Each line counts 1 worker who pre-assembles products (if necessary) and carries out the entire assembly process;
- other products – 1 fixed-position multi-model assembly line with 1 worker.

The main difference between the aluminium and the cast iron assembly lines is that the latter needs a subassembly operation for the mounting of bushings and bearings on the casing.

Within the **pre-test zone** (1 fixed-position multi-model assembly line with 1 worker), jobs coming from the assembly zone may be modified and adapted to the following test zone. Examples of adjustments may be the assembling of multiple flow pumps, such as the cast iron-aluminium double pump or the piston-gear double pump. In order to assemble multiple pumps (i.e. WIP coming from different assembly lines), the synchronization of jobs is very important at this stage.

The **test zone** (7 WSs) is fed by the assembled products and processes every item; since product and routing variability is very wide, the number of products entering the testing queue experience great variability too. The testing zone includes a total of 12 test benches: 2 benches each for WS_{10} to WS_{14}, and 1 bench for WS_9 and WS_{15}. The entire work cycle of the pump is performed at these benches and burrs are created so as to ensure better operating conditions of the end product. As at the assembly zone, testing WSs and benches are classified according to the jobs they process. In this case, however, distinctions are not so important, since most of the benches may test most of the products (at the price of significantly longer setup times).

The test phase also verifies if the performances of the products comply with standards; if this should not be the case, the products are to undergo repairs and a new test is performed. The average defect rate is 2.5% and is not dependent on product family, batch, shift or raw materials (i.e. random). Since the testing zone processes more items than the assembly zone (because of repairs) and has long CTs (up to 45 minutes), it works on three 6.5-hour shifts per day.
The **post-test zone** operates on a single 8-hour shift per day, and counts 4 WSs with 1 worker each (fixed position). Repairs and other post-testing operations are performed here, so as to identify and solve problems that may arise during the testing phase.

In the **painting and shipping zone** (2 WSs) items are painted and prepared for shipping, when needed. Before shipping, purchase orders must be fulfilled, i.e. all the items must be synchronized. This zone is directly fed by the post-test zone and works on a single 8-hour shift per day.
A close analysis of the firm suggests that the plant bottleneck is located at the test zone: indeed, these WSs appear to be the ones with the highest long term utilization (hence they work on three shifts per day), because of long CTs and the need to test every product at least once (or even twice, if repairs are needed).

3.2 Group jobs into families

The description hitherto presented has not mentioned CTs, SSDs and production rates of each workstation of the plant. All of these data, in fact, may be measured for each job at each WS, but there are far too many to manage. The number of different product codes, in fact, is roughly equal to 25,000, with an average process batch of 10 parts, while the workstations identified in this part of the plant amount to 21. For this reason, jobs were grouped into families with the same routing so as to allow the (future) simulation processes. The number of jobs was reduced from 25,000 different jobs to 195 different routings, each one passing in the same sequence through 21 WSs, though not necessarily through the same machine/assembly station at each WS.

3.3 Connect every family with an average value of CT and SSD

Each family of products was then associated with the total part number, average CT and SSD of all the product codes and part numbers belonging to it. The average CTs ($\bar{t}_{i,n}$) and SSDs ($s_{i,n}$) of every product family at each workstation were calculated as follows.

$i = 1\ to\ 195$	$[n]$	family of products F_i
$J_i = 1\ to\ N_i$	$[n]$	number of jobs belonging to F_i
$P_{J,i} = 1\ to\ N_{J,i}$	$[n]$	number of parts belonging to J_i
$WS_n\quad n = 1\ to\ 21$	$[n]$	workstations
$t\left(P_{J,i}\,;n\right)$	$[min]$	cycle time of $P_{J,i}$ at WS_n
$\bar{t}_{i,n}$	$[min]$	average CT of F_i at WS_n
$s_{i,n}$	$[min]$	sample standard deviation of F_i at WS_n

and

$$\bar{t}_{i,n} = \frac{\sum_{J_i=1}^{N_i}\sum_{P_{J,i}=1}^{N_{J,i}} t\left(P_{J,i}\,;n\right)}{\sum_{J_i=1}^{N_i} N_{J,i}} \tag{1}$$

$$s_{i,n} = \sqrt{\frac{\sum_{J_i=1}^{N_i}\sum_{P_{J,i}=1}^{N_{J,i}}\left[t\left(P_{J,i}\,;n\right) - \bar{t}_{i,n}\right]^2}{\sum_{J_i=1}^{N_i}\left(N_{J,i}\right) - 1}} \tag{2}$$

Table 2 reports the routing, $\bar{t}_{i,n}$ and $s_{i,n}$ for families from 1 to 5 according to Eq. (1) and Eq. (2) (see also Bertolini & Romagnoli, 2013). For example, a part belonging to F_3 will visit WS_3 with an average CT of 0.73 min and a SSD of 0.094, then *WS8* with an average CT of 1.39 min and a SSD of 0.176 etc. Families in Table 2 are arranged in decreasing order of the total number of parts ($\sum_{J_i=1}^{N_i} N_{J,i}$). In the same way, routings, average CTs and SSDs were calculated for the 195 families. The data reported in Table 2 suggest the following considerations:

- since values of CT and SSD are connected with families of different products (even if sharing the same routing), it may be that several products of a family experience simple operations at some WSs, and this leads to short average CTs;
- it often happens that WS_1 and WS_{21} experience simple operations (such as applying stickers or preparing small kits) and this leads to short CTs at these WSs;
- sometimes post testing operations only involve visual inspections on products;
- finally, it may be seen that, though a dominant flow direction exists between WSs (i.e. ascending order), the workflow is not unidirectional (see for example F_1).

Table 2
Average CTs and sample standard deviation for the first 5 families of products (Bertolini & Romagnoli, 2013)

Family	Workstation	Average CT [min]	Sample Standard Deviation [min]
F_1	WS_{17}	0.31	0.101
	WS_1	0.08	0.021
	WS_3	5.13	1.857
	WS_{11}	1.95	0.730
	WS_{18}	0.46	0.078
	WS_{21}	0.15	0.046
F_2	WS_1	0.60	0.071
	WS_3	5.13	0.920
	WS_{12}	3.61	0.943
	WS_{18}	0.55	0.067
	WS_{21}	0.14	0.058
F_3	WS_3	0.73	0.094
	WS_8	1.39	0.176
	WS_{14}	3.71	1.066
	WS_{18}	0.20	0.022
F_4	WS_3	5.17	1.870
	WS_{11}	1.81	0.504
	WS_{21}	0.13	0.047
F_5	WS_8	1.44	0.220
	WS_{13}	3.83	1.535
	WS_{20}	1.83	0.728

3.4 *Choose characteristics of the CONWIP system*

After a close examination of the production system, we chose to simulate the implementation of a CONWIP production plan and control with multiple-product families and the following features (see Hopp & Spearman, 2008):

- *pull-from-the-bottleneck scheme*: cards trigger strategy is based on the bottleneck status, where the bottleneck was identified at the test zone, i.e. at WS_9 to WS_{15}. This means that cards only flow from the beginning of each routing to the test zone, and the WIP level is held constant in the WSs up to and including the bottleneck;
- *path type cards*: each card defines the WS in which testing will follow (i.e. the 7 testing WSs, from WS_9 to WS_{15}), and therefore product families will be grouped according to their testing WSs. This choice leads to 7 different types of cards and offers the following advantages: (i) setup times at testing workstations are shorter because test benches are dedicated; (ii) WIP

traceability is higher, because we already know in which queue (in front of a testing workstation) the product will be waiting for processing; (iii) boxes (or posts) for CONWIP cards are smaller, though increased in number, and this improves internal logistics;
- *common unit to measure WIP*: the sum of the working capacities (in minutes) that all the jobs released will occupy at the 7 testing WSs.

Current characteristics of the shop floor and assumptions made for the simulation of the present and future states of the system are summarized in Table 3.

Table 3
Assumptions and comparison between factory conditions and the present and future state simulations

Element of comparison	Factory conditions	Simulation (AS IS)	Simulation (TO BE)
Scheduling system	MRP	MRP	Pull-from-the-bottleneck CONWIP with path type cards
Dispatching rule	EDD	EDD	$(Due\ Date - \alpha) - \dfrac{number\ of\ jobs\ in\ process}{total\ number\ of\ jobs}$
Correction on dispatching rule	Expediting	None	Value of α and percentage of jobs in process belonging to the same purchase order
WIP	Not controlled	Not controlled	Limit on the total load (i.e. sum of occupied capacities) of all jobs to the testing WSs
Part transfer	Negligible	Not considered	Not considered
Number of jobs	25,000	195 families	195 families

3.5 Define the cards release strategy

The logic behind the releasing of the cards is illustrated below.

$WS_n \quad n = 9\ to\ 15$	$[n]$	WS of the testing zone
$J_n = 1\ to\ M_n$	$[n]$	number of jobs released that will be tested at WS_n
L_n	$[min]$	maximum product *load* at WS_n (working minutes)
RC_n	$[min]$	residual capacity at WS_n
$OC_{J,n}$	$[min]$	occupied capacity by job J at WS_n
$OR_{J,n}$	$[0; 1]$	releasing of a card for job J that will be tested at WS_n
$OF_{J,n}$	$[0; 1]$	finishing of job J at WS_n

In this context, each one of the 7 testing WSs (let's say WS_{11}) will process a maximum number M_n of jobs. In the same time, WS_{11} is characterized by a *product load* (L_{11}), i.e. the process variable that indicate the maximum total amount of WIP (expressed in testing minutes at station *11*) allowed in the system for that specific working station, and by two variables, *residual capacity* (RC_{11}) and the *capacity occupied* by a general job J ($OC_{J,11}$).

Each job J_{11} released in the system will occupy an average capacity of $OC_{J,11}$ testing minutes at WS_{11}, and the residual capacity of a general WS_n is defined in Eq. (3).

$$RC_n = L_n - \sum_{j=1}^{M_n} OC_{J,n} \tag{3}$$

As Eq. (3) shows, RC_n is always less than or equal to L_n. By means of RC_n the system deals with the possibility of releasing new cards for jobs that will be processed at WS_n, as is shown below.

Initialization of the system

$$OR_{J,n} = 0$$
$$OF_{J,n} = 0$$
$$for\ n = 9\ to\ 15\ do$$
$$RC_n = L_n$$
$$J_n = 1$$
$$\quad if \quad RC_n > 0\ then\ OR_{J,n} = 1;\ J_n = J_n + 1;\ RC_n = RC_n - OC_{J,n}$$
$$\quad else\ \ OR_{J,n} = 0$$
$$\quad end\ if$$
$$next$$

Running phase

$$if \quad OF_{J,n} = 1\ \ then\ \ RC_n = RC_n + OC_{J,n}$$
$$\quad if\ RC_n > 0\ \ then\ \ OR_{J,n} = 1; J_n = J_n + 1; RC_n = RC_n - OC_{J,n}$$
$$\quad else\ \ OR_{J,n} = 0$$
$$\quad end\ if$$
$$end\ if$$

The first loop previously illustrated (initialization of the system) ends when all residual capacities are not positive, so no other card can be released in production. Afterwards, during the running phase, when the capacity of a general J_n is freed (for its test has finished), the $OC_{J,\ n}$ is released and if RC_n becomes positive, the card for a new job to be tested at WS_n will be released. In this way, we achieve a CONWIP system capable of keeping WIP under control up to the bottleneck through path type cards by measuring WIP in minutes of occupied capacity at the testing workstation (WS_n). Once released, jobs follow their routing up to the testing zone: each time a job J_n finishes its test, the value of the occupied capacity of J_n ($OC_{J,\ n}$) is added to the residual capacity of WS_n (RC_n), where n is the WS in which the testing of J_n takes place. Fig. 2 shows a diagram of this process. As can be seen in Fig. 2, another main difference between the current and the future state is the organization of the backlog list. By now, the backlog is a list of jobs to be processed sorted according to the EDD rule. In the future state, backlogs will be created for each testing WS, so as to create 7 different backlog lists associated with WS_9 to WS_{15}. Section 3.6 explains how jobs will be released into the system from the backlog list, as well as how they will be dispatched when queueing in front of a processing workstation.

3.6 Define a dispatching rule

According to the specific needs of the firm, the following goals were identified as guidelines for the scheduling problem:

- punctuality on delivery of customer orders;
- synchronization of customer orders (when belonging to the same purchase order);
- priority to *hot* jobs.

Jobs are defined as being *hot* when belonging to an important purchase order and coming from priority customers, i.e. when the delay will cause serious damage to a *hot* customer. Purchase orders are independent, and an analysis of the firm's database showed that jobs were equally parted in hot (50%) and normal (50%).

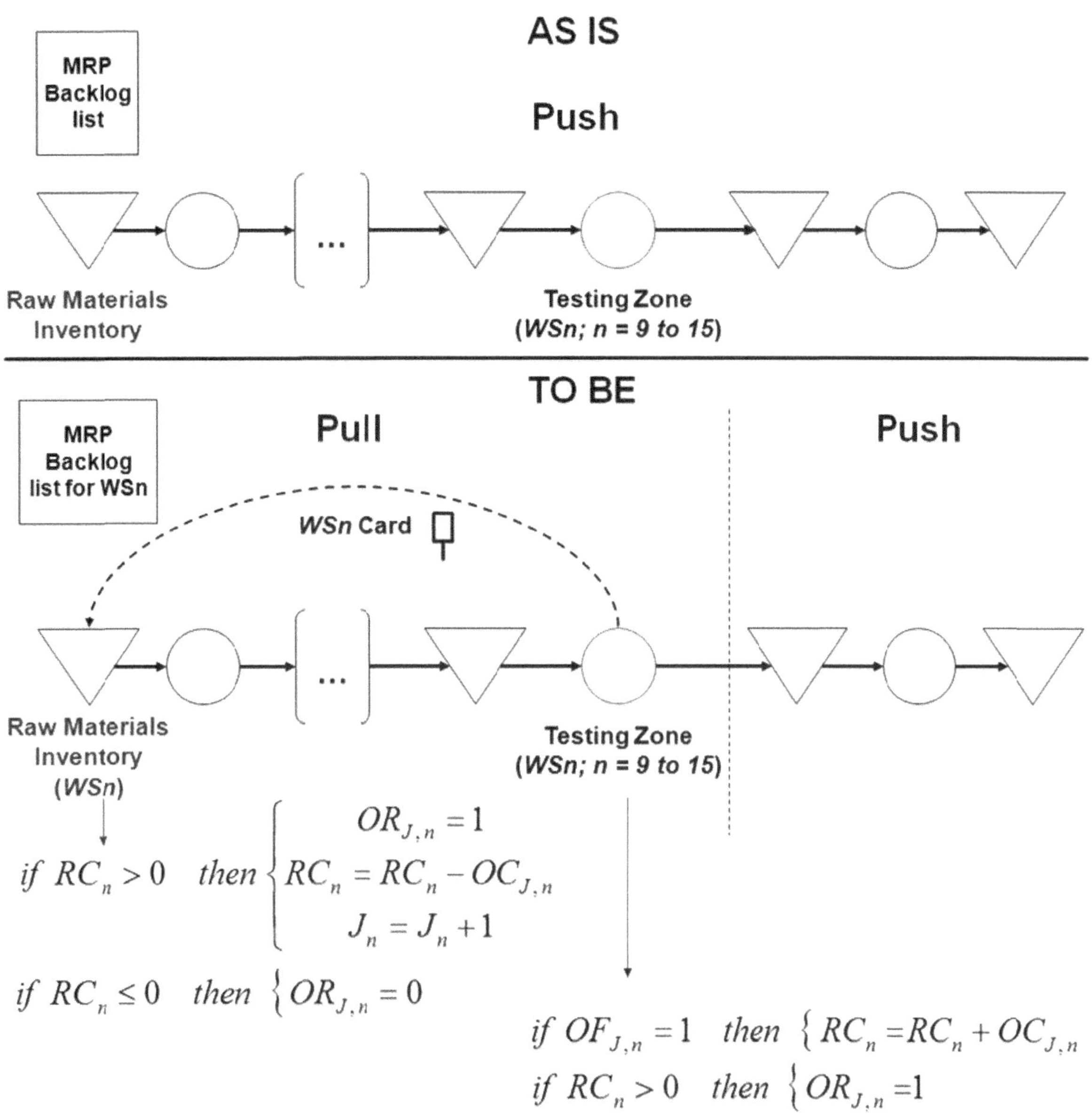

Fig. 2. cards release strategy, comparison between the current and the future state

We referred to the Pinedo approach (Pinedo, 2002) a | b | c, which indicates:

- a = machine environment, *Flexible Job-shop* with 21 stages of parallel machines (FJ_{21});
- b = processing characteristics and constraints, *Release Time* (r_j);
- c = objectives, which are (i) average job *Tardiness*, (ii) average *Staging Time* and (iii) average *Weighted Lateness* of purchase orders. The weight criterion, *w*, is set equal to 10 for *hot* jobs and 1 for *normal* jobs, and the Staging Time is equal to the wait-to-match time (i.e. the time some parts 'wait' because of lack of synchronization of purchase orders).

According to Thiagarajan and Rajendran (2005), the dispatching rule for product families shall be a mix of:

- *EDD* in order to apply a control on items punctuality;
- *NUJOB*, modified so as to manage the synchronization of different jobs belonging to the same purchase order;
- application of factor α, which indicates the advance we plan to have (measured in number of days ahead of schedule, when $\alpha > 0$) so as to consider the importance of a *hot* order.

This leads to the following dispatching rule, where the fraction considers the ratio between the number of jobs already in process and the total number of jobs belonging to the same purchase order (see also Table 3):

$$Dispatching\ rule\colon\ (Due\ Date - \alpha) - \frac{number\ of\ jobs\ in\ process}{total\ number\ of\ jobs} \tag{4}$$

3.7 Build the simulation model of the current state

The simulation model that has been used is a discrete, dynamic and stochastic *discrete-event simulation model*, implemented in an *object-oriented* software, *Simul8 (release 12).* The model of the assembling and testing area has been organized in 21 workstations, with a variable number of parallel machines, and the simulation of each workstation was made with the utilization of $\bar{t}_{i,n}$ and $s_{i,n}$, according to the product family (see Section 3.3). CTs and SSDs used in this model refer to year 2008. They are represented as normal distributions with $CT_{i,n} \sim \mathcal{N}(\bar{t}_{i,n}\,;\, s_{i,n}^2)$.

Replicating the conditions of the actual shop floor attentively, we created all the connections between job-shops and stock points, in order to simulate the routings of 195 product families F_i. For the purposes of the simulation, *Simul8* software was linked to several MS Excel worksheets, so as to allow the processing of a greater number of data (to the detriment of simulation times).

3.8 Validate the simulation model

Particular attention was paid to the validation of the current state model. The steady state was graphically identified after a start-up period of 20 working days by monitoring the reciprocal of the standard deviation of daily production rate (see Fig. 3).

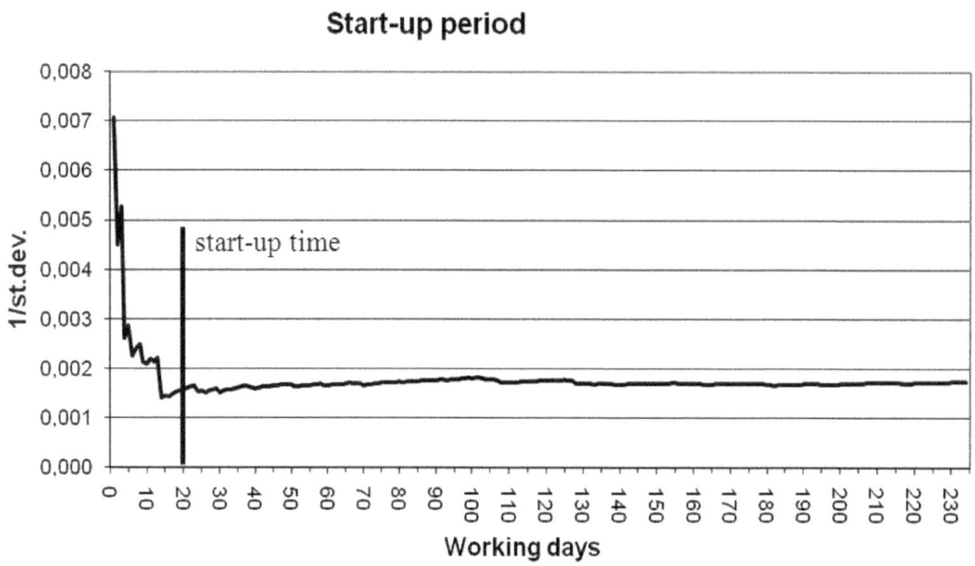

Fig. 3. start-up period definition

We then carried out 10 consecutive simulation runs of the current state, each lasting one year, and analysed their output, i.e. the average values of response variables.
Model validation was implemented by controlling the results we obtained for the following variables: (i) average daily production, (ii) average weekly WIP, (iii) average flow time and (iv) average job tardiness. Table 4 reports the comparison between data retrieved from the firm's database and the ones obtained with the simulation. Since every response variable, obtained through the simulation model, is close to the reference data from the database, the model is consistent.

Table 4
Comparison between data from the firm's database and the simulation (AS IS) for the validation of the model

		Industry database	Current state simulation	Comparison
Validation parameters	Average daily production (parts)	2,600	2,615	+ 0.6%
	Average weekly WIP (parts)	18,000	17,630	- 2.1%
	Average flow time (days)	9.2	8.7	- 5.8%
	Average job tardiness (days)	7.2	7.8	+ 8.3%

3.9 Build the simulation model of the future state

Once the validation of the current state was implemented, the future state model was built, in order to simulate a system with the characteristics described in Sections 3.4 to 3.6. Only two important process factors still need to be determined for the optimization of the future state, i.e. the maximum value of WIP allowed in the system and the value of α, the planned advance in due date for hot jobs (see Sections 3.4 and 3.5). Factors and levels at this step were chosen by mutual consent with the firm's management, and they are reported in Table 5. We have proceeded with a quick Design Of Experiment (DOE) applied to the simulation process, according to Montgomery (2001).

Table 5
Factors and levels for the future state

	Factors	**Lower Level (-)**	**Higher Level (+)**
A	α - days ahead of schedule for important orders (days)	5	10
B	WIP - sum of working minutes at testing WSs (minutes)	28,500	34,200

Table 6
Design matrix and results of the 2^2 factorial analysis

	A	B	R_1	R_2	R_3
Design point	α	WIP	*Average job Tardiness*	*Average Staging time*	*Average Weighted Lateness of purchase orders*
(1)	5	28.500	9,45	4,06	30,85
a	10	28.500	10,32	4,00	14,16
b	5	34.200	9,06	4,00	19,20
ab	10	34.200	8,36	4,25	24,63
			$A_1 = 0.09$	$A_2 = 0.10$	$A_3 = -5.63$
			$B_1 = -1.18$	$B_2 = 0.10$	$B_3 = -0.59$
			$AB_1 = -0.79$	$AB_2 = 0.15$	$AB_3 = 11.06$

Response variables are reported in Section 3.6 (i.e. objectives of the a | b | c Pinedo approach). We applied a full factorial design and made a 2^2 sampling campaign with 10 replications. Replications were made by conducting "Trials" in the simulation software (i.e. a series of runs performed with all the same parameters in the simulation except for the random numbers). Results are displayed in Table 6, where each design point is characterized by a factor-level combination and an average outcome of the three response variables (R_1; R_2 and R_3); at the same time, each response variable is related to the main effect of factor A and B and to the interaction effect (AB). Table 7 reports the ANOVA of the factors related to the three response variables.

Table 7
ANOVA of the three response variables

R1 - Average tardiness

Source of variation	*Sum of squares*	*DoF*	*Mean square*	F_0	*P-value*
A	0.08	1	0.076	0.129	0.721
B	13.80	1	13.800	23.415	0.000
AB	6.16	1	6.159	10.449	0.003
Error	21.22	36	0.589		
Total	41.25	39			

R2 - Average staging time

Source of variation	*Sum of squares*	*DoF*	*Mean square*	F_0	*P-value*
A	0.09	1	0.091	0.491	0,488
B	0.09	1	0.088	0.477	0.494
AB	0.25	1	0.251	1.357	0.252
Error	6.67	36	0.185		
Total	7.10	39			

R3 - Average weighted lateness

Source of variation	*Sum of squares*	*DoF*	*Mean square*	F_0	*P-value*
A	316.85	1	316.848	64.513	0.000
B	3.50	1	3.498	0.712	0.404
AB	1,223.12	1	1,223.125	249.039	0.000
Error	176.81	36	4.911		
Total	1,720.28	39			

As it can be seen from Table 7, response variable number 2 (average staging time) is not influenced by any of the factors (minimum p-value is 0.252). However, factor B (WIP) significantly influences average job tardiness (R_1), and factor A (alpha) influences average weighted lateness of purchase orders (R_3); furthermore, both R_1 and R_3 are influenced by the interaction between the factors (AB). If we go back to Table 6, we can deduce that B_1 (i.e. the main effect of B measured on R_1) is negative but small; that is an increase of B slightly diminishes average job lateness (which is good). On the other hand, A_3 is negative but relatively big; and this means that an increase of A significantly diminishes average weighted lateness of purchase orders. Furthermore, the interaction AB_3 is positive and big, and this means that, in order to limit R_3, the factors must not be both at the high (or low) level (i.e. no *ab* and no *(1)* combinations are recommended).

For these reasons, the best configuration was found to be design point *a*:

- WIP = 28,500 minutes
- α = 10 days

3.10 Simulate the future state

The results of the simulation and the comparison between the current and the future state of the system are reported in Fig. 4 and Table 8. Fig. 4 graphically shows the comparison in average WIP (*parts/week*), while Table 8 reports average values and standard deviations in production, WIP, flow times, job tardiness, staging times and weighted lateness.

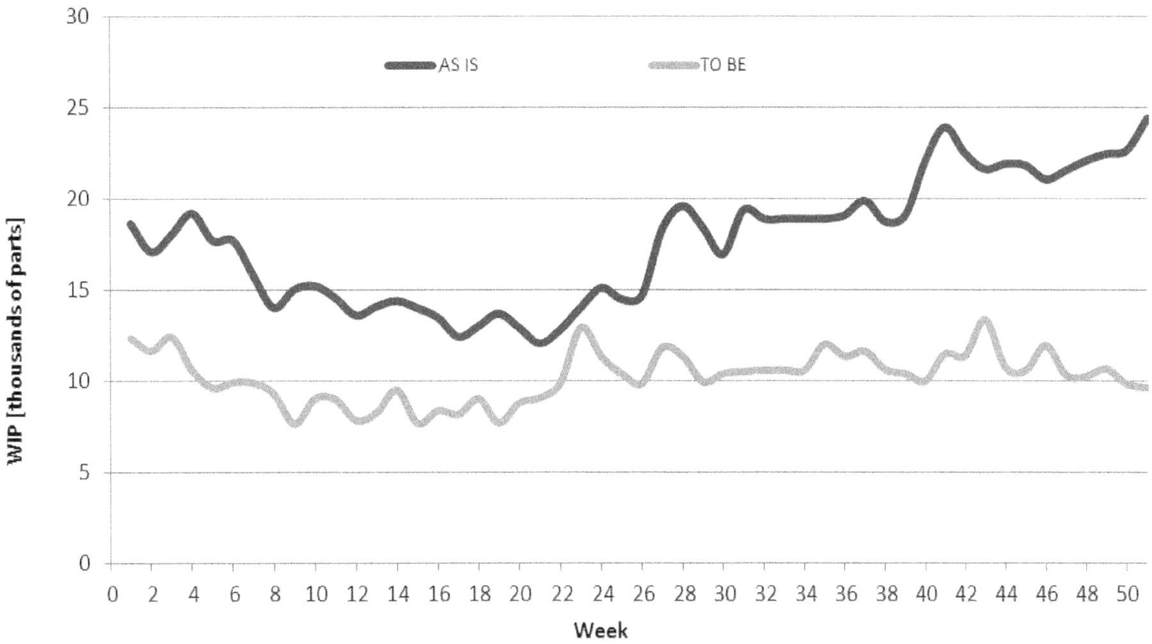

Fig. 4. Comparison between WIP average levels (thousands of parts per week)

Table 8
Comparison between the current and the future state of the system

		as is	to be	Comparison
Production [parts/day]	Average	2,615	2,561	-2.1%
	Std Dev	12.7	18.7	+47.6%
WIP [parts/week]	Average	17,630	10,238	-41.9%
	Std Dev	749	325	-56.6%
Flow time[days]	Average	8.67	5.28	-39.1%
	Std Dev	0.65	0.31	-52.3%
Job Tardiness [days]	Average	7.80	10.32	+32.3%
	Std Dev	0.66	0.63	-4.5%
Staging Times [days]	Average	7.24	4.00	-44.8%
	Std Dev	0.43	0.62	+44.2%
Weighted Lateness [days]	Average	22.00	14.16	-35.6%
	Std Dev	4.69	5.26	+12.2%

3.11 Analyse results

Let us consider the outcome of Fig. 4 and Table 8: the average level of WIP (parts/week) is greatly reduced in both its average value (- 41.9%) and its standard deviation (-56.6%); the same results could be achieved by measuring WIP in total working minutes at the testing zone, but this is quite obvious, since this parameter is directly controlled by the CONWIP system. Average flow times are also almost 40% shorter than before, and their standard deviation has decreased by more than 50%.

As a control of the dispatching rule adopted, Table 8 also reports the comparison of job Tardiness, Staging Times and Weighted Lateness. It emerges that the future state will increase average job

Tardiness (by more than 30%) without significantly modifying its standard deviation. This result is due to the importance given to hot jobs and to the synchronization of purchase orders. In fact, the application of the dispatching rule (Section 3.6) leads to significant reductions in average Staging Times and Weighted Lateness. In both cases, the increase in their standard deviations can be ascribed to the importance given to hot jobs.

Finally, the average daily production in the future state will be 2% smaller and its standard deviation will be increased, but these are relative losses. In fact, the current system is meeting demand, as often happens in MTO/ETO systems, and the demand profile is highly variable. At present, the plant experiences short order backlogs and high levels of WIP. However, as previously stated, the implementation of the simulated system will lead to lower weighted lateness and a considerable reduction of WIP (to the detriment of order backlogs). Potential consequences of these differences should be gain in market shares, economic savings and, finally, the possibility to invest in new test benches, thus increase bottleneck capacity and productivity.
A visual comparison of results is reported in Fig. 5. We can therefore conclude that the system we have simulated (future state) has fulfilled 2 objectives (out of the three defined) with a significant reduction in flow times and WIP and without significantly affecting the production rate.

Unfortunately, the firm has so far not proceeded with the implementation of the future state, so Section 3.12 is not reported.

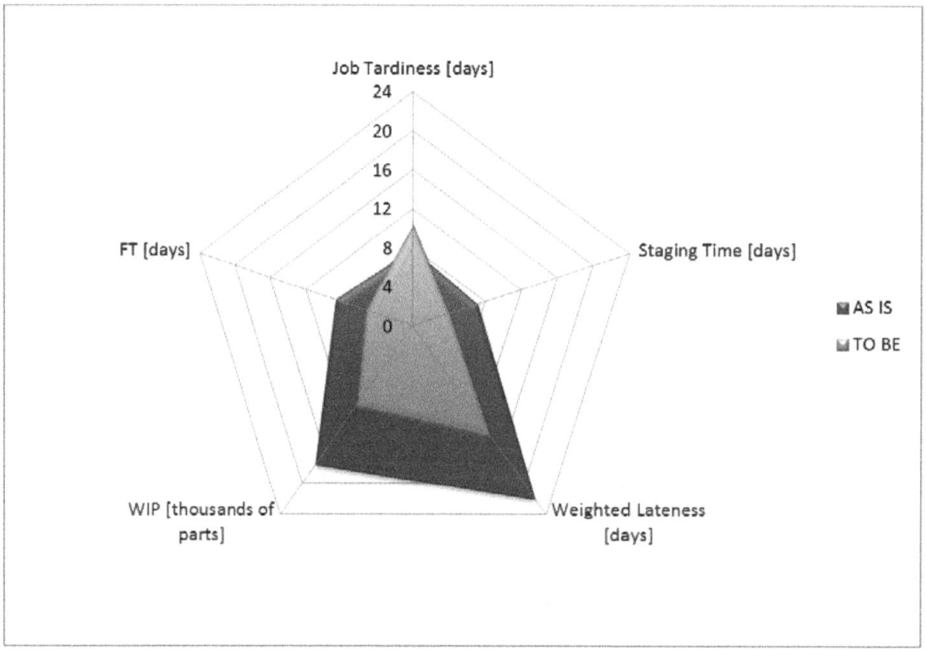

Fig. 5. comparison of average results on a radar diagram

4. Conclusions

Modern manufacturing firms need to be planned and controlled by advanced systems. Weak points of push systems were discovered from both the theoretical and practical point of view. Pull systems, on the other hand, usually achieve better results when they can be applied, though their implementation is not so immediate. The paper has presented the simulative implementation of CONWIP in an industrial case (a complex manufacturing firm). The proposed solution consists of a pull-from-the-bottleneck CONWIP with path type cards and a common unit to measure WIP (sum of working capacities occupied at the testing WSs). The approach we followed is first presented in section 2, and then applied to the industrial case (section 3).

The results we have achieved confirm the validity of CONWIP. Average WIP, measured in parts per week, decreased by over 40% and its standard deviation by over 50% without significantly affecting production rate; flow times were also reduced in both their average values and their standard deviation. The reduction in WIP and flow time standard deviations is extremely important, because it can ensure more precise and robust values to rely on when planning production. The dispatching rule we have chosen was evaluated using three indicators that express the system's capacity to meet management requirements. These indicators are (i) average tardiness, measuring order punctuality, (ii) average staging time, indicating the synchronization of parts belonging to the same purchase order, and (iii) average weighted lateness, reporting the ability to ensure punctuality of *hot* jobs. All of these indicators were measured in 10 simulation runs and reported in their average values and standard deviations. The synchronization of purchase orders was considerably increased, by reducing average staging time to 4 days (- 44%), and average weighted lateness was decreased by 36%, that is greater attention was paid to hot jobs. Nevertheless, these achievements led to the growth (+ 32%) of average job tardiness, though this result was partially expected: one of the goals of the dispatching rule we adopted is to increase the priority of hot jobs, to the detriment of normal ones and with the increase in average (non-weighted) tardiness. For all of these reasons, we can state that the proposed solution could be an excellent scheduling and control system for the shop floor of a complex manufacturing firm, being integrated in a planning system such as (classic) MRP.

Possible future developments of this study may be the optimization of model factors in a systematic way (i.e. not only relying on practical experience of the firm's management and common sense). On the other hand, especially when dealing with everyday scheduling problems of the SF, an on-line simulation may be the only way to achieve optimum results in a reasonable amount of time.

Finally, it is important to remark that the main drawback in this approach, even though it is applied to a real manufacturing environment, lies in the impossibility of providing optimum results without the comparison and evaluation of a certain number of different planning and control systems (such as synchro-MRP, workload control and drum-buffer-rope) and/or dispatching rules.

References

Adetunji, O.A.B. & Yadavalli, S.V. (2012). An integrated utilisation, scheduling and lot-sizing algorithm for pull production, *International Journal of Industrial Engineering: Theory, Applications and Practice*, 19(3).

Anavi-Isakow, S. & Golany, B. (2003). Managing multi-project environments through constant work-in-process, *International Journal of Project Management*, 21(1), 9-18.

Benton, W.C. & Shin, H. (1998). Manufacturing Planning and Control: The evolution of MRP and JIT integration, *European Journal Of Operational Research*, 110(3), 411-440.

Bertolini, M., Braglia, M., Romagnoli, G. & Zammori, F. (2013). Extending value stream mapping: the synchro-MRP case, *International Journal of Production Research*, 51(18), 5499-5519.

Bertolini, M. & Romagnoli, G. (2013). Design and simulation of multi-CONWIP into a Make-To-Order firm with General Job Shop configuration, Proceedings of the 43rd International Conference on Computers and Industrial Engineering (CIE43), 16-18 October 2013, Hong Kong, China.

Bonvik, A.M., Couch, C.E. & Gershwin, S.B. (1997). A comparison of production-line control mechanisms, *International Journal of Production Research*, 35(3), 789-804.

Braglia, M., Frosolini, M., Gabbrielli, R. & Zammori, F. (2011). CONWIP card setting in a flow-shop system with a batch production machine, *International Journal of Industrial Engineering Computations*, 2(1), 1-18.

Chong, M.Y., Prakash, J., Ng, S.L., Ramli, R. & Chin, J.F. (2013). Parallel Kanban-ConWIP system for batch production in electronics assembly, *International Journal of Industrial Engineering: Theory, Applications and Practice*, 20(7-8).

Duri, C., Frein, Y. & Lee, H.-S. (2000). Performance evaluation and design of a CONWIP system with inspections, *International Journal of Production Economics*, 64(1), 219-229.

Enns, S.T. (1995). An economic approach to job shop performance analysis, *International Journal of Production Economics*, 38(2-3), 117-131.

Flapper, S.D.P., Miltenburg, G.J. & Wijugard, J. (1991). Embedding JIT into MRP, *International Journal of Production Research*, 29(2), 329-341.

Framinan, J.M., González, P.L. & Ruiz-Usano, R. (2003). The CONWIP production control system: review and research issues, *Production Planning & Control*, 14(3), 255-265.

Framinan, J.M., González, P.L. & Ruiz-Usano, R. (2006). Dynamic card controlling in a CONWIP system, *International Journal of Production Economics*, 99(1-2), 102-116.

Geraghty, J. & Heavey, C. (2004). A comparison of Hybrid Push/Pull and CONWIP/Pull production inventory control policies, *International Journal of Production Economics*, 91(1), 75-90.

Gibson, P., Greenhalgh, G. & Kerr, R. (1995). *Manufacturing Management: Principles and concepts,* London: Chapman & Hall.

Hall, R.W., (1981). *Driving the productivity machine: Production planning and control in Japan,* Falls Church, Virginia: American Production and Inventory Control Society.

Hodgson, T.J. & Wang, D. (1991). Optimal hybrid push/pull control strategies for a parallel multistage system: Part I, *International Journal of Production Research*, 29(6), 1279-1287.

Hopp, W.J. & Spearman, M.L. (2008). *Factory Physics*, 3rd ed., Dubuque, Iowa: McGraw-Hill/Irwin.

Huang, M., Wang, D. & Ip, W.H. (1998). Simulation study of CONWIP for a cold rolling plant, *International Journal of Production Economics*, 54(3), 257-266.

Koulouriotis, D.E., Xanthopuolos, A.S. & Tourassis V.D. (2010). Simulation optimisation of pull control policies for serial manufacturing lines and assembly manufacturing systems using genetic algorithms, *International Journal of Production Research*, 48(10), 2887-2912.

Lambrecht, M.R. & Decaluwe, L. (1988). JIT and constraint theory: the issue of bottleneck management, *Production and Inventory Management Journal*, 29(3), 61-65.

Larsen, N.E. & Alting, L. (1993). Criteria for selecting a production control philosophy, *Production Planning and Control*, 4(1), 54-68.

Lengyel, A., Hatono, I. & Ueda, K. (2003). Scheduling for on-time completion in job-shops using feasibility function, *Computers & Industrial Engineering*, 45(1), 215-229.

Li, G.-W. (2010). Simulation study of coordinating layout change and quality improvement for adapting job shop manufacturing to CONWIP control, *International Journal of Production Research*, 48(3), 879-900.

Mahdavi, I., Shirazi, B. & Solimanpur, M. (2010). Development of a simulation-based decision support system for controlling stochastic flexible job shop manufacturing systems, *Simulation Modelling Practice and Theory*, 18(6), 768-786.

Monden, Y. (1998). *Toyota Production System, An Integrated Approach to Just-In-Time*, 3rd ed., Norcross, Georgia: Chapman & Hall.

Montgomery, D.C. (2001). *Design and analysis of experiments*, 5th ed., New York, New York: John Wiley & Sons.

Ohno, T. (1988). *Toyota Production System: Beyond Large-Scale Production*, New York, New York: Productivity Press.

Özbayrak, M., Papadopoulou, T.C. & Samaras, E. (2006). A flexible and adaptable planning and control system for an MTO supply chain system, *Robotics and Computer-Integrated Manufacturing*, 22(5-6), 557-565.

Pettersen, J.A. & Segerstedt, A. (2009). Restricted work-in-process: A study of differences between Kanban and CONWIP, *International Journal of Production Economics*, 118(1), 199-207.

Pinedo, M. (2002). *Scheduling: theory, algorithms and systems*, 2nd ed., Upper Saddle River, New Jersey: Prentice Hall.

Ramsay, M.L., Brown, S. & Tabibzade, K. (1990). Push, pull and squeeze shop floor control with computer simulation, *Industrial Engineering*, 22(2), 39-45.

Renna, P. (2010). Dynamic control card in a production system controlled by CONWIP approach, *Jordan Journal of Mechanical and Industrial Engineering,* 4(4), 425-432.

Roderick, L.M., Toland, J. & Rodriguez, F.P. (1994). A simulation study of CONWIP versus MRP at Westinghouse, *Computer & Industrial Engineering*, 26(2), 237-242.

Rubiano Ovalle, O. & Crespo Marquez, A. (2003). Exploring the utilization of a CONWIP system for supply chain management. A comparison with fully integrated supply chains. *International Journal of Production Economics*, 83(2), 195-215.

Scholl, A. (1999). Balancing and sequencing of assembly lines, 2nd ed., Munich: Physica Verlag.

Sharma, S. & Agrawal, N. (2009). Selection of a pull production control policy under different demand situations for a manufacturing system by AHP-algorithm, *Computer & Operations Research*, 36(5), 1622-1632.

Spearman, M. L., Woodruff, D. L. & Hopp, W. J. (1990). CONWIP: a pull alternative to kanban, *International Journal Of Production Research*, 28(5), 879-894.

Stevenson, M., Hendry, L.C. & Kingsman, B.G. (2005). A review of production planning and control: the applicability of key concepts to the make-to-order industry, *International Journal of Production Research*, 43(5), 869-898.

Sultana, I & Ahmed, I. (2014). A state of art review on optimization techniques in just in time. *Uncertain Supply Chain Management*, 2(1), 15-26.

Thiagarajan, S. & Rajendran, C. (2005). Scheduling in dynamic assembly job-shops to minimize the sum of weighted earliness, weighted tardiness and weighted flowtime of jobs, *Computers & Industrial Engineering*, 49(4), 463-503.

Villa, A. & Watanabe, T. (1993). Production management: beyond the dichotomy between push and pull, *Computer Integrated Manufacturing Systems*, 6(1), 53-63.

Vollmann, T.E., Berry, W.L. & Whybarck, D.C. (1997). *Manufacturing planning and control systems*, 4th ed., New York, New York: McGraw-Hill.

Yang, T., Fu, H.P. & Yang, K.Y. (2007). An evolutionary-simulation approach for the optimization of multi-constant work-in-process strategy – A case study, *International Journal of Production Economics*, 107(1), 104-114.

14

Machining parameter optimization in turning process for sustainable manufacturing

S. G. Dambhare[a*], S. J. Deshmukh[b] and A. B. Borade[c]

[a]*Mechanical Engineering Department, PVPIT Pune, M.S., India*
[b]*Mechanical Engineering Department, PRMITR Badnera Amaravati, M.S., India*
[c]*Mechanical Engineering Department, JDIET Yavatmal, M.S., India*

CHRONICLE

Keywords:
Machining
Turning
Sustainability
ANOVA
RSM

ABSTRACT

There is an increase in awareness about sustainable manufacturing process. Manufacturing industries are backbone of a country's economy. Although it is important but there is a great concern about consumption of resources and waste creation. The primary aim of this study was to explore sustainability concern in turning process in an Indian machining industry. The effect of cutting parameters, Speed/Feed/Depth of Cut, the machining environment, Dry/MQL/Wet, and the type of cutting tool on sustainability factors under study were observed. Analysis of Variance (ANOVA) was used to analyse the data obtained from experimentation in a small scale machining industry. The process is modelled mathematically using response surface methodology (RSM).The economic and environmental aspect like surface roughness, material removal rate and energy consumption were considered as sustainability factors. The model helps to understand the effect of the cutting parameters and conditions on surface finish, energy consumption, and material removal rate. The process was optimized for minimum power consumption considering environmental concern as prime importance. Studies suggest that the cutting environment and tool type influenced on the power consumption during turning process. Extended form of the proposed model could be useful to predict the environmental impact due to machining process, which would bring environmental concern into conventional machining.

1. Introduction

In recent era, manufacturing industry is focusing their capabilities towards achieving sustainable products through sustainable manufacturing. This is an effect of increased awareness amongst the manufacturer and the users (Averam et al., 2011). The associated countries are being compelled to reduce negative environmental impact due to manufacturing process. We must understand that the cost of environment is higher than any other objective (personal or of nation) for our better future (CPCB, 2010). Machining industry is the most energy consuming and waste generating industry. How a manufacturing process can be used so that the emissions would be on lower side and would provide high productivity is a question each industry is facing (Tan et al., 2011). Sustainability is no longer a choice but rather it has become a necessity.

The sustianability approach is based on three pillars namely economic, environment and social aspects, well known as tripple bottom line approach. The factors involved in machining proceess can be categorized in three groups viz. Economic Indicator, Environmental Indicator and Social Indicator. Fig.1 illustrates the approach used in sustainable manufacturing. The US department of commerce defines the sustainable manufaturing as

Sustainable manufacturing is the creation of manufactured products through economically-sound processes that minimize negative environmental impacts while conserving energy and natural resources. Sustainable manufacturing also enhances employee, community, and product safety.

Sustainability related aspects like quality of product, energy consumption, emissions and other environmental aspects are getting an integrated part at operational level and decision making in manufacturing (Shao et al., 2010). As the concept of product sustainability is getting important amongst organization, its assessment is becoming more challenging. There appears to be no universal methodology for sustainability assessment of products or processes. This is because of problems in quantifying the parameters commonly known as indicators of sustainability pertaining to manufacturing process or manufactured product (Jawahir et al., 2006).

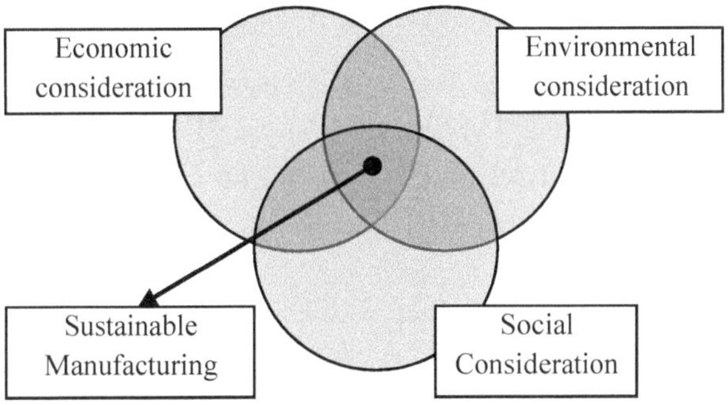

Fig. 1. Three pillers of sustianbility (Aurich et. al., 2013)

Machining process is one of the most used production processes to shape the products. It is the most power consuming and waste generating operations amogst all becuse of its profound use. The machining industries of all sizes have capability of saving money and improve their environmental performance (Kopac & Pusavec, 2009). Therefore, sustainability assessment of machining process is major area of focus for most researchers. First, it is important to define the parameters involved in machining process and categorize those in three aspects of sustainability. Fig. 2 shows the sustainability parametrs for a typical machining process and sustainability assessment model of a machining process.

The cause and response are considerded from a small scale industry point of view taking into account limitation in their measurement during operations. The input parameters identified from available literature include work piece material, work piece geometry, tool material, tool geometry, process parameters; namely speed, feed and depth of cut, and the machining environment dry, minimum quantity lubrication (MQL) and wet machining. The response parameters taken for assessment are catagorized into three aspects of sustainability. The machining cost, surface roughness, material removal rate, tool life were considered as economic aspect, energy consumption, tool temperature, CO_2 equivalant, water use, waste disposal as environmental aspect while noise, vibration, illness rate, absenteeism and job satisfaction were considered as social aspects.

2. Literature Review

Various methodologies have been used by researchers to understand the sustainability of a machining process. A three dimensional system approach (Yuan et al., 2012) highlighted sustainability issues of manufacturing from pollution prevention point of view. Three key components of process; namely technology, energy and material were considered for the study. Supported case study shows the effective use of methodology in a nano-scale manufacturing unit. Li et al.(2012) presented eco efficiency approach for evaluating energy consumption as well as the resource utilization in manufacturing process supported by a case study of grinding process. Munoz and Sheng(1995) focused on waste streams of machining process considering process mechanics, tool wear, metal working fluid loss, chip waste and energy consumption. Jiang et al.(2012) described a new method for environmental assessment of manufacturing process for entire process plan. The weights for process plan parameters were obtained using Analytical hierarchy processing (AHP) approach.

Life cycle Assessment (LCA) is widely considered for understanding the environmental impact of a manufacturing process. It is an authoritative instrument to analyze the manufacturing process. Narita et al. (2006) developed a methodology to assess environmental burden of a machining process using LCA methodology. The model developed provides equivalent CO_2 for the process using the energy consumption, metal working fluid used, chip generated and tool used in the process. Branker et al.(2011) presented new economic model based on LCA methodology. Theoretical and experimental results were used to validate the model for carbon emission and cost sensitivity. Use of LCA methodology demands the knowledge of LCA study. Many of the industries may not have recourses available for the same. Also the availability of environmental data of the country is a major concern.

Large numbers of parameters namely economic, environmental and social need to be evaluated for sustainability assessment of a machining process (Weiser et al., 2008; Jiang et al., 2012). In any analysis it is important to define the boundary of the study first (Smith et al., 2012). Multi criteria decision making approaches like Analytical Hierarchy Process (AHP), Analytical Network Process (ANP), Technique for Order Preference by Similarity to an Ideal Solution (TOPSIS) etc. can be applied for the assessment. The outcome of such analysis will be based on the judgment of the decision maker, which in turn, depends on his/her experience and knowledge.

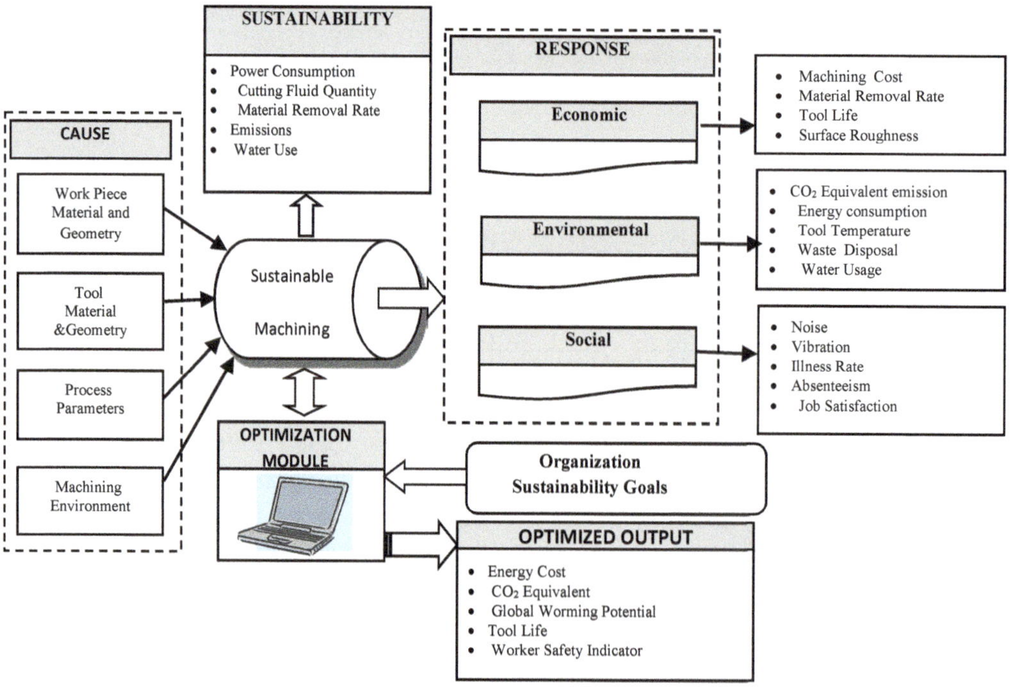

Fig. 2. Sustainability assessment model

Indian Machining industry scenario is slightly different from other developed countries. Because of profound availability of skilled manpower, conventional machining is still being used by large organizations. Majority of workers found in such small scale industry do not have proper training and hence it is very difficult to help them understand the concepts of mathematical model and outcomes. While most methods are being developed and applied for CNC machining, sustainability assessment of a conventional machining process at small scale industry is hardly sufficient attention. Thus there is a need of developing a simple model, which could be easy to understand and could be easily incorporated for small scale industries working on conventional machining (Smith et al., 2012).

Process parameter and cutting environment plays important role in end result of a machining process. Dahmus and Gutowski (2004) presented a system-level environmental analysis of machining process. They considered the environmental impact due to machining process and associated material preparation process and suggested that further focus in this area is needed. An economic and environmental impact aspect of machining highly depends on these parameters. Muthukrishnan and Davim(2009) studied the effect of machining parameters on surface roughness of Al-SiC using coarse grade Polycrystalline diamond (PCD) inserts. Artificial Neural Network (ANN) and Analysis of Variance (ANOVA) were used to validate the results. Rajemi et al. (2010) developed a model for optimizing energy footprint of machined product. They identified critical parameters in minimizing energy use and hence reducing energy cost and environmental impact. Table 1 shows the typical indicators used by various researchers for sustainability assessment purpose. The sustainability issues presented in this paper emphasises on economic and environmental aspect of machining process based on energy use, material removal rate, and surface finish of product manufactured. Here attempt has been made to avoid complexity in the model, which would allow any person working in the conventional machining industry to understand the process impacts easily.

Table 1
Typical Indicators used by various researchers for sustainability analysis

Economic Indicators	Environmental Indicators	Social Indicators
• Energy Cost	• Air quality	• Customer Satisfaction
• Investment	• Carbon Footprint	• Employee Health
• Labour Cost	• Chip mass	• Employment creation
• Machining cost	• Coolant usage	• Employee Safety
• Material consumption cost	• Coolant Mist Produced	• Health related absenteeism rate
• Machining time	• CO_2Emissions	• Human energy
• Productivity	• CO_2Equivalant	• Injury Rate
• Product quality	• Energy Use	• Job satisfaction
• Production quantity	• GHG Emissions	• Labour Turnover
• Profit	• Global warming potential	• Mist / dust level
• Rate of Return	• Heat generation	• Skill Improvement
• Surface roughness	• Land pollution	
• Tool wear	• Natural resource depletion	
• Tool life	• Noise level	
	• Tool Temperature	
	• Water quality	
	• Water Usage	

3. Selection of study parameters

There are literally a large number of variables involved in the sustainability analysis of a machining process and the model is targeted for the small scale industries. Therefore, it is necessary to reduce the number of input/output parameters to avoid complexity. The literature review suggested that major influencing parameters in a machining process are work piece material, cutting environment (Dry / Wet / MQL), Type of Tool, and the process parameters i.e. cutting speed, feed & depth of cut. But all the mentioned factors are hardly considered together for the study. In this paper three important output parameters; namely surface roughness, material removal rate and power consumption are considered. The system boundary for this study is depicted in Fig. 3.

4. Material and Method

4.1 Experimental Conditions

Three machines in different industries were selected to accomplish the experimentation work and gathering the necessary data. Workers having different skill sets and educational levels were chosen for absorbing the variation in the process modeling. Experimentation was performed on three conventional medium duty lathe machines with varying capacity and motor power. AISI 1040 carbon steel commonly known as EN-8 was selected for the case study because of its wide engineering applications. Three types of tools namely Brazed ceramic tool, Insert with Titanium Nitride (TiN) Coating and Insert with Titanium Aluminum Nitride (TiAlN) coating, which are most commonly used in machining industry were selected for study purpose. The details are listed in Table 2. The geometry of the component was kept constant throughout experimentation

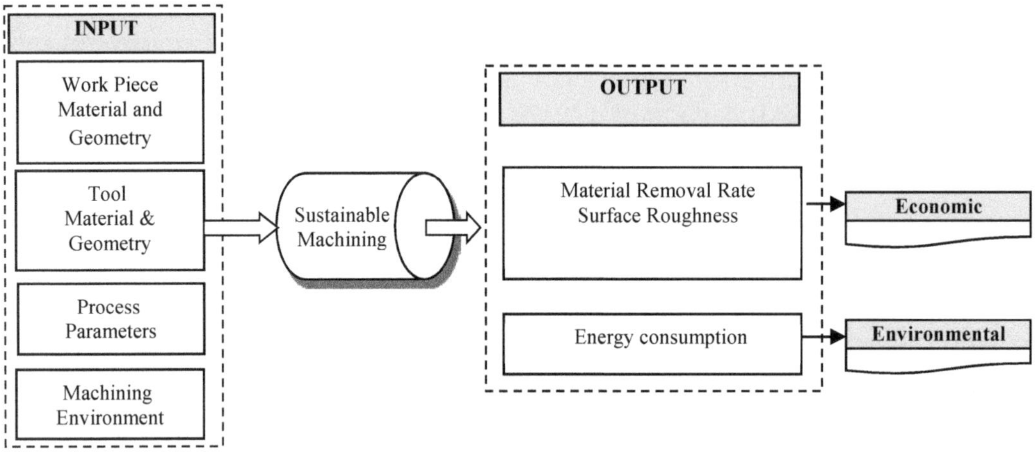

Fig. 3. System boundary for study

Experimental design is widely used in many engineering applications. In factorial design, as the number of input parameters increases, the experiment to be conducted rapidly increases. L27 orthogonal array is a systematic and effective method, which provides better results with reduced number of experiments when all factors are considered at three levels. To compensate for the measurement errors, each experiment is being replicated three times and average value of variable is considered for experimentation and analysis. Table 2 shows details of the machine, cutting tool, tool holder and cutting fluid used during the experimentation. The input parameters were varied at three levels. The variable values at various levels used during experimentation are listed in table 3. The experiments were carried out in accordance with 3 level L27 orthogonal array using design of experiment technique. Table 4 shows the material composition of AISI 1040 used for the study

Table 2
Details of the machine, tool , tool holder and cutting fluid

SN	Particular	1	2	3
1	Lathe Motor Power	1.5 HP	2.0 HP	3.0 HP
2	Cutting Tool Insert	CNMG 120412 TF (AlTiN coated) CNMG 120412 MP (TiN coated)		
3	Tool Holder	PCLNR/L 1616H 12-M SANDVIK		
4	Cutting fluid	Servocut-S (Manufactured by Indian Oil)		
5	Cutting Fluid flow rate (Wet)	5 % cutting fluid mixed with water and flow rate of 15 lit /hr.		
6	Compressor Air Pressure (MQL)	Air at 4 bar & 5% cutting fluid mixed with water & cutting fluid flow rate of 90 ml/ hr.		

Table 3
Machining parameters

SN	Parameter	Level 1	Level 2	Level 3
1	Cutting Speed (m/min)	21.93	33.73	50.55
2	Feed (mm/rev)	0.1658	0.1855	0.2107
3	Depth of Cut (mm)	0.5	0.75	1.0
4	Environment	Dry	MQL	Wet
5	Tool	Brazed tip	TiN coated Insert	TiAlN coated Insert

Table 4
Material Composition of AISI 1040 (EN-8)

C %	Si %	Mn %	S %	P %	Cr %	Mo %	Ni %
0.36 - 0.44	0.10 - 0.40	0.60 - 1.00	0.05 max	0.05max	-	-	-

4.2 Experimental procedure and measurements

The work piece geometry as shown in Fig.4 was provided to the worker. All the machines selected were having four spindle speeds available to use. The most suited and used speeds were selected for the experimentation. The depth of cut was selected as 0.5 mm, 0.75 mm and 1 mm. The worker was allowed to choose feed by their experience. They were asked to vary the feed as slow, medium and fast. Since the experimentation was carried out as a manual operation there was variation in the feed hence we decided to perform three replicates of each experiment and take average value to minimize the measurement errors. The orthogonal array L 27 was followed for the experimentations. The experimentations were carried out in three shifts i.e. Morning, afternoon and evening.

The raw work piece material of 30 mm diameter was cut into pieces of 43 mm length. All the pieces were weighed with the help of weighing scale and the dimensions were measured using digital Vernier caliper. This was necessary to find the amount of material removed during the machining accurately as it could be difficult to collect and measure the weight of the chips produced during operation. Each raw work piece was coded to facilitate the after analysis of finished product. To determine the spindle speed noncontact tachometer was used. To maintain constant flow rate of the cutting fluid coolant pump was set to deliver flow rate of 15 lit / hour. The flow rate was selected based on worker experience and was maintained constant throughout experimentation. To deliver air and cutting fluid mixture during minimal quantity lubrication condition spray gun and the compressor was used with a coolant flow rate of 90 ml /hr at 4 bar air pressure. The variation in current during operation and idle running was measured using the clamp meter and Power required was calculated using Eq. (1), where $\cos\Phi$ represents power factor value. Lathe motor being inductive load, in this study it is considered as 0.7 (Bureau of Indian Standard IS:7752, 2007). The power consumed by the coolant pump during wet machining was also measured in the same way. The Power required by the compressor was measured for the amount of time the compressor was switched on for compressing the air. It was observed that the total cycle time of charging and discharging was approximately 34 minutes. Out of that the compressor was using the electricity for 5 minutes to get the required air pressure. Once the pressure was reached to 6 bar the power supply was cut off. The amount of electricity used during this cycle was distributed over the machining time of the work piece and added in total power consumption.

$$P = \sqrt{3}V_L \times I_L \times \cos \emptyset \qquad (1)$$

The surface roughness of the machined component was measured using Handy-surf E35-B surface roughness tester shown in Fig. 5 with a cut-off length of 0.8 mm and sampling length 5 mm. Average surface roughness(Ra) was calculated by taking average of three readings obtained at three different points of machined surface. After machining was over the finished component was measured for dimensional accuracy and weighed to determine the amount of material removed. The tolerance for dimensional accuracy was maintained at ± 0.05 mm. It could have been difficult to measure timings accurately during the process hence we decided to take video shooting of all experiments for later analysis.

4.3 Data Analysis

The experimentations were conducted on three machines to compensate for variation in data. Data was collected for 27 experiments with three replications on each machine. We recorded 243 data samples in all. For analysis purpose the similar experimental conditions on all machines were grouped and average value of input / output variable was used for the analysis. The data was analyzed using Minitab 16 software. This software provides excellent tools for statistical analysis of the data. MS-Excel was also used for calculation and plotting. Fig 6 depicts the finished components obtained from the experimentation.

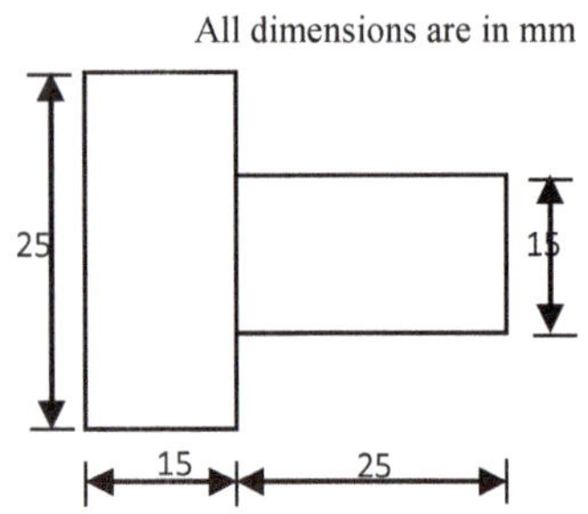

Fig. 4. Work piece geometry

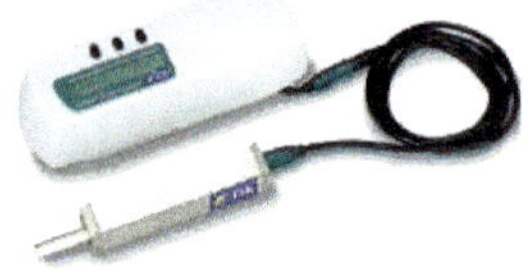

Fig.5. Handy surf E35 B

Fig. 6. Manufactured Work piece (Dambhare et al., 2014)

5. Results and Discussion

5.1 Taguchi analysis

Most analysis presented in literature review section used Taguchi technique to find the value of response variable. Taguchi methodology provides results using fewer experimental runs than other techniques. A small number of experiments can be used to develop a model although a bigger number of experiments will provide more accurate results (Lakshminarayanan & Balasubramanian, 2009). The results obtained may be not optimal, but when these results are implemented, process is improved (Carmita Camposeco-Negrete, 2013). The purpose of this study was to investigate factors influencing sustainability issues in machining industries. The Taguchi analysis was performed using Minitab 16 software to understand the influencing parameters on responses.

Table 5

Response Table for Signal to Noise Ratios for Surface Roughness (Smaller is better)

Level	Environment	Tool type	Cutting speed	Feed	Depth of cut
1	-14.08	-15.56	-13.48	-12.82	-12.61
2	-11.29	-11.03	-12.86	-12.12	-12.40
3	-12.57	-11.33	-11.59	-12.99	-12.93
Delta	2.79	4.53	1.88	0.87	0.53
Rank	2	1	3	4	5

Table 6

Response Table for Signal to Noise Ratios for Material Removal Rate (Larger is better)

Level	Environment	Tool type	Cutting speed	Feed	Depth of cut
1	30.61	29.93	30.21	30.04	29.84
2	30.46	31.52	30.24	30.84	31.33
3	31.67	31.29	32.29	31.86	31.57
Delta	1.20	1.59	2.08	1.82	1.73
Rank	5	4	1	2	3

Table 7

Response Table for Signal to Noise Ratios for Power required for machining (Smaller is better)

Level	Environment	Tool type	Cutting speed	Feed	Depth of cut
1	35.54	28.78	28.95	29.27	28.93
2	24.81	30.25	29.60	29.52	30.31
3	28.89	30.22	30.69	30.45	30.00
Delta	10.73	1.47	1.74	1.18	1.38
Rank	1	3	2	5	4

Table 5 shows the results for signal to noise ratio of surface roughness (Ra) verses the input parameters. Smaller is better criteria was selected for analysis. The ranking depict that tool type, machining environment and cutting speed are ranked 1, 2 & 3 as influencing parameters on surface finish. Table 6 suggests that material removal rate (MRR) depends on cutting speed, feed and depth of cut while tool type and machining environment also contributes to certain extent. Table 7 shows the signal to noise ratio for power required for machining (P). Cutting environment is significant parameter for power consumption compared to rest as during wet conditions the pump power and during MQL condition the compressor power is added while calculating total power. The S/N ratio shows close relationship of the Input parameters on the responses.

5.2 Response Surface Method

In this study all the variables are quantifiable hence we decided to use response surface methodology which is a statistical technique to analyze a number of independent variables influencing the response (Muthukrishnan & Davim, 2009). Response Surface Methodology (RSM) is a set of techniques used in the empirical study of relationships between one or more responses and a group of variables (Cornell, 1990). In RSM second order polynomial equation used to represent response Y is given as Eq. (2),

$$Y = b_0 + \sum b_i x_i + \sum b_{ii} x_i^2 + \sum b_{ij} x_i x_j + e_r \tag{2}$$

Here, the polynomial is being developed for five influencing variables on responses Ra, MRR and P.

5.2.1 RSM model for surface roughness Ra

Table 7 shows the ANOVA results of the model. The data was provided in coded form for input variables. The 'p – value' in the last column represents the influence of the terms. For 95% confidence level the p-value less than 0.05 we reject the null hypothesis that parameter does not affect the response in other words it indicates significant influence of the parameter. Lower p value of the regression model shows that the model is significant. It can be inferred from table 8 that machining Environment and Type of tool influences the surface roughness while feed contributes to certain extent. Eq. 3 represents the RSM model for surface roughness (Ra). The values $R^2 = 93.63$ % and R^2(adj) = 82 % obtained for model indicates high significance of the model.

$$\text{Surface Roughness (Ra)} = 24.2191 - 7.7294Env - 7.8198Tt + 1.1085V_c - 3.4217f - 1.7592a + 1.2396Env^2 + 1.4083Tt^2 - 0.5085\times V_c^2 + 0.4096f^2 + 0.2773a^2 + 0.4118(Env\times Tt) + 0.4262(Env\times f) + 0.0760(Env\times a) + 0.0525(TT\times f) - 0.0052(Tt\times a) + 0.1317(V_c\times f) + 0.2659(f\times a) \tag{3}$$

Table 8

Analysis of Variance results for Surface Roughness Ra (Significant terms Only)

Source	DF	Seq SS	Adj SS	Adj MS	F	p
Regression	17	69.5398	69.5398	4.0906	7.78	0.002
Linear	5	41.0942	29.5827	5.9165	11.26	0.001
Env	1	4.2018	10.4528	10.452	19.89	0.002
Tt	1	28.9701	12.0953	12.095	23.02	0.001
Square	5	22.6427	23.1037	4.6207	8.79	0.003
Env × Env	1	9.2192	9.2192	9.2192	17.55	0.002
Tt × Tt	1	11.8994	11.8994	11.899	22.65	0.001
f×a	1	0.4241	0.4241	0.4241	0.81	0.392
Residual Error	9	4.7291	4.7291	0.5255		
Total	26	74.2689				
Std. Deviation		0.72488		R-Sq =		93.63%
Press		50.0944		R-Sq(adj) =		81.60%

Fig. 7 validates the model developed for surface roughness (Ra). Correlation factor between experimental and calculated value was found to be 0.9676 which indicates model holds good for predicting the Ra value.

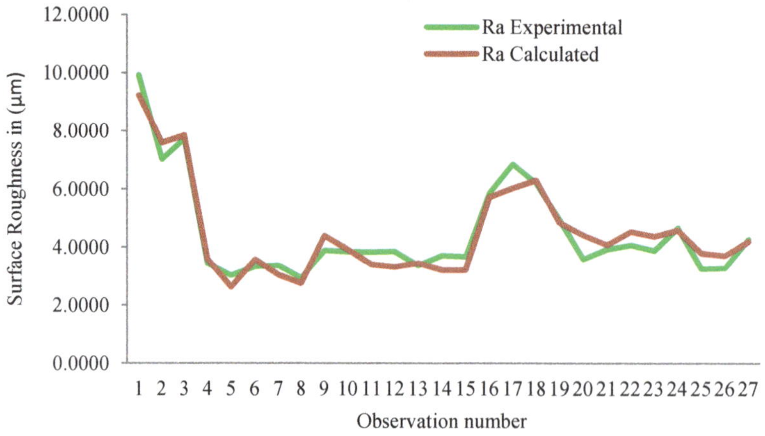

Fig. 7. Validation of Model for Surface Roughness

5.2.2 RSM model for material removal rate (MRR)

Table 9 shows ANOVA results for MRR model. The p-value for almost all input parameters except cutting environment is less than 0.05 which indicates that the terms Tool Type, and Process parameters speed, feed and depth of cut have strong influence on the MRR . The model is also significant as per the results shown. The values of $R^2 = 97.22$ % and $R^2(adj) = 92$ % demonstrate close significance of the model. Eq. 4 represents the model obtained for MRR using RSM.

Table 9

Analysis of Variance results for Material Removal Rate (MRR) (Significant terms Only)

Source	DF	Seq SS	Adj SS	Adj MS	F	p
Regression	17	1463.0	1463.0	86.061	18.52	0.000
Linear	5	949.74	439.28	87.856	18.90	0.000
Tt	1	103.84	236.54	236.54	50.90	0.000
Vc	1	343.26	232.33	232.33	49.99	0.000
f	1	241.34	36.65	36.645	7.89	0.020
a	1	198.85	71.74	71.737	15.44	0.003
Square	5	309.86	386.88	77.377	16.65	0.000
Env × Env	1	52.35	52.35	52.347	11.26	0.008
Tt × Tt	1	75.27	75.27	75.269	16.20	0.003
Vc × Vc	1	141.60	218.63	218.62	47.04	0.000
a×a	1	38.56	38.56	38.560	8.30	0.018
Interaction	7	203.44	203.44	29.063	6.25	0.007
Env×Tt	1	54.81	93.99	93.991	20.22	0.001
Env×f	1	2.42	30.92	30.916	6.65	0.030
Env×a	1	11.87	40.97	40.969	8.82	0.016
Tt×a	1	1.76	47.44	47.442	10.21	0.011
Vc×f	1	70.79	70.79	70.795	15.23	0.004
f×a	1	39.78	39.78	39.784	8.56	0.017
Residual Error	9	41.83	41.83	4.647		
Total	26	1504.8				
Std. Deviation		2.15579		R-Sq =		97.22%
Press		359.963		R-Sq(adj) =		91.97%

$$MRR = 18.4114 + 3.1327Env + 34.5809Tt - 45.8682V_c - 16.0796f + 20.8425a + 2.9537Env^2 - 3.5419Tt^2 + 10.4553V_c^2 + 0.5890f^2 - 2.5351a^2 - 5.5973(Env\times Tt) + 2.0722(Env\times f) - 3.0173(Env\times) - 0.1615(Tt\times f) - 3.2469(Tt\times a) + 4.2070(V_c\times f) + 2.5750(f\times a) \quad (4)$$

Fig. 8 justifies the trueness of the model with correlation coefficient of 0.9860. The experimental and calculated values are closely matching which proves soundness of the model.

5.2.3 RSM model for power required during machining (P)

Table 10 indicates the ANOVA results for power required during machining. Small p-value for the regression suggest model is significant. Power required for machining by large depends on machining environment, tool type and cutting speed while feed and depth of cut has no significance. Eq. 5

represents the RSM model for power required during machining. The values of $R^2 = 98.05\%$ and R^2(adj) = 94.36 % reveal significance of the model.

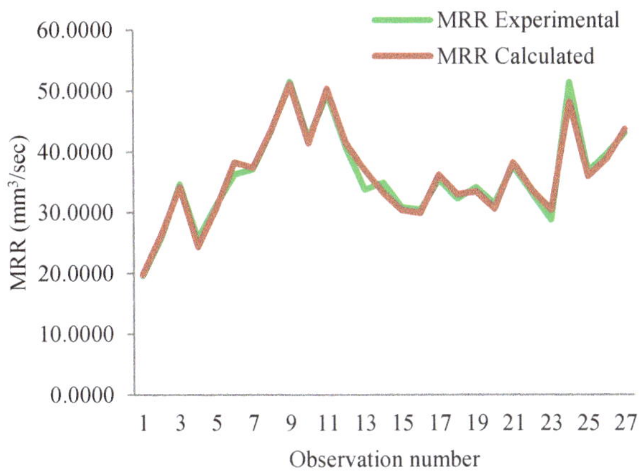

Fig. 8. Validation of Model for MRR

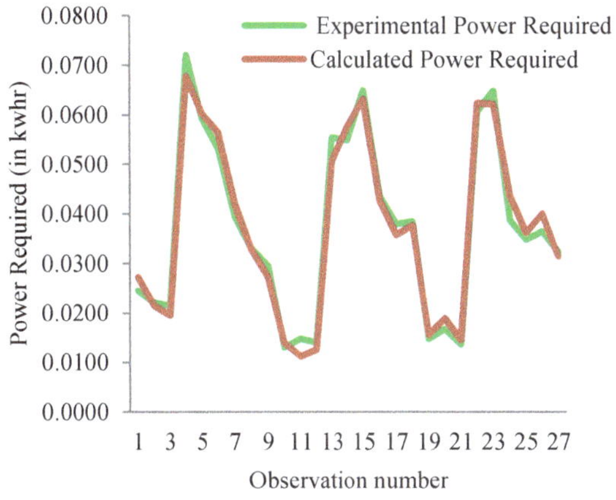

Fig. 9. Validation of Model for Power Required for Machining (P)

Table 10

Analysis of Variance results for Power Required (P) (Significant terms Only)

Source	DF	Seq SS	Adj SS	Adj MS	F	p
Regression	17	0.008437	0.008437	0.000496	26.60	0.000
Linear	5	0.002049	0.005179	0.001036	55.52	0.000
Env	1	0.001625	0.003381	0.003381	181.24	0.000
Tt	1	0.000050	0.000098	0.000098	5.27	0.047
Vc	1	0.000085	0.000122	0.000122	6.56	0.031
a	1	0.000137	0.000066	0.000066	3.54	0.093
Square	5	0.006110	0.006170	0.001234	66.15	0.000
Env × Env	1	0.005971	0.005971	0.005971	320.09	0.000
Vc × Vc	1	0.000036	0.000096	0.000096	5.16	0.049
Residual Error	9	0.000168	0.000168	0.000019		
Total	26	0.008605				
Std. Deviation		2.15579			R-Sq	= 97.22 %
Press		359.963		R-Sq(adj) = 91.97 %		

$$P = -0.071746 + 0.130744Env - 0.022287Tt + 0.033281Vc + 0.010216f - 0.020002a - 0.31457\,Env^2 + 0.000806\,Tt^2 - 0.006936Vc^2 - 0.000858f^2 + 0.003954a^2 + 0.004491(Env\times Tt) - 0.002110(Env\times f) + 0.000092(Env\times a) + 0.002302(Tt\times f) + 0.001798(Tt\times a) - 0.003856(Vc\times f) - 0.001179(f\times a) \quad (5)$$

Fig.9 reveals the close relationship between experimental and calculated values with correlation factor of 0.9901.

5.2.4 Response optimization for sustainability

The objective was to optimize the influencing parameters to improve sustainability of a machining process. The goal was set to keep power consumption to minimum, surface roughness to minimum and to maximize material removal rate. Relative importance was provided accordingly as shown in Table 11.

Table 11

Parameter conditions for Optimization

	Goal	Lower	Target	Upper	Weight	Importance
Ra	Min	2.9500	2.9500	6.4500	1	2
MRR	Max	35.5000	51.4500	51.450	1	3
P	Min	0.0129	0.0129	0.0424	1	1

Table 12
Global Solution

Parameter	Opt. Value
Environment	1
Tool type	2.57576
Cutting Speed	3
Feed	3
Depth of cut	1.67016

Table 13
Predicted Responses

Parameter	Output	Desirability
Ra	2.7623	1.000000
MRR	53.6737	1.000000
P	0.0040	1.000000

Table 12 shows the global solution obtained by performing the RSM optimization using Minitab 16. Dry environment with TiAlN coated tool, cutting speed = 50.55 m/min, feed=0.2107 mm/rev and depth of cut =1 are the optimized values for given conditions. The values in fraction are rounded off to next higher level. Table 13 shows the predicted values of the responses for the optimized solution.

6. Conclusion

Conventional machining was selected for the study purpose. Sustainability issues related to Economic and Environmental aspect in the form of surface roughness, material removal rate and power consumption were studied. Experiments were conducted with varying conditions for speed feed, depth of cut, machining environment and cutting tool type. Taguchi analysis was performed to understand the ranking of factors affecting the response. The process was modelled using Response surface methodology (RSM). ANOVA results were obtained to understand the significance of the model developed.

Study has revealed that surface roughness by large is influenced by cutting environment and tool used. Material removal rate is influenced by tool type, cutting velocity, feed and depth of cut while power required for machining depends on cutting environment, tool type cutting velocity and depth of cut. The experimental and the results obtained from model are closely related. The results found are in line with the previous studies by various researchers.

The results were optimized from sustainability point of view providing importance to power consumption and to keep it to minimum. The outcome of the model facilitate for setting machining parameters to accomplish the objective. Future work will cover more critical analysis of input parameters from overall sustainability point of view and to assess sustainability of conventional turning process.

Acknowledgements

The authors are grateful to M/S Jisbon Industries, Pune, Maharashtra, India and M/S Kiran Industries Pune, Maharashtra, India for extending their facilities to carry out the experimentation work. The authors also acknowledge the help rendered by Mr. Girish Bhaladhare and Mr. Subhash Aher during the experimentation and analysis phase.

References

Aurich, J. C., Linke, B., Hauschild, M., Carrella, M., & Kirsch, B. (2013). Sustainability of abrasive processes. *CIRP Annals-Manufacturing Technology,62*(2), 653-672.

Avram, O., Stroud, I., & Xirouchakis, P. (2011). A multi-criteria decision method for sustainability assessment of the use phase of machine tool systems. *The International Journal of Advanced Manufacturing Technology, 53*(5-8), 811-828.

Bureau of Indian Standards. (2007), Indian standard guide for improvement of power factor in consumers' installation part I low and medium supply voltages, retrived from https://law.resource.org/pub/in/bis/S05/is.7752.1.1975.html

Branker, K., Jeswiet, J., & Kim, I. Y. (2011). Greenhouse gases emitted in manufacturing a product—a new economic model. *CIRP Annals-Manufacturing Technology, 60*(1), 53-56.

Camposeco-Negrete, C. (2013). Optimization of cutting parameters for minimizing energy consumption in turning of AISI 6061 T6 using Taguchi methodology and ANOVA. *Journal of Cleaner Production, 53*, 195-203.

CPCB. (2010), Evaluation of Central Pollution Control Board, Report submitted to Ministry of Environment and Forest Government of India, by Indian Institute of Management, Lucknow February 2010.

Dahmus, J. B. and Gutowski, T. G. (2004). An environmental analysis of machining, *Proceedings of IMECE2004*, November 13-19, 2004, Anaheim, California USA, 01 -10.

Dambhare, S.G., Deshmukh, S.J., Digalwar, A.K., Borade, A.B. and Phate, M.R. (2014) Sustainability issues in turning process: A study in Indian machining Industry, *Procedia CIRP*, 12th GCSM held at Johor Bahru, Malaysia during 22-24 Sept 2014.

Jawahir, I. S., Wanigarathne, P. C., & Wang, X. (2006). Product design and manufacturing processes for sustainability. *Chapter 12 Mechanical Engineers' Handbook*, 414-443.

Jiang, Z., Zhang, H., & Sutherland, J. W. (2012). Development of an environmental performance assessment method for manufacturing process plans. *The International Journal of Advanced Manufacturing Technology*, *58*(5-8), 783-790.

Jiang, Z., Zhang, H., Yan, W., Zhou, M., & Li, G. (2012). A method for evaluating environmental performance of machining systems. *International Journal of Computer Integrated Manufacturing*, *25*(6), 488-495.

Kopac, J., & Pusavec, F. Concepts of sustainable machining processes. In*Proceedings of the 13th international research/expert conference 'Trends in the Development of Machinery and Associated Technology'TMT 2009*.

Lakshminarayanan, A.K. and Balasubramanian, V. (2009). Comparison of RSM with ANN in predicting tensile strength of friction stir welded AA7039 aluminium alloy joints, *Transactions of Met. Soc. China*, 19, 9 – 18.

Li, W., Winter, M., Kara, S., & Herrmann, C. (2012). Eco-efficiency of manufacturing processes: A grinding case. *CIRP Annals-Manufacturing Technology*, *61*(1), 59-62.

Munoz, A. A., & Sheng, P. (1995). An analytical approach for determining the environmental impact of machining processes. *Journal of Materials Processing Technology*, *53*(3), 736-758.

Muthukrishnan, N., & Davim, J. P. (2009). Optimization of machining parameters of Al/SiC-MMC with ANOVA and ANN analysis. *Journal of materials processing technology*, *209*(1), 225-232.

Narita, H., Kawamura, H., Norihisa, T., Chen, L. Y., Fujimoto, H., & Hasebe, T. (2006). Development of prediction system for environmental burden for machine tool operation. *JSME International Journal Series C*, *49*(4), 1188-1195.

Rajemi, M. F., Mativenga, P. T., & Aramcharoen, A. (2010). Sustainable machining: selection of optimum turning conditions based on minimum energy considerations. *Journal of Cleaner Production*, *18*(10), 1059-1065.

Shao, G., Kibira, D., & Lyons, K. (2010, January). A virtual machining model for sustainability analysis. In *ASME 2010 International Design Engineering Technical Conferences and Computers and Information in Engineering Conference* (pp. 875-883). American Society of Mechanical Engineers.

Smith, L., & Ball, P. (2012). Steps towards sustainable manufacturing through modelling material, energy and waste flows. *International Journal of Production Economics*, *140*(1), 227-238.

Tan, X. C., Wang, Y. Y., Gu, B. H., Mu, Z. K., & Yang, C. (2011). Improved methods for production manufacturing processes in environmentally benign manufacturing. *Energies*, *4*(9), 1391-1409.

Reich-Weiser, C., Vijayaraghavan, A., & Dornfeld, D. A. (2008, January). Metrics for sustainable manufacturing. In *ASME 2008 International Manufacturing Science and Engineering Conference collocated with the 3rd JSME/ASME International Conference on Materials and Processing* (pp. 327-335). American Society of Mechanical Engineers.

Yuan, C., Zhai, Q., & Dornfeld, D. (2012). A three dimensional system approach for environmentally sustainable manufacturing. *CIRP Annals-Manufacturing Technology*, *61*(1), 39-42.

15

ANN modeling of kerf taper angle in CO_2 laser cutting and optimization of cutting parameters using Monte Carlo method

Miloš Madić[a*], Miroslav Radovanović[a] and Marin Gostimirović[b]

[a]*Faculty of Mechanical Engineering, University of Niš, A. Medvedeva 14, Niš, Serbia*
[b]*Faculty of Technical Science, University of Novi Sad, Trg Dositeja Obradovića 6, Serbia*

CHRONICLE

Keywords:
CO_2 laser cutting
Kerf taper
Modeling
Optimization
Artificial neural network
Monte Carlo method

ABSTRACT

In this paper, an attempt has been made to develop a mathematical model in order to study the relationship between laser cutting parameters such as laser power, cutting speed, assist gas pressure and focus position, and kerf taper angle obtained in CO_2 laser cutting of AISI 304 stainless steel. To this aim, a single hidden layer artificial neural network (ANN) trained with gradient descent with momentum algorithm was used. To obtain an experimental database for the ANN training, laser cutting experiment was planned as per Taguchi's L_{27} orthogonal array with three levels for each of the cutting parameters. Statistically assessed as adequate, ANN model was then used to investigate the effect of the laser cutting parameters on the kerf taper angle by generating 2D and 3D plots. It was observed that the kerf taper angle was highly sensitive to the selected laser cutting parameters, as well as their interactions. In addition to modeling, by applying the Monte Carlo method on the developed kerf taper angle ANN model, the near optimal laser cutting parameter settings, which minimize kerf taper angle, were determined.

1. Introduction

Laser cutting is a thermal energy based advanced machining process which has many applications in industry where a variety of components in large numbers are required to be machined with high quality and close tolerance at low costs. The wide spectrum of industrial application of the laser cutting is due to its convenience of operation, high precision, small heat-affected zone (HAZ), minimum deformity, low cost, high product quality, high cutting speed, low level of noise, flexibility, ease of automation, etc. (Madić & Radovanović, 2013).

Due to many advantages that offer, laser cutting technology is area of continuous research and development. A number of analytic, numerical and experimental modeling studies have been carried out in order to analyze the laser cutting process, and some of the findings and possibilities of this cutting technology are summarized in a comprehensive review paper (Dubey & Yadava, 2008).

* Corresponding author.

Of particular interest to companies using laser cutting are the maximization of the productivity and the subsequent quality of components made by the laser cutting process (Rajaram et al., 2003). There are different quality characteristics used for assessing the laser cut quality, wherein kerf quality characteristics, surface roughness, and size of the HAZ were mostly investigated. It has been reported in literature that the optimum parameter settings for one quality characteristic may deteriorate other quality characteristics. With a limited theoretical and practical background to assist in systematical selection, these parameters are usually set by previous experience in a time consuming trial and error procedure. However, this trial-and-error approach is extremely costly in terms of time and labor (Ciurana et al., 2009).

Hence, it is of great importance to exactly quantify the relationship between cut quality characteristics and laser cutting conditions through mathematical modeling and subsequently determine the near optimal laser cutting parameter settings via application of optimization algorithms. In the field of laser cutting, most of the time the optimal cutting conditions are determined using the Taguchi method (Prajapati et al., 2013) and by coupling response surface models (Sivarao et al., 2013) and ANNs models with different optimization and metaheuristic algorithms such as particle swarm optimization (Ciurana et al., 2009), genetic algorithm (GA) (Tsai et al., 2008), and simulated annealing (Chaki & Ghosal, 2011). The open literature reveals several research attempts based on ANNs such as for modeling and optimization of laser micromachining process (Ciurana et al., 2009; Biswas et al., 2010; Dhara et al., 2008; Dhupal et al., 2007), selection of optimal laser cutting parameters through integration of ANNs with GA (Tsai et al., 2008; Ghoreishi & Nakhjavani 2008), development of a prediction model through integration with the Taguchi method (Yang et al., 2012) and parametric modeling and optimization of lasox cutting (Chaki & Ghosal, 2011). The ANNs are the learning algorithms and mathematical models, which imitate the information processing capability of human brain and can be applied to non-linear and complex data, even if the data are imprecise and noisy (Raja et al., 2012). The ability of ANNs to capture any complex input-output relationships from limited data is very valuable in manufacturing processes, where huge experimental data for the process modeling is difficult and expensive to obtain (Karnik et al., 2008).

This paper aims at understanding the relations between the laser cutting parameters (laser power, cutting speed, assist gas pressure and focus position) and kerf taper angle obtained in CO_2 laser cutting of stainless steel. Back propagation (BP) ANN trained with gradient descent with momentum algorithm was applied to construct a mathematical model for the kerf taper angle. To obtain data for ANN training, the laser cutting experiment was conducted according to Taguchi's L_{27} orthogonal array (OA) experimental layout plan. Statistically assessed as adequate, the ANN model was then used to investigate the effect of the laser cutting parameters on the kerf taper angle by generating 2D and 3D plots. In addition to modeling, the kerf taper angle mathematical model was optimized. Although the aforementioned metaheuristic algorithms can provide acceptable results, its effective application, requires a substantial amount of parameter tuning. On the other hand, the Monte Carlo method has been proven efficient in optimization, but its application in solving machining optimization problems has been given less attention (Madić et al., 2014). Furthermore, to the authors' knowledge, little work has been reported in the literature on optimization through integration of ANN models and the Monte Carlo method. Hence in this paper the near optimal laser cutting parameter settings that minimize kerf taper angle were identified by exploring the solution space of the developed ANN kerf taper model using the Monte Carlo method.

2. Experimental work

In this study, AISI 304 stainless steel was chosen as the work material because of its extensive use in the aerospace, automotive and food industry. The chemical composition is given in Table 1. The sheet dimensions were 500 x 500 mm with thickness of 3 mm.

Table 1
Nominal chemical composition of AISI 304 stainless steel

Cr	Ni	C	Mn	Si	S	Fe
18.9	9.22	0.07	1.64	0.5	0.006	Balance

The experiment was performed using a ByVention 3015 CO_2 laser cutting machine with a nominal power of 2.2 kW at a wavelength of 10.6 μm, operating in CW mode. The cuts were performed with a Gaussian distribution beam mode (TEM_{00}). In consideration of the numerous parameters that influence the cutting process, some of the process parameters were kept constant throughout the experimentation. During conducting experiment trials, laser beam was focused through a lens of focal length of 127 mm and the distance between workpiece and nozzle was controlled at 1 mm. The nitrogen gas with 99.95 % purity was used as assist gas and it was passed through a conical shape nozzle (HK20) with nozzle diameter of 2 mm.

On the other hand, the laser cutting parameters such as the laser power (P), cutting speed (v), assist gas pressure (p), and focus position (f) were taken as controllable input parameters. Since it was assumed that the effects of laser cutting parameters on the kerf taper angle were complex and non-linear, the experiment was set up with parameters with more levels. The numerical values of control parameters at different levels are shown in Table 2. The value range for each parameter was chosen such that full cut for each parameter combination was achieved and by considering manufacturer's recommendation for parameter settings.

Table 2
Cutting parameters and their levels

Cutting parameter	Unit	Level		
		1	2	3
Laser power, P	kW	1.6	1.8	2
Cutting speed, v	m/min	2	2.5	3
Assist gas pressure, p	bar	9	10.5	12
Focus position, f	mm	-2.5	-1.5	-0.5

The experiment trials were conducted according to Taguchi's L_{27} orthogonal array design matrix (Table 3). The L_{27} chosen consisted of 13 columns and 27 rows (experimental trials). Control parameters, laser power, cutting speed, assist gas pressure and focus position were assigned to columns 1, 2, 5 and 9, respectively.

In this study, kerf taper angle was selected as the laser cut quality characteristic for modeling and optimization. Kerf taper angle is one of the most important cut quality characteristics in laser cutting since it determines the geometrical accuracy of the finished parts. Due to converging-diverging shape of laser beam profile the kerf taper angle always exists during laser cutting. The kerf taper angle (K_t) was calculated using the following formula:

$$K_t(^\circ) = \left(\frac{K_w - K_b}{2 \cdot d} \right) \cdot \frac{180}{\pi} \tag{1}$$

where d is the workpiece thickness, and K_w and K_b represent the top and bottom kerf width, respectively. The kerf widths at top and bottom of each machined workpiece were measured at three equally distanced locations along the length of the cut by means of optical microscope (Leitz, Germany). The average kerf taper angle values corresponding to each experimental trial are listed in Table 3.

Table 3
L_{27} matrix for the experiment and kerf taper angle experimental results

Exp. trial	P (kW)	v (m/min)	p (bar)	f (mm)	K_t (°)
1	1.6	2	9	-2.5	1.18
2	1.6	2	10.5	-1.5	-0.6
3	1.6	2	12	-0.5	-6.02
4	1.6	2.5	9	-1.5	0.83
5	1.6	2.5	10.5	-0.5	-3.63
6	1.6	2.5	12	-2.5	1.46
7	1.6	3	9	-0.5	-4.07
8	1.6	3	10.5	-2.5	2.07
9	1.6	3	12	-1.5	0.51
10	1.8	2	9	-1.5	-4.46
11	1.8	2	10.5	-0.5	-4.93
12	1.8	2	12	-2.5	1.34
13	1.8	2.5	9	-0.5	-5.63
14	1.8	2.5	10.5	-2.5	1.27
15	1.8	2.5	12	-1.5	-3.12
16	1.8	3	9	-2.5	1.5
17	1.8	3	10.5	-1.5	-3.09
18	1.8	3	12	-0.5	-4.01
19	2	2	9	-0.5	-5.06
20	2	2	10.5	-2.5	1.15
21	2	2	12	-1.5	-5.57
22	2	2.5	9	-2.5	0.35
23	2	2.5	10.5	-1.5	-5.25
24	2	2.5	12	-0.5	-5.95
25	2	3	9	-1.5	-4.84
26	2	3	10.5	-0.5	-6.49
27	2	3	12	-2.5	1.12

3. Kert taper angle mathematical model

Modeling studies in laser cutting are the scientific ways to study the system behaviors and help us to get a better understanding of this complex process (Syn et al., 2011). A mathematical model of a system is the relationship between input and output parameters in terms of mathematical equations (Dubey & Yadava 2008). Although the regression models are very promising for practical applications, they are of limited applicatability and reliability in laser cutting modeling. Therefore, to establish a mathematical relationship between the kerf taper angle and the laser cutting parameters, a multilayer perceptron type ANN was selected.

3.1. ANN architecture and training

Four neurons at the input layer (for each of the cutting parameter), one neuron at the output layer for calculating kerf taper and only one hidden layer were used to define ANN architecture. The determine the number of hidden neurons the following equation was considered (Madić & Radovanović, 2013):

$$T = m(n + k + 1) + k \qquad (2)$$

where n is the number of input neurons, m is the number of hidden neurons and k is the number output neurons and T is the total number of weights and biases. Regarding previous equation and the number of available experimental data, 4-4-1 ANN architecture was selected for kerf taper modeling. Nonlinear and linear activation (transfer) functions were used in hidden layer and output layer, respectively. These transfer functions were used since it was assumed that there existed a nonlinear relationship between input and output process parameters. In order to develop robust and accurate mathematical model, the ANN training process was repeated several times using different initial weights. 10000 iterations was set as termination criterion for ANN training while learning rate and momentum were set at 0.1 and 0.9, respectively. The mean squared error at the end of the training process was found to be 0.00509298.

3.2. ANN model function and validation

After ANN training, the set of weights and biases of the developed ANN model were saved (Table 4), upon which explicit mathematical model for the kerf taper angle was developed.

Table 4
The weights and biases of the developed ANN model for the kerf taper angle

W_1				W_2	B_1	B_2
0.897	-1.197	-0.579	-0.356	0.131	-2.543	1.035
-0.649	-1.764	-1.406	3.705	-0.871	1.796	
-0.094	-1.460	1.494	1.446	0.112	-0.001	
-2.501	-0.831	-1.477	-0.430	0.897	-3.043	

W_1: weights between input and hidden layer; W_2: weights between hidden and output layer; B_1: biases of the hidden neurons; B_2: bias of the output neuron

The mathematical equation for the prediction of the kerf taper angle is as follows:

$$K_t = 4.28 \cdot \left(\left(\left[\frac{2}{1+e^{-2(X \cdot W_1 + B_1)}} - 1 \right] \cdot W_2 + B_2 \right) + 1 \right) - 6.49 \quad (3)$$

where X is the column vector which contains normalized values of P, v, p and f. To test the prediction capability of the developed model, the trained ANN was initially tested by presenting 27 input data patterns, which were employed for the training purpose. Furthermore, 4 new experiment trials were conducted with cutting parameter levels which did not belong to the training data set (Table 5).

Table 5
Experiment trials for testing the ANN

Test no.	P (kW)	v (m/min)	p (bar)	f (mm)	Experimentally measured	ANN predicted
1	2	2.5	10.5	-2.5	1.18	1.15
2	1.8	2	12	-1.5	-5.25	-5.31
3	2	2.5	12	-1.5	-0.8	-0.81
4	1.8	3	10.5	-2.5	1.43	1.08

The effectiveness of training of the ANN model and its ability to predict correct values for unseen data can be given in a form of regression analysis where correlation coefficient R determines the correspondence between the ANN outputs and targets (Bajić et al., 2008). The R-values of 0.987 and 0.999 were obtained for the training and testing of the ANN model, respectively. These values indicate that the ANN model has high prediction accuracy and that it also has good generalization ability. Thus, the ANN model can be used to analyze the effects of the laser cutting parameters on the kerf taper, as well as to serve as a fitness function for the optimization of laser cutting parameters.

4. Effect of laser cutting parameters on the kerf taper angle

4.1. Main effect plots

Initially, the effect of the laser cutting parameters on the kerf taper angle was analyzed by changing one parameter at a time, while keeping all other parameters constant at low, center and high level. The effect of the laser cutting parameters on the kerf taper angle is given in Fig. 1.

From Fig. 1a it can be seen that when all other parameters are kept at high level, an increase in the laser power results in a small linear decrease in the kerf taper angle. On the other hand, the effect of the laser power causes a non-linear change in the kerf taper angle. When all parameters are on low level, laser power in the (1.6 – 1.8 kW) range results in about ± 2° change in the kerf taper angle with the minimal value when using laser power of about 1.75 kW.

From Figs. 1b and 1c it is seen that an increase in the cutting speed and assist gas pressure results in nonlinear increase in the kerf taper angle and small linear increase when all other parameters are kept at high level. Further, as it could be seen, when all other parameters are at low level, increase in the cutting speed or assist gas pressure results in positive kerf taper angle. On the other hand, kerf taper angle has negative values.

In the case of the focus position (Fig. 1d), when all other parameters are kept at low level, the effect of the focus position has negligible influence on kerf taper angle change. On the other hand, the focus position has strong influence on kerf taper angle which can be related by nonlinear functional dependence. When all other parameters are kept at middle level, focusing the laser beam deep into the bulk of material is beneficial for obtaining small kerf taper angle. However, when using laser power of 2 kW, cutting speed of 3 m/min and assist gas pressure of 12 bar, focusing the laser beam up to the -1 mm results in small kerf taper angle.

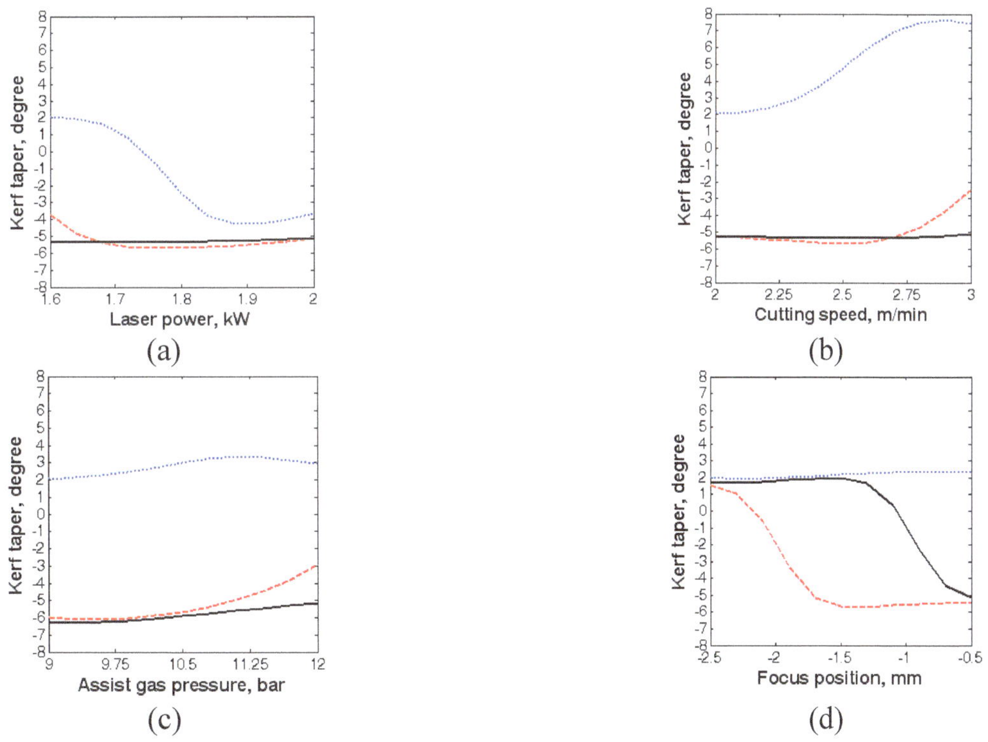

Fig. 1. Effect of laser cutting parameters on kerf taper

4.2. Interaction effect plots

The results from Fig. 1 indicate that there are certain interactions between the laser cutting parameters and the kerf taper angle. In order to determine the interaction effects of the cutting parameters on the kerf taper angle, 3D surface plots were generated considering two parameters at a time, while the third and fourth parameter were kept constant at level 2 (Fig. 2).

From Fig. 2a, it can be seen that the minimal kerf taper angle would be obtained either by using high laser power (P= 2 kW) in combination with high cutting speed (v= 3 m/min) or by using low laser power (P= 1.6 kW) in combination with low cutting speed (v= 2 m/min). A similar effect of the laser power can be observed in interaction with the assist gas pressure as shown in Fig. 2b. From Fig. 2c it can be seen that the minimal kerf taper angle is obtained by focusing the laser beam deep into the bulk of material (about f= -2 mm) independently from the laser power level. In the case of cutting speed and assist gas pressure interaction (Fig. 2d), it can be seen that higher cutting speed in conjunction with higher assist gas pressure is beneficial for minimizing the kerf taper angle. As shown in Fig. 2e, the interaction between the focus position and cutting speed results in a linear change in the kerf taper angle, and this is valid when focusing the laser beam beneath the half of the material thickness for all cutting speed values. A similar conclusion can be drawn from Fig. 2f in the case of focus position and assist gas pressure interaction.

Note that the plots in Fig. 1 were generated by keeping the three parameters constant at low, center and high level. Similarly the 3D surface plots in Fig. 2 were generated considering two laser cutting parameters at a time, while the third and fourth parameter were kept constant at center level. However, finding an optimal set of the laser cutting parameter values to meet the desired kerf taper angle, calls for the parameter optimization in four-dimensional laser cutting parameter hyperspace.

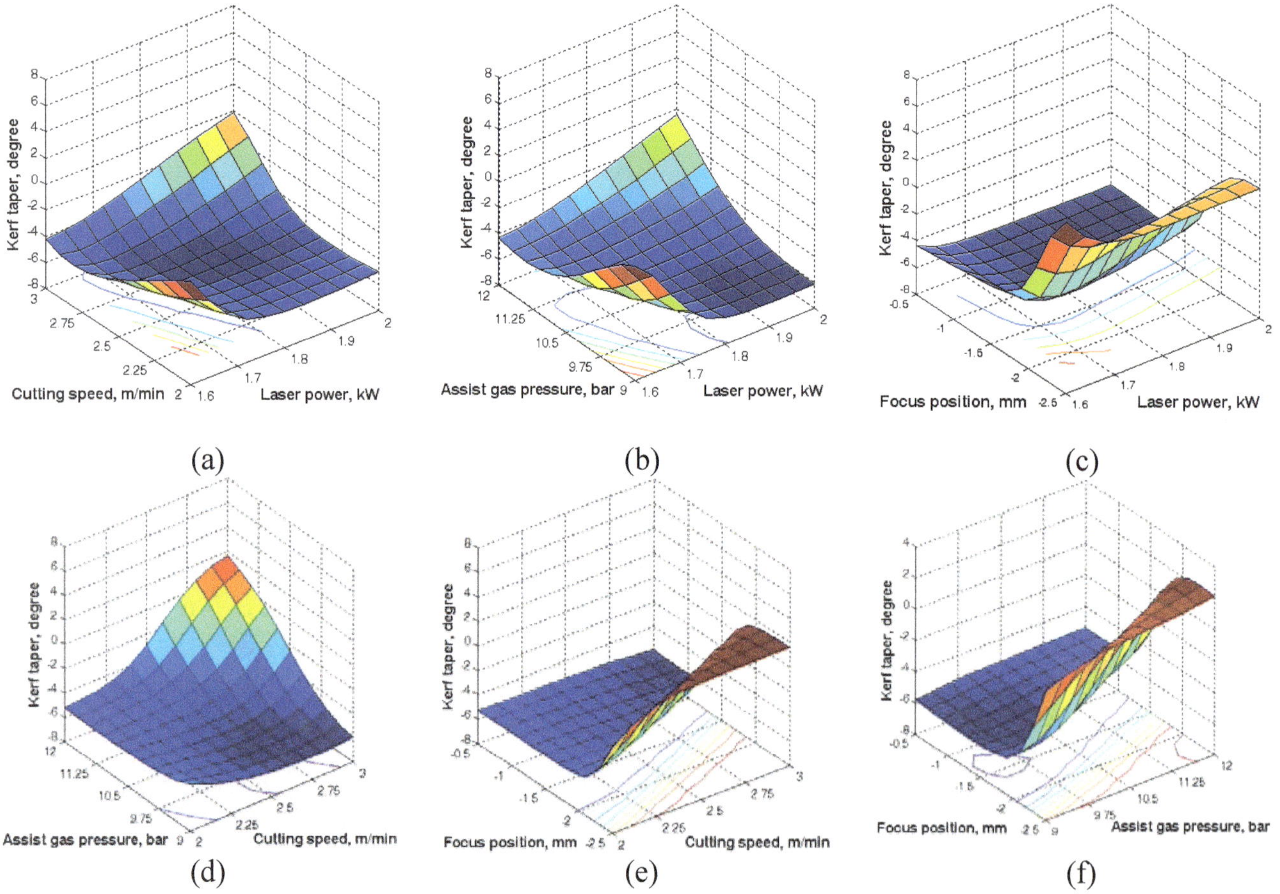

Fig. 2. Effect of laser cutting parameter interactions on kerf taper

5. Optimization methodology

For effective utilization of the laser cutting processes, it is very important to find out the optimal combinations of their process parameters to achieve enhanced machining performance with high dimensional accuracy (Samanta & Chakraborty, 2011). There are numerous methods and algorithms which provide acceptable results in laser cutting optimization. However, each of these techniques has its own advantages and disadvantages and often requires fine tuning of one or more parameters which can be difficult to use for engineers who are not experts in optimization theory and artificial intelligence.

The Monte Carlo method is an attractive and relatively simple technique which utilizes random numbers in simulation algorithm. The Monte Carlo method can be considered as a very general mathematical tool for solving a great variety of problems. The salient feature is that it offers a parameter free optimization procedure at low computation time and high effectiveness. The Monte Carlo optimization procedure was implemented in MS Excel which confirms the easiness of implementation. The details about the Monte Carlo, optimization problem formulation and results are discussed bellow.

5.1. Optimization problem formulation

The goal of the optimization process in this study is to determine the optimal laser cutting parameter values at which the minimum kerf taper angle (K_t) is obtained. An appropriate selection of cutting conditions should increase the product quality through minimizing kerf taper angle, increase the laser productivity and decrease the operation costs. Laser productivity and operation costs are in direct relation with the cutting speed and assist gas pressure, respectively. Therefore, the optimization problem can be formulated as:

$$\begin{aligned}
&\text{Find: } P_{opt}, v_{opt}, p_{opt}, f_{opt} \\
&\text{to minimize: } K_t = \mathrm{f}(P, v, p, f) \\
&\text{subject to: } 1.6\ \mathrm{kW} \le P \le 2\ \mathrm{kW} \\
&\qquad -2.5\mathrm{mm} \le f \le -0.5\ \mathrm{mm} \\
&\qquad v = v_{\max} = 3\ \mathrm{m/min} \\
&\qquad p = p_{\min} = 9\ \mathrm{bar}
\end{aligned} \tag{4}$$

For calculating kerf taper angle the ANN function given in Eq. 3 was employed.

5.2. Optimization solution with Monte Carlo

The optimization problem in Eq. 4 was solved with the Monte Carlo simulation using the following procedure. The first step in the optimization procedure is the generation of random numbers $r_{i,j}$ uniformly distributed in the range [0,1] using the function *rand*. To satisfy the limitations of the laser cutting parameters values, random numbers $r_{i,j}$ were used to generate random numbers $q_{i,j}$ uniformly distributed into the range of interest for each of the laser cutting parameter $[q_i^{min}, q_i^{max}]$. This was accomplished using the following equation (Madić et al., 2014):

$$q_{i,j} = q_i^{min} + r_{i,j} \cdot \left(q_i^{max} - q_i^{min}\right) \tag{5}$$

Therefore, for each laser cutting parameter, the randomized values which were uniformly distributed in the interval of interest were generated and subsequently the kerf taper angle was calculated using the ANN mathematical model. In the optimization procedure only 225 Monte Carlo simulations were performed since the acceptable near optimal solution was identified. The results of Monte Carlo simulation runs are plotted in Fig. 3. Each point in the graph corresponds to a particular input laser cutting parameter settings.

As shown in Fig. 3b, in the 168th Monte Carlo simulation run, the minimal value of kerf taper angle $K_{t\,(min)} = 0.1°$ was observed. This solution corresponds to the following combination of the laser cutting parameters: P=1.6 kW, v=3 m/min, p=9 bar and f= -1.966 mm. This was considered as an acceptable solution. Otherwise, the number of simulation runs would be increased.

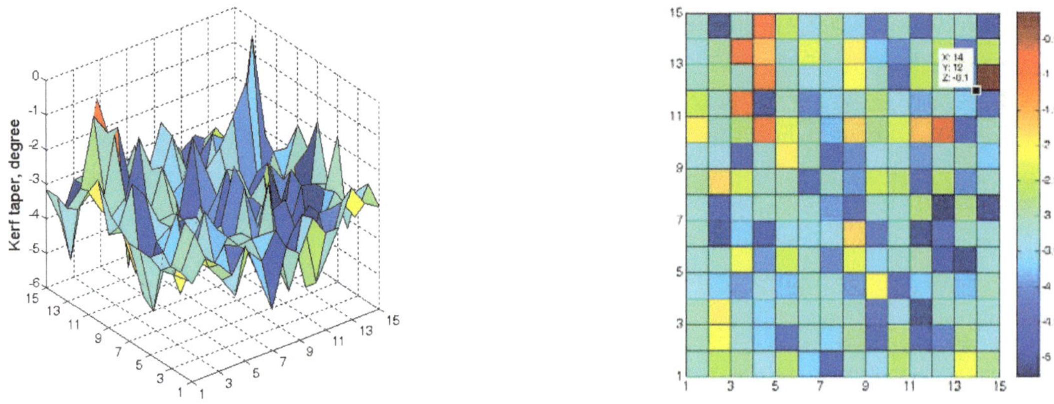

(a) (b)

Fig. 3. Monte Carlo simulation runs (a) 3D view, (b) top view with identified near optimal solution

6. Conclusion

In this paper, ANN based mathematical model is developed in order to relate the laser cutting parameters i.e. laser power, cutting speed, assist gas pressure and focus position, and kerf taper angle in CO_2 laser cutting of stainless steel. Experiment trials were performed according to Taguchi's L_{27} experimental design and the obtained data was used for ANN training. Statistical results indicate that the ANN model can predict the kerf taper with good accuracy. By applying the Monte Carlo method, the optimal laser cutting parameter settings, which minimize kerf taper angle were determined. It was found that focusing the laser beam in about 2/3 of material thickness under the low assist gas pressure (9 bar) at combination of low laser power (1.6 kW) and high cutting speed (3 m/min), produced an acceptable kerf taper angle. The conclusions drawn can be summarized by the following points:

- kerf taper angle is highly sensitive to the selected laser cutting parameters, as well as their interactions,
- change in a given laser cutting parameter value can result in negative or positive kerf taper angle, depending of the values of other parameters,
- the influence of a given laser cutting parameter must be considered through interactions with other laser cutting parameters,
- the combined ANN-Monte Carlo approach is found to be very simple, easy to implement and efficient for the optimization of CO_2 laser cutting process.

Acknowledgements

This work was carried out within the project TR 35034 financially supported by the Ministry of Education and Science of the Republic of Serbia.

References

Bajić, D., Lela, B., & Živković, D. (2008). Modeling of machined surface roughness and optimization of cutting parameters in face milling. *Metalurgija*, *47*(4), 331-334.

Biswas, R., Kuar, A. S., Biswas, S. K., & Mitra, S. (2010). Artificial neural network modelling of Nd: YAG laser microdrilling on titanium nitride—alumina composite. *Proceedings of the Institution of Mechanical Engineers, Part B: Journal of Engineering Manufacture*, *224*(3), 473-482.

Chaki, S., & Ghosal, S. (2011). Application of an optimized SA-ANN hybrid model for parametric modeling and optimization of LASOX cutting of mild steel. *Production Engineering*, *5*(3), 251-262.

Ciurana, J., Arias, G., & Ozel, T. (2009). Neural network modeling and particle swarm optimization (PSO) of process parameters in pulsed laser micromachining of hardened AISI H13 steel. *Materials and Manufacturing Processes*, *24*(3), 358-368.

Dhara, S. K., Kuar, A. S., & Mitra, S. (2008). An artificial neural network approach on parametric optimization of laser micro-machining of die-steel. *The International Journal of Advanced Manufacturing Technology*, *39*(1-2), 39-46.

Dhupal, D., Doloi, B., & Bhattacharyya, B. (2007). Optimization of process parameters of Nd: YAG laser microgrooving of Al2TiO5 ceramic material by response surface methodology and artificial neural network algorithm. *Proceedings of the Institution of Mechanical Engineers, Part B: Journal of Engineering Manufacture*, *221*(8), 1341-1350.

Dubey, A. K., & Yadava, V. (2008). Laser beam machining—a review. *International Journal of Machine Tools and Manufacture*, *48*(6), 609-628.

Ghoreishi, M., & Nakhjavani, O. B. (2008). Optimisation of effective factors in geometrical specifications of laser percussion drilled holes. *Journal of materials processing Technology*, *196*(1), 303-310.

Karnik, S. R., Gaitonde, V. N., Rubio, J. C., Correia, A. E., Abrão, A. M., & Davim, J. P. (2008). Delamination analysis in high speed drilling of carbon fiber reinforced plastics (CFRP) using artificial neural network model. *Materials & Design*, *29*(9), 1768-1776.

Madić, M., & Radovanović, M. (2013). Application of RCGA-ANN approach for modeling kerf width and surface roughness in CO2 laser cutting of mild steel. *Journal of the Brazilian Society of Mechanical Sciences and Engineering*, *35*(2), 103-110.

Madić, M., Kovačević, M., & Radovanović, M. (2014). Application of multi-stage Monte Carlo method for solving machining optimization problems. *International Journal of Industrial Engineering Computations*, *5*(4), 647-659.

Prajapati, B. D., Patel, R. J., & Khatri, B. C. (2013). Parametric investigation of CO_2 laser cutting of mild steel and hardox-400 material. *IJETAE*, *3*(4), 204-208.

Raja, T. V., Ruhil, A. P., & Gandhi, R. S. (2012). Comparison of connectionist and multiple regression approaches for prediction of body weight of goats. *Neural Computing and Applications*, *21*(1), 119-124.

Rajaram, N., Sheikh-Ahmad, J., & Cheraghi, S. H. (2003). CO_2 laser cut quality of 4130 steel. *International Journal of Machine Tools and Manufacture*, *43*(4), 351-358.

Samanta, S., & Chakraborty, S. (2011). Parametric optimization of some non-traditional machining processes using artificial bee colony algorithm. *Engineering Applications of Artificial Intelligence*, *24*(6), 946-957.

Sha, W., & Edwards, K. L. (2007). The use of artificial neural networks in materials science based research. *Materials & design*, *28*(6), 1747-1752.

Sivarao, S., Robert, M., & Samsudin, A. R. (2013). RSM Modelling and Optimization of CO2 Laser Machining of Industrial PVC Foam. *International Review on Modelling and Simulations*, *6*(4), 1339-1343.

Syn, C. Z., Mokhtar, M., Feng, C. J., & Manurung, Y. H. (2011). Approach to prediction of laser cutting quality by employing fuzzy expert system. *Expert Systems with applications*, *38*(6), 7558-7568.

Tsai, M. J., Li, C. H., & Chen, C. C. (2008). Optimal laser-cutting parameters for QFN packages by utilizing artificial neural networks and genetic algorithm. *Journal of Materials Processing Technology*, *208*(1), 270-283.

Yang, C. B., Deng, C. S., & Chiang, H. L. (2012). Combining the Taguchi method with artificial neural network to construct a prediction model of a CO2 laser cutting experiment. *The International Journal of Advanced Manufacturing Technology*, *59*(9-12), 1103-1111.

Permissions

All chapters in this book were first published in IJIEC, by Growing Science; hereby published with permission under the Creative Commons Attribution License or equivalent. Every chapter published in this book has been scrutinized by our experts. Their significance has been extensively debated. The topics covered herein carry significant findings which will fuel the growth of the discipline. They may even be implemented as practical applications or may be referred to as a beginning point for another development.

The contributors of this book come from diverse backgrounds, making this book a truly international effort. This book will bring forth new frontiers with its revolutionizing research information and detailed analysis of the nascent developments around the world.

We would like to thank all the contributing authors for lending their expertise to make the book truly unique. They have played a crucial role in the development of this book. Without their invaluable contributions this book wouldn't have been possible. They have made vital efforts to compile up to date information on the varied aspects of this subject to make this book a valuable addition to the collection of many professionals and students.

This book was conceptualized with the vision of imparting up-to-date information and advanced data in this field. To ensure the same, a matchless editorial board was set up. Every individual on the board went through rigorous rounds of assessment to prove their worth. After which they invested a large part of their time researching and compiling the most relevant data for our readers.

The editorial board has been involved in producing this book since its inception. They have spent rigorous hours researching and exploring the diverse topics which have resulted in the successful publishing of this book. They have passed on their knowledge of decades through this book. To expedite this challenging task, the publisher supported the team at every step. A small team of assistant editors was also appointed to further simplify the editing procedure and attain best results for the readers.

Apart from the editorial board, the designing team has also invested a significant amount of their time in understanding the subject and creating the most relevant covers. They scrutinized every image to scout for the most suitable representation of the subject and create an appropriate cover for the book.

The publishing team has been an ardent support to the editorial, designing and production team. Their endless efforts to recruit the best for this project, has resulted in the accomplishment of this book. They are a veteran in the field of academics and their pool of knowledge is as vast as their experience in printing. Their expertise and guidance has proved useful at every step. Their uncompromising quality standards have made this book an exceptional effort. Their encouragement from time to time has been an inspiration for everyone.

The publisher and the editorial board hope that this book will prove to be a valuable piece of knowledge for researchers, students, practitioners and scholars across the globe.

List of Contributors

Roya Soltani
School of Industrial Engineering, Iran University of Science and Technology, Tehran, Iran

Arun Kumar Parida
School of Mechanical Engineering, KIIT, Bhubaneswar, Odisha, India

Rajesh Kumar Bhuyan
School of Mechanical Engineering, KIIT, Bhubaneswar, Odisha, India

Bharat Chandra Routara
School of Mechanical Engineering, KIIT, Bhubaneswar, Odisha, India

Masoud Rabbani
School of Industrial Engineering, College of Engineering, University of Tehran, Tehran, Iran

Nadia Pourmohammad Zia
School of Industrial Engineering, College of Engineering, University of Tehran, Tehran, Iran

Hamed Rafiei
School of Industrial Engineering, College of Engineering, University of Tehran, Tehran, Iran

Mortaza Honarmande Azimi
Department of Management, College of Management, Economics & Accounting, Tabriz Branch, Islamic Azad University, Tabriz, Iran

Houshang Taghizadeh
Associate Professor, Department of Management, College of Management, Economics & Accounting, Tabriz Branch, Islamic Azad University, Tabriz, Iran

Nasser Fegh-hi Farahmand
Associate Professor, Department of Management, College of Management, Economics & Accounting, Tabriz Branch, Islamic Azad University, Tabriz, Iran

Jafar Pourmahmoud
Assistant Professor, Department of Applied Mathematics, Azarbaijan Shahid Madani University, Tabriz, Iran

Rajan Jindal
M.Tech. Scholar, Department of Mechanical Engineering, Yamuna Institute of Engineering & Technology, Gadholi, Distt. Yamuna Nagar, 133103, India

Deepak Choudhary
Assistant Professor, Department of Mechanical Engineering, Yamuna Institute of Engineering & Technology, Gadholi, Distt. Yamuna Nagar, 133103, India

Yuri Mauergauz
Sophus Group, Moscow 119296, Russian Federation

Sujan Piya
Department of Mechanical and Industrial Engineering, College of Engineering, Sultan Qaboos University, AL-Khod 123, Muscat, Sultanate of Oman

Miloš Madić
Faculty of Mechanical Engineering, University of Niš, A. Medvedeva 14, Niš, Serbia

Marko Kovačević
Faculty of Electronic Engineering, University of Niš, A. Medvedeva 14, Niš, Serbia

Miroslav Radovanović
Faculty of Mechanical Engineering, University of Niš, A. Medvedeva 14, Niš, Serbia

Suryakant B. Chandgude
K. K. Wagh Institute of Engineering Education and Research, Nashik-422003, Maharashtra, India

Padmakar J. Pawar
K. K. Wagh Institute of Engineering Education and Research, Nashik-422003, Maharashtra, India

Mudigonda Sadaiah
Dr. Babasaheb Ambedkar Technological University, Vidyavihar, Lonere- 402103, Maharashtra, India

Rinto Yusriski
Department of Industrial Engineering and Management, Institut Teknologi Bandung, Bandung 40132, Indonesia

Sukoyo Sukoyo
Department of Industrial Engineering and Management, Institut Teknologi Bandung, Bandung 40132, Indonesia

T. M. A Ari Samadhi
Department of Industrial Engineering and Management, Institut Teknologi Bandung, Bandung 40132, Indonesia

Abdul Hakim Halim
Department of Industrial Engineering and Management, Institut Teknologi Bandung, Bandung 40132, Indonesia

Mohammad Shahrazad
Department of Electrical Engineering, Brunel University, London, UK

Amir Hessam Alikhanzadeh
Department of Electrical Engineering, Brunel University, London, UK

André Serra e Santos
Department of Production and Systems - School of Engineering - University of Minho (EEUM/UM), Portugal

Ana Maria Madureira
GECAD - Knowledge Engineering and Decision Support Research Center - School of Engineering - Polytechnic of Porto (ISEP/IPP), Portugal

Maria Leonilde R. Varela
Department of Production and Systems - School of Engineering - University of Minho (EEUM/UM), Portugal

Giovanni Romagnoli
Department of Industrial Engineering, University of Parma, Italy

S. G. Dambhare
Mechanical Engineering Department, PVPIT Pune, M.S., India

S. J. Deshmukh
Mechanical Engineering Department, PRMITR Badnera Amaravati, M.S., India

A. B. Borade
Mechanical Engineering Department, JDIET Yavatmal, M.S., India

Miloš Madić
Faculty of Mechanical Engineering, University of Niš, A. Medvedeva 14, Niš, Serbia

Miroslav Radovanović
Faculty of Mechanical Engineering, University of Niš, A. Medvedeva 14, Niš, Serbia

Marin Gostimirović
Faculty of Technical Science, University of Novi Sad, Trg Dositeja Obradovića 6, Serbia